Advances in Hydrogen Energy - Production, Storage, and Utilization

Edited by Muhammad Bilal Tahir, Muhammad Shahid Rafique, Muhammad Rafique and Muhammad Sagir

Published in London, United Kingdom

Advances in Hydrogen Energy - Production, Storage, and Utilization
http://dx.doi.org/10.5772/intechopen.1008574
Edited by Muhammad Bilal Tahir, Muhammad Shahid Rafique, Muhammad Rafique and Muhammad Sagir

Contributors
Ashton Swartbooi, Bilal Ahmed, Claudelle Sybilline Anensong Djadock, Ghulam Nabi, Hugo M. Oliveira, Jalaal A. Hayes, João Piteira, Karuppiah Veerathangam, Marcela de Oliveira Brahim Cortez, Muhammad Bilal Tahir, Muhammad Rafique, Nancy Anna Anasthasiya Anthuvan, Noemí Cristina Silva de Souza, Relebohile Mckete, Renata Pereira Lopes Moreira, Saban Pusat, Shawn Gouws, Tahir Iqbal, William G. Coors, Yasin Karagoz, Zohreh Akbari

First published in London, United Kingdom, 2026 by IntechOpen

IntechOpen is the global imprint of INTECHOPEN LIMITED, registered in England and Wales, registration number: 11086078, 167-169 Great Portland Street, London, W1W 5PF, United Kingdom

For EU product safety concerns: IN TECH d.o.o., Prolaz Marije Krucifikse Kozulić 3, 51000 Rijeka, Croatia, info@intechopen.com or visit our website at intechopen.com.

British Library Cataloguing-in-Publication Data
A catalogue record for this book is available from the British Library

Advances in Hydrogen Energy - Production, Storage, and Utilization
Edited by Muhammad Bilal Tahir, Muhammad Shahid Rafique, Muhammad Rafique and Muhammad Sagir
p. cm.
Print ISBN 978-1-83634-794-1
Online ISBN 978-1-83634-793-4
eBook (PDF) ISBN 978-1-83634-795-8

Meet the editors

Dr. Muhammad Bilal Tahir is a physicist and academic leader serving at the Institute of Physics, KFUEIT, Rahim Yar Khan. His research interests span nanomaterials, renewable energy technologies, hydrogen systems, and advanced materials for environmental applications. He has contributed extensively to scientific literature and has led several national and international academic initiatives. Known for his commitment to academic development and collaborative research, he has played a key role in strengthening interdisciplinary programs in physics and engineering. Dr. Tahir is also actively involved in curriculum development, scientific outreach, and mentoring young researchers.

Prof. Dr. Muhammad Shahid Rafique is a distinguished physicist at the Department of Physics, Main Campus, UET Lahore. His expertise includes laser–material interactions, thin films, advanced materials, and energy applications. With numerous publications, funded projects, and postgraduate supervisions, he is recognized for his contributions to research and higher education in Pakistan. He has represented his institution in various international forums and remains committed to fostering scientific innovation, capacity building, and high-quality academic training.

Dr. Muhammad Rafique serves in the Department of Physics at the University of Sahiwal, where he is actively engaged in teaching, research, and student mentorship. His research interests include materials science, renewable energy systems, and experimental physics. He has contributed to several publications and academic activities, advancing physics education and collaborative research. Dr. Rafique is known for his dedication to academic excellence and his efforts to promote emerging scientific fields within his institution.

Prof. Dr. Muhammad Sagir is a senior faculty member at the Institute of Chemical and Environmental Engineering, KFUEIT, Rahim Yar Khan. His work focuses on chemical process design, environmental engineering, energy systems, and sustainable technologies. With a strong academic and research background, he has contributed significantly to teaching, industrial collaboration, and applied research projects. Prof. Sagir is committed to advancing innovative solutions in chemical and environmental engineering and plays an active role in training the next generation of engineers and researchers.

Contents

Preface

Hydrogen energy stands today at the forefront of global efforts to transition toward cleaner, more resilient, and sustainable energy systems. As nations confront the escalating challenges of climate change, resource limitations, and the need for diversified energy portfolios, hydrogen has emerged as a versatile and robust solution. Its capacity for carbon-free utilization, long-term storage, and integration across industrial, transportation, and power-generation sectors makes it a cornerstone of future low-carbon strategies. This edited volume, *Advances in Hydrogen Energy – Production, Storage, and Utilization*, has been assembled to present a clear, comprehensive, and up-to-date account of ongoing scientific and technological developments across the hydrogen value chain.

The chapters in this book collectively explore the full spectrum of hydrogen technologies in a coherent and structured manner. Beginning with foundational and emerging production routes, the initial chapters outline key advancements in electrolysis, thermochemical cycles, renewable-driven processes, and biological systems. These contributions highlight the steady progress toward higher efficiency, reduced environmental impact, and cost-effective scalability. Following this, the volume presents a detailed discussion on hydrogen storage, one of the most critical challenges in the widespread adoption of hydrogen-based systems. Chapters dedicated to advanced solid-state materials, nanostructured adsorbents, liquid carriers, metal hydrides, and high-pressure containment technologies showcase the remarkable strides being made through material innovation and system-level optimization. This section underscores how improvements in storage directly influence the feasibility and safety of downstream applications. The later chapters turn to utilization technologies, ranging from fuel cells and industrial combustion systems to transportation solutions, microgrid integration, and hybrid energy systems. Together, these analyses provide a holistic view of how hydrogen can decarbonize diverse sectors while simultaneously supporting energy security and economic development. Complementary discussions on techno-economic analyses, safety protocols, regulatory frameworks, and future policy directions ensure that the volume offers not only scientific insights but also practical considerations relevant to stakeholders and decision-makers. Figures and tables included throughout the volume have been specially curated to help readers visualize processes, compare technological pathways, and interpret key findings with clarity. These visual components are designed to enhance comprehension and support both teaching and research activities.

We acknowledge, with deep appreciation, the valuable efforts of all contributing authors who offered their expertise and timely revisions throughout this project. We are also grateful to the reviewers, whose constructive assessments played an essential role in ensuring academic rigor. Our sincere thanks are extended to Ms. Ivana Boskovic, Publishing Process Manager and Team Leader at IntechOpen, whose professional guidance, coordination, and support were instrumental in bringing this edited book to completion.

On a personal note, we extend heartfelt gratitude to our beloved family members, Hafiza Zunaira Bilal, Arfa Bilal, Zainab Chaudhary, Mohammad Musa, and Bariza Fatima, whose encouragement, patience, and unwavering support enabled us to complete this work within the required timelines. Their motivation has been a constant source of strength throughout the editorial process.

We hope that this edited volume will serve as a meaningful resource for researchers, engineers, students, and professionals working in the rapidly advancing field of hydrogen energy. It is our aspiration that the knowledge presented here will inspire innovation, foster collaboration, and contribute to the global pursuit of sustainable energy solutions.

Dr. Muhammad Bilal Tahir
Institute of Physics, KFUEIT,
Rahim Yar Khan, Pakistan

Dr. Muhammad Shahid Rafique
Professor,
Department of Physics,
Main Campus UET Lahore,
Lahore, Pakistan

Dr. Muhammad Rafique
Department of Physics,
University of Sahiwal,
Sahiwal, Pakistan

Dr. Muhammad Sagir
Professor,
Institute of Chemical and Environmental Engineering, KFUEIT,
Rahim Yar Khan, Pakistan

Section 1

Advances in Hydrogen Energy: Production

Chapter 1

Catalytic Hydrogen Combustion

Zohreh Akbari

Abstract

The intermittency associated with renewable energy sources can be mitigated through the use of an energy carrier that can be produced and stored during periods of energy abundance. Hydrogen (H_2), as a carbon-free energy carrier, possesses the highest mass energy density relative to other fuels, making it a promising candidate. The chemical energy of H_2 can be harnessed through oxidation via a highly exothermic reaction. Nonetheless, the broad flammability range of H_2 in air, along with flame instability and elevated temperatures, presents challenges to conventional H_2 combustion. Catalytic H_2 combustion (CHC) involves the recombination of oxygen (O_2) and H_2 on the surface of a catalyst. With CHC, the temperature can be easily controlled, thereby minimizing the risks associated with flashbacks and the emissions of nitrogen oxides (NO_x). CHC can be used directly for heat and power generation or to improve safety in H_2-related applications by preventing the release of surplus H_2 into the atmosphere at concentrations exceeding the lower flammability limit of H_2 in air. Although the state-of-the-art catalysts for the CHC reaction are based on precious metals, ongoing research aims to discover alternative catalysts, understand the CHC mechanism over different catalysts, and develop new approaches to stabilize the H_2 flame. This chapter reviews the efforts toward discovery and investigation of catalysts for the CHC reaction, the mechanism and kinetics involved, as well as applications of CHC for heat and power generation. Finally, the challenges associated with CHC and future research directions are discussed in the conclusion section.

Keywords: catalytic H_2 combustion, hydrogen, heterogeneous catalysis, catalysts, reaction kinetics, reaction mechanism

1. Introduction

The projection of energy demands indicates an increase by 2050, with a significant contribution from renewable energies. However, the intermittency associated with renewable energy must be addressed by a reliable and accessible energy vector. Among all possible energy vectors, H_2 has the highest mass energy density and is carbon-free. Moreover, H_2 has an inherently closed cycle [1]. In other words, if we use renewable electricity to electrolyze water and produce H_2, it can be directly used or stored for later applications. To release the chemical energy of H_2, we can use it in a direct combustion engine or in a fuel cell. In both scenarios, water is the only byproduct, which can be recycled in an electrolyzer.

In conventional H_2 combustion, temperature control is challenging, as it exceeds the auto-ignition temperature of H_2 in air (585°C). The flame temperature of H_2 in air exceeds 2100°C, which leads to NO_x emissions. Additionally, in conventional combustion, efficiency is low, and there is a risk of flashbacks, raising safety concerns with this technique.

In contrast, with catalytic H_2 combustion (CHC), we can regulate the temperature, prevent flame formation, and prevent reaching the auto-ignition temperature of H_2 in air. This way, there is no risk of NO_x emissions. Furthermore, by controlling the H_2 and air feed direction, we can avoid gas premixing and thereby mitigate the risk of flashbacks.

The H_2 flammability and explosion limits in an oxidizer gas mixture of nitrogen (N_2) and O_2 at an ambient pressure of 1 bar are shown in **Figure 1**. H_2 has a lower and upper flammability limit of 4 and 75 vol% in air, respectively. This means in any H_2-sector, the H_2 content of the effluent gas should be kept below 4 vol%.

Since 2020, numerous groups have focused on various aspects of the CHC process, and the trend in the number of publications, as well as citations, shows significant growth, as presented in **Figure 2**, which represents these data as of August 2025.

Historically, in 1816, Humphry Davy introduced the concept of flameless combustion of methane (CH_4) on Pt. A few years later, in 1823, Johann Wolfgang Döbereiner discovered that in the presence of air, H_2 reacts with O_2 to form condensed water on the surface of finely powdered Pt at temperatures below RT [3]. Since then, CHC has been used mainly in three applications: (1) heat generation, (2) power generation, and (3) safety, to reduce the H_2 concentration in the effluent gas below 4 vol%. These applications will be discussed in Section 5.

During the CHC process, H_2O and heat are released due to H_2 and O_2 recombining in a highly exothermic reaction on the surface of a catalyst (Eq. (1)).

$$H_2 + \frac{1}{2}O_2 \rightarrow H_2O \; \Delta H^f_{H_2O(g)} = -241.82 \; {}^{kJ}/_{mol} \quad (1)$$

$\Delta H^f_{H_2O(g)}$ is the heat of formation of water vapor at room temperature (RT). The negative enthalpy of the CHC reaction makes it thermodynamically favorable.

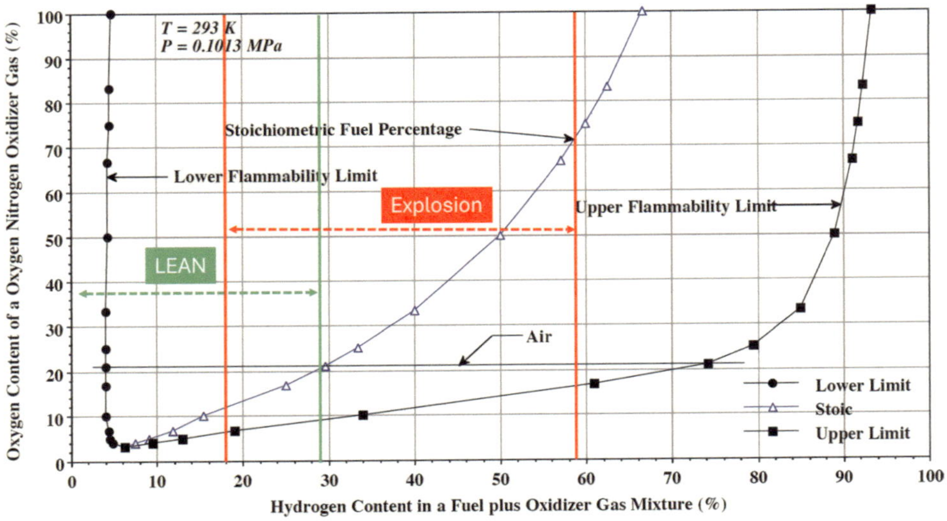

Figure 1.
The flammability and explosion limits of H_2 in an oxidizer gas (N_2 + O_2) at 1 bar [2].

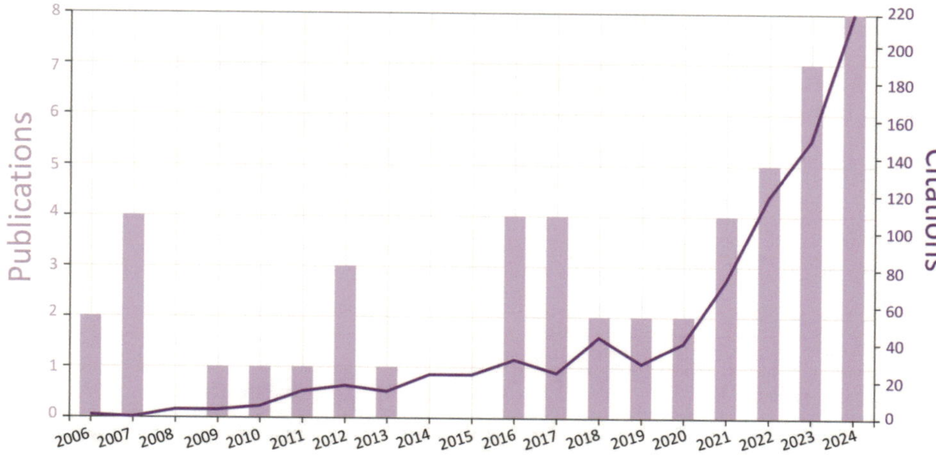

Figure 2.
The number of publications and citations by August 2025 for the keyword "catalytic hydrogen combustion", obtained from Web of Science.

The state-of-the-art catalysts for the CHC process are based on Pt and Pd metals. These catalysts can initiate the CHC at temperatures below RT, making them suitable for specific low-temperature applications. Therefore, till now, the majority of mechanistic and kinetic studies has been focused on these catalysts.

However, due to the scarcity of Pt and Pd resources, along with their high costs and low stability at elevated temperatures, researchers have concentrated on partially or entirely replacing these metals.

This chapter discusses first the catalysts for the CHC reaction and the catalytic assessment methods, such as FTIR, IR thermography, mass spectrometry, H_2 sensors, and gas chromatography, used to evaluate their catalytic activity. It then explores the mechanism and kinetics of the CHC reaction across different catalysts. Finally, the various applications of the CHC reaction and the challenges and prospects in the field are reviewed.

2. Catalysts for CHC

Although CHC catalysis was initiated with the Pt catalysts, various catalyst classes have been studied for decades. These include pristine metals [4, 5], (complex) oxides [6–8], and supported metal catalysts. Supported metal catalysts are particularly attractive because of their high surface area, which provides well-dispersed active sites, enhances catalyst utilization, benefits from the synergy between metal and support, and offers scalability and reproducibility. This has made supported metal catalysts the most favored class in catalysis and, consequently, in CHC catalysis.

Table 1 summarizes the supported monometallic catalysts studied in the field of CHC reaction. As shown, the majority of the focus is on the Pt- and Pd-based catalysts. Bi-metallic catalysts have also been widely studied in the field. However, most of these studies focus on the partial substitution of the Pt or Pd catalysts [19, 20, 32].

Numerous studies have been dedicated to reducing the Pt and Pd content in catalysts through various strategies such as nanoparticle (NP) utilization, partial substitution with other transition metals (TMs), doping of Pt inside the support material

Support	Catalyst	Reducibility	Water affinity	Ref
Al_2O_3	Pt	Non-reducible	Hydrophilic	[9–12]
	Pd			[10, 13–15]
	Rh			[16]
	Ru			[17, 18]
	Cu			[19–23]
	Co			[18]
	Ni			
	Mo			
Anodized Al oxide (AAO)	Pt	Non-reducible	Hydrophilic	[24–27]
ZrO_2	Cu	Non-reducible	Hydrophilic	[21, 23]
SiC	Pt	Non-reducible	Hydrophilic	[9, 28, 29]
SiO_2	Pt	Non-reducible	Hydrophobic	[9, 12, 30]
	Pd			[31]
	Cu			[21]
TiO_2	Pt	Reducible	Hydrophilic	[9, 32]
	Pd			[32]
	Au			[33]
SnO_2	Cu	Reducible	Hydrophilic	[21]
CeO_2	Pt	Reducible	Hydrophobic	[34]
	Pd			[32]
Co_3O_4-ZrO_2	Pt	mixed	—	[35]
	Pd			[35]
	Ni			[36]
	La			[36]
Ti-mesh	Pt	Reducible	Hydrophilic	[37]
Stainless steel-mesh	Pt	—	—	[38]
Ni-foam	Pt	—	—	[39]

Table 1.
Supported mono metal catalyst in the CHC field.

[40, 41], and the employment of single-atom catalysts. Nonetheless, the selection of an appropriate catalyst must consider the specific application in which the catalyst will be utilized. If activation at room temperature is not required, then the use of Pt- and Pd-based catalysts may be deemed unnecessary.

Furthermore, a significant challenge associated with Pt- and Pd-based catalysts is their deactivation caused by water at temperatures below the water boiling point. Particularly at high H_2 concentrations, water droplets tend to occupy the catalysts' active sites, leading to the cessation of the reaction [42, 43]. Consequently, the selection of these catalysts is contingent upon the specific requirements of the application.

One of the promising catalysts in the CHC field is the Ru-based catalyst, which demonstrates superior stability relative to other TM oxides [18]. The supported Ru catalyst activates the CHC reaction at approximately 100°C, thereby effectively addressing issues related to water-induced deactivation.

Although Ru is a cheaper metal compared to Pt and Pd, it remains classified within the precious metal groups (PGMs). Consequently, its efficient utilization is of paramount importance. Akbari et al. [17] investigated the influence of Ru NPs size on the catalytic performance of Ru-γAl_2O_3 under the CHC reaction. Their findings indicated that nanoparticles measuring approximately 0.77 nm, corresponding to a loading of 1 wt% Ru, exhibited the highest CHC catalytic activity. This enhancement was attributed to the optimal balance between Ru dispersion and metal-support interaction (MSI) and the resulting oxidation state of Ru within the NPs.

A similar study was also conducted on Cu-Al_2O_3 catalysts, highlighting the significance of NP size engineering in catalyst design to attain optimal CHC activity and catalyst utilization [22].

3. Catalytic activity measurement

The catalytic activity of CHC can be quantitatively evaluated using various parameters, such as H_2 conversion rate, the reaction onset temperature, the temperature at which complete H_2 conversion occurs, and the temperature interval between the onset and full conversion.

State-of-the-art Pt- and Pd-based catalysts initiate the CHC reaction at temperatures below ambient conditions [44]. Complete H_2 conversion can be achieved at temperatures under 100°C, depending on the catalyst loading and the size of the catalyst NPs.

In a typical experiment to examine catalytic activity, the catalyst is placed in a plug-flow reactor, then the reactive gas contacts the catalyst bed, and the reaction occurs depending on the reactor temperature. To determine the H_2 conversion, various analytical techniques such as mass spectrometry [17, 18], gas chromatography [29, 45], and an H_2 sensor may be employed to quantify the H_2 content in the exhaust gases. Subsequently, by knowing the initial and final H_2 concentrations, H_2 conversion and its rate can be calculated using Eqs. (2) and (3), respectively. Moreover, the reaction turnover frequency (TOF) can be calculated using Eq. (4).

$$H_2\ conversion\ (X_{H_2}) = \frac{H_{2,in} - \frac{N_{2,in}}{N_{2,out}} \times H_{2,out}}{H_{2,in}} \tag{2}$$

$$H_2\ conversion\ rate\ (r) = \frac{X_{H_2} \times flow\ rate}{catalyst\ amount} \tag{3}$$

$$TOF = r \times \frac{Molecular\ weight\ of\ the\ metal}{Dispersion \times metal\ percentage} \tag{4}$$

A typical H_2 conversion versus temperature curve exhibits an S-shaped profile, as illustrated in **Figure 3**. At lower temperatures, where the H_2 conversion remains minimal, the CHC reaction is predominantly controlled by kinetic factors. Under these conditions, the activation energy barrier must be overcome for the reaction to proceed. Therefore, the kinetic information can only be obtained in this region of the curve.

As the temperature increases over time and provides sufficient activation energy, the reaction transitions to being limited by diffusion processes. Specifically, this

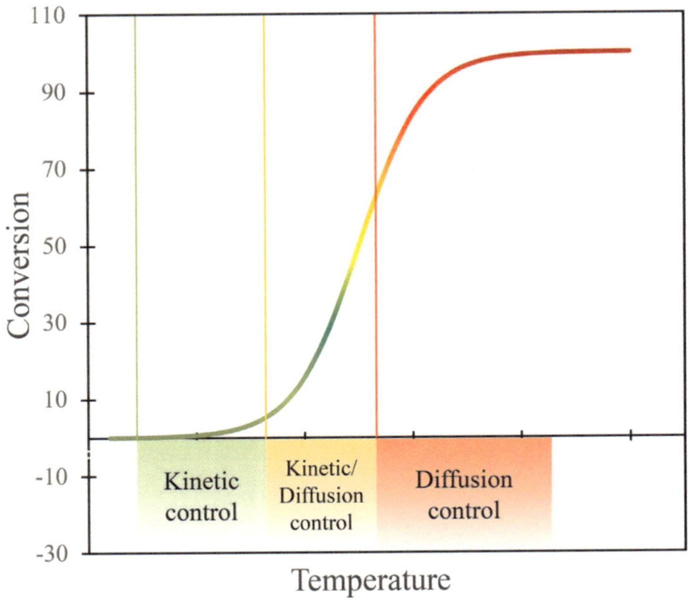

Figure 3.
Typical H_2 conversion curve vs. temperature under CHC reaction with three characteristic regions: kinetic control, mixed control, and diffusion control regions.

involves the diffusion of reactants from the gas stream to the catalyst surface, as well as the diffusion of the resulting products away from the surface into the exhaust gas. From this point forward, elevating the temperature no longer influences the H_2 conversion rate.

It is important to note that should the temperature attain the autoignition point of H_2 in air (585°C), homogeneous (open flame) combustion will occur [46].

The catalytic activity of CHC can also be evaluated by monitoring changes in specific properties of either the reactants (H_2 and O_2) or the products (H_2O and heat). Techniques based on infrared (IR) radiation can be employed to observe temperature variations (by IR thermography [42]) or the evolution of the water adsorption band (IR spectroscopy). However, because H_2 and O_2 are IR-inactive, their consumption cannot be directly measured using IR methods. Common IR techniques used for such analyses include Fourier-Transform Infrared (FTIR) spectroscopy [29].

These methods primarily demonstrate the catalyst's immediate response during CHC. To ensure a comprehensive understanding, it is also essential to investigate the development of catalytic activity over time and evaluate its stability in CHC, as briefly discussed in the following section.

4. Stability of the catalytic activity

The decline in catalytic activity occurs when the metal catalyst is exposed to high temperatures resulting from the exothermic H_2 oxidation, particularly above 500°C. This decline is primarily due to the agglomeration of metal particles. The presence of water can accelerate this deactivation. Limited studies have [47] investigated catalyst durability under the CHC reaction conditions, and these are summarized in **Table 2**. For example, the Pd-Ni-Al_2O_3 catalyst demonstrated promising stability after

DOI: http://dx.doi.org/10.5772/intechopen.1014178

Catalyst	Particle Size (nm)	Flow rate (ml/min)	Gas Mixture	Time (h)	Temperature (°C)	Conversion before test	Conversion after test	Ref.
0.27% Pt-SiC	11	200	1 vol% H_2 in air	1	420	100%	100%	[29]
2%Pd-Al_2O_3	1.9	25	0.5 vol% H_2 in air	70	25	59 %	70 %	[48]
2%Pd90-Ni10-Al_2O_3	2.6					64 %	65 %	
2%Pd80-Ni20-Al_2O_3	2.9					64 %	63 %	
2%Pd70-Ni30-Al_2O_3	3.7					64 %	60 %	
2% Pd-Al_2O_3	-	7	7.2 vol% H_2 in air	1	22	72 %	65 %	[30]
2% Pt-Al_2O_3	-					72 %	68 %	
2%Pd-Pt (1:1)-Al_2O_3	-					78 %	78 %	
2%Pd-Pt (2:1)-SiO_2	11					92 %	92 %	
2%Pd-Pt (1:1)-SiO_2	10					74 %	78 %	
2%Pd-Pt (1:2)-SiO_2	9					70 %	68 %	
CeO_2	-	100	2.7 vol% H_2	12	500	90 %	90 %	[49]
$CeVO_4$	-		1.35 vol% O_2			98 %	95 %	
Fe_2O_3	-		(H_2:O_2=2:1)			100 %	85 %	
$FeVO_4$	-					100 %	95 %	
1 wt% Ru-γAl_2O_3	0.8	20	H_2:O_2:N_2 = 4:2:96	45	240	100 %	100 %	[18]
10 wt% Co-γAl_2O_3	5.9					100 %	99 %	
Pd-Cu- Al_2O_3	5.7	200	3 vol% H_2 in air	40	600	100 %	100 %	[20]
Cu- Al_2O_3	5.4					100 %	100 %	
Co_3O_4−ZrO_2	9.1	100	H_2:O_2:N_2 = 5:10:85	50	200	32 %	35 %	[36]
Ni- Co_3O_4−ZrO_2	11.6					55 %	51 %	
La- Co_3O_4−ZrO_2	8.9					40 %	40 %	

Table 2.
Stability studies for different catalysts and different CHC conditions.

70 hours, comparable to its monometallic counterpart. This suggests that Pd-Ni-Al_2O_3 could serve [48] as a cost-effective alternative to Pd-Al_2O_3, given its lower price.

While most studies have focused on precious metals, Co- and Cu-based catalysts demonstrate notable stability. Co-Al_2O_3 exhibited only a 1% decrease in H_2 conversion over 45 hours, from 100–99% at 240°C, and its activity was fully restored after a regeneration step involving reduction, returning to 100% H_2 conversion.

Characterization of the catalysts before and after the CHC reaction provides essential insights into parameters including phase transformation, metal dispersion, oxidation state of the metal catalyst, particle size, and distribution.

5. CHC kinetics and mechanism

5.1 CHC kinetics

Kinetic studies offer insights into various aspects of chemical reactions, including the determination of the rate-limiting step, activation energy, and the influence of reactant concentrations on the overall reaction rate.

The CHC reaction exhibits rapid kinetics. As previously noted, kinetic analyses should be conducted exclusively within the kinetic control region of the conversion versus temperature curve, provided that temperature varies with time.

To identify the kinetic control region, several approaches can be employed, including verifying the Weisz-Prater criteria to ensure the absence of mass transfer and the Anderson and Mears criteria to ensure the absence of heat transfer limitations (see **Table 3**) [50], and confirming the conditions of the differential reactor [51].

In essence, the initial step in a kinetic study involves experimentally establishing the kinetic control region, followed by validation using the previously outlined criteria. The Arrhenius kinetic model serves as a useful tool for experimentally identifying this region, as it provides a foundation for analysis. In this model $TOF \propto A \ \exp\left(\frac{-E_a}{RT}\right)$, where A represents the frequency factor, which shows the frequency of the successful collisions, E_a indicates the activation energy, T is the temperature, and R denotes the universal gas constant. Based on this model, where the slope of TOF vs. 1000/T is zero (i.e., constant E_a), kinetics controls the reaction [18, 52].

It is important to note that for a reliable kinetic study, it is essential to collect as many data points as possible within the kinetic control region. Therefore, employing a mass spectrometer with multi-ion detection (MID) mode is considered the most appropriate analytical method.

Criteria	**Equation**
Weisz-Prater: No Internal Mass Transfer	$\frac{r_{obs} R_p^2}{C_S D_{TA}^e} < 1, or < \frac{1}{n}$
Weisz-Prater: No External Mass Transfer	$\frac{r_{obs} R_p^2}{C_B k_c} < \frac{0.15}{n}$
Anderson: No Internal Heat Transfer	$\frac{\lvert\Delta H_r\rvert r_{obs} R_p^2}{\lambda^e T_s} < \frac{0.75 R T_s}{E_a}$
Mears: No External Heat Transfer	$\frac{\lvert\Delta H_r\rvert r_{obs} R_p^2}{h_t T_B} < \frac{0.15 R T_B}{E_a}$

Table 3.
Criteria to avoid mass and heat transfer limitations [50].

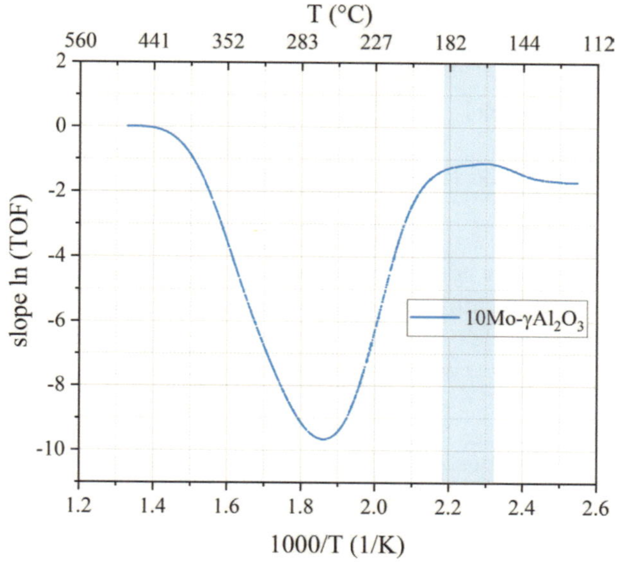

Figure 4.
Plot of slope of TOF vs. 1000/T for the CHC reaction over a 10 wt% Mo-γAl_2O_3 catalyst. The CHC reaction was conducted at 4 vol% H_2 with a H_2:O_2 molar ratio of 2:1, a flow rate of 20 mL/min, and a catalyst weight of 200 mg. The highlighted section is the kinetic control region for this catalyst under the conditions.

For instance, **Figure 4** illustrates the plot of slope TOF – 1000/T for Mo-γAl_2O_3 catalyst under the CHC reaction. The 20°C highlighted section shows the kinetic control region where the slope of TOF vs. 1000/T is zero. Interested readers are directed to reference [18] for a detailed example of rate determination or activation energy extraction under actual CHC reaction conditions. Nevertheless,

The calculation of a reliable and correct TOF is crucial for kinetic studies. Nonetheless, accurately determining the true dispersion of active sites is complicated by numerous dynamic factors affecting the catalyst during CHC reaction conditions. A primary source of error in TOF calculations stems from incorrect assumptions regarding the shape of NPs. This error can be minimized by considering the shape and the crystal structure of the NPs. An incorrect assumption about the shape can lead to over- or underestimation of catalyst dispersion and, subsequently, TOF [53].

Differential reactor condition is a steady-state approximation in which the H_2 conversion is below 10% and therefore any concentration gradient through the reactor bed can be neglected. In this condition, the reaction rate can be calculated using Eq. (5):

$$r = \frac{F_{A0} - F_{Ae}}{\Delta W} = \frac{X}{\frac{\Delta W}{F_{A0}}} \tag{5}$$

where F_{A0} and F_{Ae} are the inlet and outlet flow rates, respectively, and ΔW represents the mass of the catalyst.

Under this condition, the reaction rate has a linear relation with the H_2 conversion (X) and $\frac{\Delta W}{F_{A0}}$ [17, 35]. Therefore, any deviation from this linear relationship indicates a transition from the kinetic control region to the mass diffusion control region.

Once the kinetic control is identified and validated, the kinetic parameters are derived using an appropriate kinetic model. Several models have been documented in the literature, with two models being predominantly utilized to determine the kinetic parameters of the CHC reaction over various catalysts. These models are as follows:

1. Arrhenius kinetic model [17, 23]

2. Langmuir-Hinshelwood (L-H) model [19]

Mostly, the L-H models are applied to supported Pt and Pd catalysts, where both O_2 and H_2 are dissociatively adsorbed on the catalyst. Shinde et al. [21] proposed a modified L-H model to account for the effects of different supports.

A comprehensive overview of the reported activation energies for various catalysts and CHC reaction conditions is presented in **Table 4**.

Catalyst	Amount of catalyst	Gas composition	GHSV (h^{-1})	T100% (°C)	Ea (kJ/mol)	ref
$Co_3O_4 - ZrO_2$	—	H_2:O_2:N_2 = 5:10:85	47,750	~300	34.3	[38]
Pt-Co_3O_4-ZrO_2	2 at%	H_2:O_2:N_2 = 5:10:85	47,750	~100	39.3	
Pd-$Co_3O_4 - ZrO_2$	2 at%	H_2:O_2:N_2 = 5:10:85	47,750	~100	13.4	
Pt-γAl_2O_3	1 wt%	H_2:O_2:N_2 = 4:2:96	6000	60	31.7	[18]
Ru-γAl_2O_3	1 wt%	H_2:O_2:N_2 = 4:2:96	6000	235	24.7	
Co-γAl_2O_3	10 wt%	H_2:O_2:N_2 = 4:2:96	6000	250	19.3	
Ni-γAl_2O_3	10 wt%	H_2:O_2:N_2 = 4:2:96	6000	313	17.7	
Mo-γAl_2O_3	10 wt%	H_2:O_2:N_2 = 4:2:96	6000	455	10.4	
Pd-Al_2O_3	0.92 wt%	H_2: air	—	>180	36	[10]
Pt-Al_2O_3	1.78 wt%	H_2: air	—	>130	41	
Pt-Al_2O_3	0.38 wt%	H_2:O_2:Ar = 0.04:1:96	40,260	280	22.3	[11]
$Ce_{0.9}Ni_{0.1}O_{2-\delta}$	—	H_2:O_2: N_2 = 2.7:2.7:94.6	48,000	~250	49	[54]
$Ce_{0.88}Pd_{0.02}Ni_{0.1}O_{2-\delta}$	—	H_2:O_2: N_2 = 2.7:2.7:94.6	48,000	~190	34	
Pd-Al_2O_3	0.5 wt%	H_2: air = 1.5: 98.5	—	>90	73.77	[55]
Pt-SiC	0.27 wt%	H_2: air = 1: 99	—	60	35	[29]
Pt-SiC	0.5 wt%	H_2: air = 1: 99	—	>70	19	[9]
Pt-SiO_2	0.5 wt%	H_2: air = 1: 99	—	>30	33	
Pt-Al_2O_3	0.5 wt%	H_2: air = 1: 99	—	>50	29	
$Sn_{0.95}Cu_{0.05}O_{2-\delta}$	—	H_2:O_2: N_2 = 2.7:2.7:94.6	42,310	340	27.6	[9]
$Ce_{0.95}Cu_{0.05}O_{2-\delta}$	—	H_2:O_2: N_2 = 2.7:2.7:94.6	42,310	420	31.5	
$Al_{1.95}Cu_{0.05}O_{3-\delta}$	—	H_2:O_2: N_2 = 2.7:2.7:94.6	42,310	500	38.8	
$Zr_{0.95}Cu_{0.05}O_{3-\delta}$	—	H_2:O_2: N_2 = 2.7:2.7:94.6	42,310	480	56	

Catalyst	Amount of catalyst	Gas composition	GHSV (h^{-1})	T100% (°C)	Ea (kJ/mol)	ref
Pd-Cu- Al_2O_3	0.75 wt% Pd 0.25 wt% Cu	H_2: air = 1: 99	—	40	22	[19]
$Ti_{0.97}Pd_{0.03}O_{2-\delta}$-cordierite	—	H_2:O_2:Ar = 1:0.5:98.5	—	>140	32.6	[56]
Pd-Al_2O_3	2 wt%	H_2: air = 0.5: 99.5	—	>170	45	[15]
PdNi-Al_2O_3	2 wt% Pd 0.12 wt% Ni	H_2: air = 0.5: 99.5	—	>170	39	
PdNi-Al_2O_3	1.99 wt% Pd 0.27 wt% Ni	H_2: air = 0.5: 99.5	—	>170	41	
PdNi-Al_2O_3	1.99 wt% Pd 0.47 wt% Ni	H_2: air = 0.5: 99.5	—	>170	45	
$Ce_{0.98}Pt_{0.02}O_{2-\delta}$	2 at%	2 μmol H_2 in air	220,000	30	8.4	[40]
$Ti_{0.98}Pd_{0.02}O_{2-\delta}$	2 at%	2 μmol H_2 in air	220,000	30	59.8	
Pt-Al_2O_3	1.12 at%	2 μmol H_2 in air	220,000	30	33.5	
Ru-γAl_2O_3	0.5 wt%	H_2:O_2:N_2 = 4:2:96	6000	300	24.8	[17]
Ru-γAl_2O_3	1 wt%	H_2:O_2:N_2 = 4:2:96	6000	240	17.2	
Ru-γAl_2O_3	2 wt%	H_2:O_2:N_2 = 4:2:96	6000	180	22.1	
Ru-γAl_2O_3	5 wt%	H_2:O_2:N_2 = 4:2:96	6000	165	26.7	
Ru-γAl_2O_3	10 wt%	H_2:O_2:N_2 = 4:2:96	6000	150	29.9	

Table 4.
Activation energy reported for different catalysts and under different CHC reaction conditions.

Kinetic investigations have been employed to elucidate the reaction pathways and identify the rate-determining steps for catalysts based on Pt and Pd [57]. Although these studies offer valuable insights into the overall catalytic mechanism, a comprehensive understanding of the CHC reaction mechanism across various catalysts necessitates the utilization of advanced analytical techniques and *in situ* investigations.

5.2 CHC mechanism

Given the rapid kinetics of the CHC process, understanding the CHC mechanism is crucial for optimizing reaction parameters. The overall CHC reaction can be summarized as follows:

1. Reactants diffuse from the gas stream toward the active sites,

2. Adsorption of reactants on the active sites,

3. Chemical reaction (catalytic combustion),

4. Desorption of products from the active sites, and

5. Back-diffusion of products (water) to the gas stream.

The widely accepted mechanism of the H_2 and O_2 recombination reaction on the Pt and Pd catalysts is dissociative adsorption of O_2 and H_2 followed by Eqs. (6) and (7) [57].

$$O_{ads} + H_{ads} \rightarrow OH_{ads} \quad (6)$$

$$2\,OH_{ads} + H_{ads} \rightarrow H_2O_{ads} \rightarrow H_2O_{gas} \quad (7)$$

To elucidate the function of metal sites and support interactions, Singh et al. [38] utilized H_2 temperature-programmed reduction (H_2-TPR). Their findings indicated that, in a supported Pd-Co_3O_4-ZrO_2 catalyst, both H_2 and O_2 molecules adsorb dissociatively onto the Pd sites, subsequently recombining to form hydroxyl (OH) intermediates, which ultimately lead to water formation. Conversely, in the supported Pt-Co_3O_4-ZrO_2 catalyst, only H_2 dissociatively adsorbs onto the Pt sites, resulting in the formation of OH intermediates (see **Figure 5**). The study further demonstrated that the support material plays an active role in the catalytic process, with oxygen vacancies serving as active sites for O_2 adsorption and Co sites facilitating H_2 dissociation. Subsequently, hydrogen and oxygen species combine to form OH groups, which are key intermediates in the formation of water.

Further evidence of support for participation was provided by Deshpande et al. [41], who combined the FTIR and X-ray photoelectron spectroscopy (XPS) to investigate the CHC mechanism on the surface of Pt- and Pd-doped ZrO_2 and TiO_2. Their findings indicate that surface OH groups in ZrO_2 catalysts play a crucial role in the CHC process, whereas, in the case of TiO_2, lattice oxygen ions also participate actively in the reaction.

This distinction becomes particularly clear when comparing reducible and non-reducible supports (**Table 1**). In the case of reducible support materials, the presence of redox couples such as Ce^{4+}/Ce^{3+} in CeO_2 enables the lattice oxygen to interact with hydrogen species or molecular H_2 directly from the gas stream to produce water. The oxygen vacancies generated through the participation of lattice oxygen in the CHC

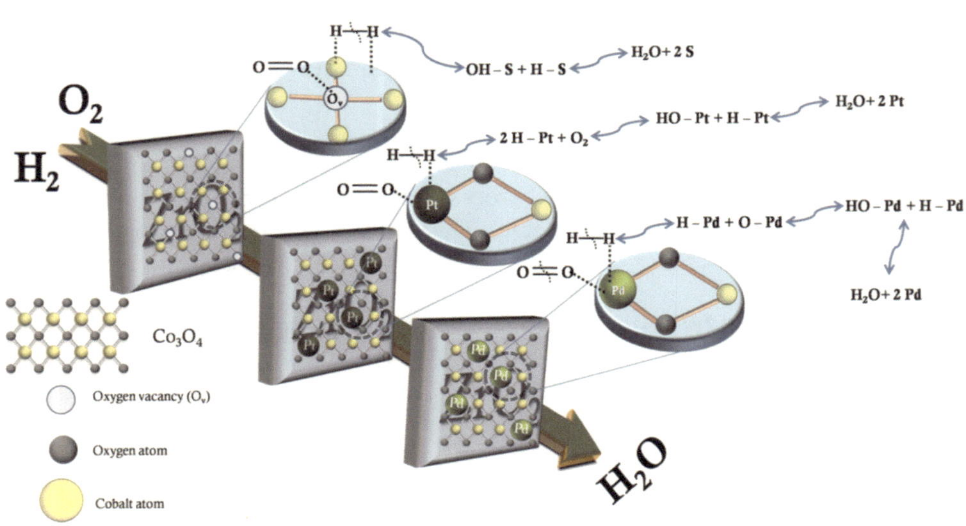

Figure 5.
CHC mechanism on undoped, Pt-doped, and Pd-doped ZrO_2-Co_3O_4 composite [35].

reaction are subsequently reoccupied by oxygen atoms via dissociative adsorption of O_2 molecules [54]. It is also conceivable that the metal site functions as an active site for the dissociative adsorption. For example, iron sites in $FeVO_4$ catalysts are believed to promote H_2 dissociation, leading to enhanced catalytic activity compared to the pure vanadate support [49]. In contrast, the non-reducible support materials do not directly participate in the CHC reaction. However, the support materials influence the stability of the catalyst nanoparticles [12].

Infrared techniques, particularly IR-based methods such as Diffuse Reflectance Infrared Fourier Transform Spectroscopy (DRIFTS), provide direct evidence of support effects and their impact on intermediate formation [58]. Zhang et al. [23] demonstrated that in reducible support materials, oxygen vacancies serve as active sites for the dissociative adsorption of O_2 and H_2. Utilizing DRIFTS, they established that these formed OH groups are critical intermediates in the CHC reaction over Cu catalyst. Subsequent investigations revealed that even in non-reducible support materials such as Al_2O_3, the OH groups remain essential reaction intermediates. In more recent research, Akbari et al. [17] identified that in supported Ru catalysts, the OH groups also function as key intermediates. Density Functional Theory (DFT) studies further corroborated these findings, indicating that the rate-limiting step in the CHC reaction over the Ru catalysts is the formation of water.

It is important to recognize that although the involvement of OH intermediates has been confirmed for Ru, Cu, Pt, Pd, and Rh, the detailed mechanisms of the CHC reaction steps for other catalysts remain to be thoroughly investigated. Typically, transition metals exhibit a very low energy barrier for the adsorption of O_2 and H_2, rendering the metal-oxygen bond a critical factor in catalytic activity. When the metal-oxygen or metal-hydrogen bond is stronger, the catalyst generally demonstrates reduced reactivity in the CHC process. In fact, it has been demonstrated that for TMs such as Co, Ni, and Mo, oxidation occurs upon exposure to reactive gases, which accounts for their comparatively sluggish CHC activity relative to noble metal catalysts [18].

6. Application

CHC has garnered significant interest across various applications (see **Figure 6**). These applications can be classified into three primary categories.

1. Safety: To prevent the release of unreacted or leaked H_2 into the atmosphere, passive autocatalytic recombiners (PARs) based on CHC have been developed [10, 59]. Additionally, Zahng et al. [60] have fabricated a CHC-based H_2 sensor specifically designed for H_2 leakage detection.

   CHC is also used for gas purification. In fuel cells, it is crucial to eliminate residual H_2 from the exhaust stream. Therefore, the exhaust can be directed to an H_2 burner and a heat recovery steam generator, producing steam that can be utilized in additional processes [34, 61].

2. Heat generation: Due to the exothermic nature of the CHC reaction, CHC can serve as a source of heat for both domestic and industrial applications. To address indoor air pollution, Jonin et al. [62] developed an H_2-fueled radiative catalytic

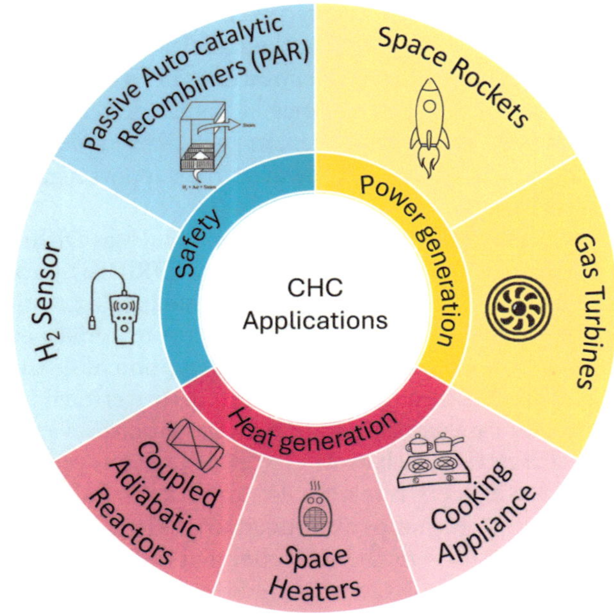

Figure 6.
Applications of CHC.

cooking stove. Additionally, in space propulsion, CHC has been employed to achieve the ignition temperature in homogeneous gas combustion processes [34].

A cooking stove, or burner, is composed of four primary components: (a) a H_2 diffuser, (b) a main catalytic combustion zone, (c) an air diffuser, and (d) a secondary catalytic combustion zone [63].

These stoves operate with H_2 concentrations exceeding the lower flammability limit of H_2 in air (4% vol H_2) and function under lean-fuel conditions. Consequently, careful design is crucial to prevent hazardous premixing of O_2 and H_2 [28]. The maximum power output achievable varies with the H_2 input; for example, a H_2 flow rate of 15 Nl/min yields a power output of 2.7 kW [64].

In another example, the Italian company GIACOMINI claims the large-scale manufacturing of H_2 catalytic heaters, marketed under the trade name H_2ydroGEM, with a power output of 5 kW. These heaters operate with a H_2 concentration of less than 4 vol%. The maximum internal temperature of the GIACOMINI boiler is approximately 350°C.

3. Power generation: To reduce NO_x emissions, CHC can be integrated into gas turbine engines for power generation [65, 66]. More recently, studies have demonstrated that employing CHC within a micro-combustor enhances wall-temperature uniformity in micro-thermophotovoltaic systems [67].

While numerous companies worldwide manufacture catalytic heaters that operate with gases other than H_2 (such as natural gas and propane), **Table 5** provides a list of companies and their commercially available products, including burners, boilers, and heaters, that utilize CHC technology for various applications.

Company	Country	Product
Giacomini	Italy	H_2ydroGEM
Neohysens (neo hydrogen sensors GmbH)	Germany	H2 Oxi-cat
EBZ GmbH	Germany	Catalytic Burner
Infragas	United Kingdom	Infragas® catalytic converters
Worcester-Bosch	United Kingdom	Hydrogen-fired boiler
Nationwide Boiler Inc.	United States	CataStak™ SCR boiler
TUBITAK	Turkey	Hydrogen-burning Catalyst

Table 5.
List of companies and their commercially available products using the CHC technology for H_2 burners, boilers, and heaters.

7. Conclusions

In summary, the CHC process is a highly exothermic reaction to recombine H_2 and O_2. By lowering the recombination temperature of H_2 and O_2, CHC provides an alternative to traditional homogeneous combustion of H_2 and O_2. This approach mitigates issues commonly associated with conventional combustion, such as the risk of flashbacks and NO_x emissions.

Although CHC has been a well-established area of research, the mechanisms and kinetics are predominantly understood in the context of state-of-the-art Pt- and Pd-based catalysts. Despite the high costs and limited availability of these catalysts, research efforts remain open to discovering novel catalyst materials that offer high TOF and elevated H_2 conversion rates at relatively low temperatures. In this context, bimetallic catalysts based on Ni, Co, and Cu are the most promising ones.

Further investigations using *in situ* characterization methods are essential to elucidate the mechanism of the CHC reaction over supported TM catalysts. As H_2 represents a critical fuel in the pursuit of decarbonization, it is anticipated that emerging applications will be developed and industrialized in the near future. This progression will likely foster greater integration of the CHC process into safety protocols, heat production, and power generation.

Moreover, although several designs for boilers and domestic heating appliances have been proposed, ongoing research is critical to devising safe, efficient, and economically viable solutions.

Acknowledgements

The author thanks Prof. Andreas Züttel and acknowledges the financial support of EPFL, and EMPA. Open access publishing was facilitated by Ecole polytechnique federale de Lausanne via the Consortium of Swiss Academic Libraries. The author acknowledges the usage of Grammarly for language polishing. There is no AI created content.

Author details

Zohreh Akbari
Laboratory of Materials for Renewable Energy (LMER), École Polytechnique Fédérale de Lausanne (EPFL), Sion, Valais, Switzerland

*Address all correspondence to: zohreh.akbari@epfl.ch; zohreh.akbari.phd@gmail.com

References

[1] Züttel A, Remhof A, Borgschulte A, Friedrichs O. Hydrogen: The future energy carrier. Philosophical Transactions of the Royal Society A: Mathematical, Physical and Engineering Sciences. 2010;**368**:3329-3342. DOI: 10.1098/rsta.2010.0113

[2] Green MA. Hydrogen safety issues compared to safety issues with methane and propane. AIP Conference Proceedings. 2006;**823**:319-326. DOI: 10.1063/1.2202431

[3] Kauffman GB. From triads to catalysis: Johann Wolfgang Döbereiner (1780–1849) on the 150th anniversary of his death. The Chemical Educator. 1999;**4**:186-197. DOI: 10.1007/s00897990326a

[4] Gland JL, Fisher GB, Kollin EB. The hydrogen-oxygen reaction over the Pt (111) surface: Transient titration of adsorbed oxygen with hydrogen. Journal of Catalysis. 1982;**77**:263-278. DOI: 10.1016/0021-9517(82)90167-1

[5] Fakheri A, Buckius RO, Masel RI. Combustion of hydrogen-rich mixtures on a nickel oxide catalyst. Combustion and Flame. 1988;**72**:259-269. DOI: 10.1016/0010-2180(88)90126-5

[6] Eom H-J, Jang JH, Lee D-W, Kim S, Lee K-Y. Catalytic combustion of hydrogen over La1−xSrxCoO3−δ + Co3O4 and LaMn1−xCuxO3+δ under simulated MCFC anode off-gas conditions. Journal of Molecular Catalysis A: Chemical. 2011;**349**:48-54. DOI: 10.1016/j.molcata.2011.08.017

[7] Buciuman F-C, Patcas F, Menezo J-C, Barbier J, Hahn T, Lintz H-G. Catalytic properties of La0.8A0.2MnO3 (a = Sr, Ba, K, Cs) and LaMn0.8B0.2O3 (B = Ni, Zn, Cu) perovskites. Applied Catalysis B: Environmental. 2002;**35**:175-183. DOI: 10.1016/S0926-3373(01)00250-8

[8] Haruta M, Sano H. Catalytic combustion of hydrogen I—Its role in hydrogen utilization system and screening of catalyst materials. International Journal of Hydrogen Energy. 1981;**6**:601-608. DOI: 10.1016/0360-3199(81)90025-2

[9] Arzac GM, Montes O, Fernández A. Pt-impregnated catalysts on powdery SiC and other commercial supports for the combustion of hydrogen under oxidant conditions. Applied Catalysis B: Environmental. 2017;**201**:391-399. DOI: 10.1016/j.apcatb.2016.08.042

[10] Morfin F, Sabroux J-C, Renouprez A. Catalytic combustion of hydrogen for mitigating hydrogen risk in case of a severe accident in a nuclear power plant: Study of catalysts poisoning in a representative atmosphere. Applied Catalysis B: Environmental. 2004;**47**:47-58. DOI: 10.1016/j.apcatb.2003.07.001

[11] Joshi SY, Ren Y, Harold MP, Balakotaiah V. Determination of kinetics and controlling regimes for H2 oxidation on Pt/Al2O3 monolithic catalyst using high space velocity experiments. Applied Catalysis B: Environmental. 2011;**102**: 484-495. DOI: 10.1016/j.apcatb.2010.12.030

[12] Liu R, Zhu Y, Sui Z, Wang H, Li P, Zhou X. Support effects on catalytic performance for selective combustion of hydrogen in the presence of propene. Fuel Processing Technology. 2013;**108**: 82-88. DOI: 10.1016/j.fuproc.2012.05.027

[13] Yan P, Xiao J, Yu W, Zhao C, Yu X, Zhao S, et al. Superhydrophobic Pd@CeO2/Al2O3 catalyst coating for

hydrogen mitigation systems in nuclear power plants. International Journal of Hydrogen Energy. 2019;**44**:21746-21758. DOI: 10.1016/j.ijhydene.2019.06.143

[14] Yu W, Yu X, Tu S-T, Tian P. Superhydrophobic Pt–Pd/Al2O3 catalyst coating for hydrogen mitigation system of nuclear power plant. International Journal of Hydrogen Energy. 2017;**42**: 14829-14840. DOI: 10.1016/j.ijhydene.2017.04.102

[15] Łomot D, Karpiński Z. Hydrogen oxidation over alumina-supported palladium–nickel catalysts. Research on Chemical Intermediates. 2015;**41**:9171-9179. DOI: 10.1007/s11164-015-1935-3

[16] Karakaya C, Deutschmann O. Kinetics of hydrogen oxidation on Rh/Al2O3 catalysts studied in a stagnation-flow reactor. Chemical Engineering Science. 2013;**89**:171-184. DOI: 10.1016/j.ces.2012.11.004

[17] Akbari Z, Alizadeh Kiapi MR, My Pham TH, Lombardo L, Fairen-Jimenez D, Zuttel A. Engineering Ru nanoparticle size and metal–support interactions for enhanced catalytic hydrogen combustion. J Mater Chem A. 2025;**13**: 20372-20382. DOI: 10.1039/D5TA03057A

[18] Akbari Z, Mensi MD, Oveisi E, Ruffiner J, Luo W, Koolen CD, et al. Catalytic hydrogen combustion over supported metal catalysts; comparison and kinetic study. International Journal of Hydrogen Energy. 2025;**141**:384-393. DOI: 10.1016/j.ijhydene.2024.08.300

[19] Kim J, Tahmasebi A, Lee JM, Lee S, Jeon C-H, Yu J. Low-temperature catalytic hydrogen combustion over Pd-Cu/Al2O3: Catalyst optimization and rate law determination. Korean Journal of Chemical Engineering. 2023;**40**:1317-1330. DOI: 10.1007/s11814-023-1437-8

[20] Kim J, Tahmasebi A, Khoshk Rish S, Jeon C-H, Yu J. Reactivity of Pd–Cu/Al2O3 and Cu/Al2O3 during catalytic hydrogen combustion – In-situ mechanistic study and rate law determination. Journal of the Energy Institute. 2023;**109**:101297. DOI: 10.1016/j.joei.2023.101297

[21] Shinde VM, Madras G. Kinetic studies of ionic substituted copper catalysts for catalytic hydrogen combustion. Catalysis Today. 2012;**198**: 270-279. DOI: 10.1016/j.cattod.2012.02.019

[22] Zhang X, Sang K, Chen W, Zhang J, Qian G, Chen D, et al. Mechanistic and kinetics insights into Cu size effects on catalytic hydrogen combustion. Chemical Engineering Journal. 2024;**485**: 149875. DOI: 10.1016/j.cej.2024.149875

[23] Zhang X, Fei N, Sang K, Chen W, Qian G, Zhang J, et al. Mechanistic insights into the support effects of Cu catalysts in hydrogen combustion. Cell Reports Physical Science. 2024:5. DOI: 10.1016/j.xcrp.2024.102014

[24] Kozhukhova AE, du Preez SP, Malakhov AA, Bessarabov DG. A thermally conductive Pt/AAO catalyst for hydrogen passive autocatalytic recombination. Catalysts. 2021;**11**:491. DOI: 10.3390/catal11040491

[25] Kozhukhova AE, du Preez SP, Bessarabov D. Preparation of highly active and thermally conductive platinum nanoparticle/Ce–Zr–Y mixed oxide/AAO Washcoat catalyst for catalytic hydrogen combustion technologies. ACS Applied Nano Materials. 2022;**5**:8161-8174. DOI: 10.1021/acsanm.2c01255

[26] Kozhukhova AE, du Preez SP, Bessarabov DG. Preparation of Pt/Ce–Zr–Y mixed oxide/anodized aluminium

oxide catalysts for hydrogen passive autocatalytic recombination. International Journal of Hydrogen Energy. 2022;**47**:12726-12738. DOI: 10.1016/j.ijhydene.2022.01.246

[27] Kozhukhova AE, du Preez SP, Shuro I, Bessarabov DG. Development of a low purity aluminum alloy (Al6082) anodization process and its application as a platinum-based catalyst in catalytic hydrogen combustion. Surface and Coatings Technology. 2020;**404**:126483. DOI: 10.1016/j.surfcoat.2020.126483

[28] Vogt U, Fumey B, Bielmann M, Siong V, Gallandat N, Züttel A. Catalytic hydrogen combustion on porous SiC ceramics. European Fuel Cell Forum. 2017:45-54

[29] Fernández A, Arzac GM, Vogt UF, Hosoglu F, Borgschulte A, Jiménez de Haro MC, et al. Investigation of a Pt containing washcoat on SiC foam for hydrogen combustion applications. Applied Catalysis B: Environmental. 2016;**180**:336-343. DOI: 10.1016/j.apcatb.2015.06.040

[30] Lalik E, Kosydar R, Tokarz-Sobieraj R, Witko M, Szumełda T, Kołodziej M, et al. Humidity induced deactivation of Al2O3 and SiO2 supported Pd, Pt, Pd-Pt catalysts in H2 + O2 recombination reaction: The catalytic, microcalorimetric and DFT studies. Applied Catalysis A: General. 2015;**501**:27-40. DOI: 10.1016/j.apcata.2015.04.029

[31] Lalik E, Drelinkiewicz A, Kosydar R, Tokarz-Sobieraj R, Witko M, Szumełda T, et al. A role of Au-content in performance of Pd-Au/SiO2 and Pd-Au/Al2O3 catalyst in the hydrogen and oxygen recombination reaction. The microcalorimetric and DFT studies. Applied Catalysis A: General. 2016;**517**:196-210. DOI: 10.1016/j.apcata.2016.03.004

[32] Kim GJ, Shin JH, Chang HS. Study on the role of Pt and Pd in Pt–Pd/TiO2 bimetallic catalyst for H2 oxidation at room temperature. International Journal of Hydrogen Energy. 2020;**45**:17276-17286. DOI: 10.1016/j.ijhydene.2020.03.062

[33] Dyrbeck H, Hammer N, Rønning M, Blekkan EA. Catalytic oxidation of hydrogen over Au/TiO2 catalysts. Topics in Catalysis. 2007;**45**:21-24. DOI: 10.1007/s11244-007-0234-7

[34] Deshpande PA, Madras G. Catalytic hydrogen combustion for treatment of combustible gases from fuel cell processors. Applied Catalysis B: Environmental. 2010;**100**:481-490. DOI: 10.1016/j.apcatb.2010.08.026

[35] Singh SA, Vishwanath K, Madras G. Role of hydrogen and oxygen activation over Pt and Pd-doped composites for catalytic hydrogen combustion. ACS Applied Materials & Interfaces. 2017;**9**:19380-19388. DOI: 10.1021/acsami.6b08019

[36] Varun Y, Sreedhar I, Singh SA. Role of Ni, La impregnation and substitution in Co3O4-ZrO2 catalysts for catalytic hydrogen combustion. Journal of Environmental Chemical Engineering. 2022;**10**:108384. DOI: 10.1016/j.jece.2022.108384

[37] du Preez SP, Jones DR, Warwick MEA, Falch A, Sekoai PT, Mota das Neves Quaresma C, et al. Thermally stable Pt/Ti mesh catalyst for catalytic hydrogen combustion. International Journal of Hydrogen Energy. 2020;**45**:16851-16864. DOI: 10.1016/j.ijhydene.2020.04.112

[38] du Preez SP, Jones DR, Bessarabov DG, Falch A, Mota das Neves Quaresma C, Dunnill CW. Development of a Pt/stainless steel mesh catalyst and its

application in catalytic hydrogen combustion. International Journal of Hydrogen Energy. 2019;**44**:27094-27106. DOI: 10.1016/j.ijhydene.2019.08.168

[39] Jo S, Jin J, Kwon S. The preparation of a metal foam support of Pt/Al2O3 for combustion of hydrogen. Catalysis Today. 2010;**155**:45-50. DOI: 10.1016/j.cattod.2009.04.021

[40] Sharma S, Hegde MS. Ti0.99Pd0.01O2−δ: A new Pt-free catalyst for high rates of H2+O2 recombination with high CO tolerant capacity. ChemPhysChem. 2009;**10**:637-640. DOI: 10.1002/cphc.200800768

[41] Deshpande AP, Madras G. Noble metal ionic sites for catalytichydrogen combustion: Spectroscopic insights. Physical Chemistry Chemical Physics. 2011;**13**:708-718. DOI: 10.1039/C0CP01110J

[42] Choi W, Kwon S, Dong SH. Combustion characteristics of hydrogen–air premixed gas in a sub-millimeter scale catalytic combustor. International Journal of Hydrogen Energy. 2008;**33**:2400-2408. DOI: 10.1016/j.ijhydene.2008.02.070

[43] Son IH, Lane AM, Johnson DT. The study of the deactivation of water-pretreated Pt/γ-Al2O3 for low-temperature selective CO oxidation in hydrogen. Journal of Power Sources. 2003;**124**:415-419. DOI: 10.1016/S0378-7753(03)00807-3

[44] Lu H, Cao M, Xie Z, Shu Q, Zhang Q. Highly dispersed Pt/γ-Al2O3/Al structured mesh-type catalysts prepared by competitive adsorption method applied in catalytic hydrogen combustion: Start-up performance and mass transfer enhancement. Catalysis Letters. 2025;**155**:217. DOI: 10.1007/s10562-025-05063-z

[45] Abel N, Bargiela P, Eternot M, Mesbah A, Cadete Santos Aires Francisco J, Gomez C, et al. Upgrading of the old Pt@Al2 O3 using Al(OH)3 as alumina precursors with remarkable catalytic hydrogen combustion performance at room temperature. Energy & Fuels. 2025;**39**:11898-11911. DOI: 10.1021/acs.energyfuels.5c01041

[46] Rudy W, Teodorczyk A, Wen J. Self-ignition of hydrogen–nitrogen mixtures during high-pressure release into air. International Journal of Hydrogen Energy. 2017;**42**:7340-7352. DOI: 10.1016/j.ijhydene.2016.06.051

[47] du Preez SP, Kozhukhova AE, Bessarabov DG. Catalytic hydrogen combustion using platinum supported on anodized aluminium oxide adhered to metallic aluminium. South African Journal of Chemistry. 2022;**76**:31-37. DOI: 10.17159/0379-4350/2022/v76a06

[48] Łomot D, Karpiński Z. Catalytic activity of Pd-Ni in the oxidation of hydrogen for the safety of catalytic activity of Pd-Ni in the oxidation of hydrogen for the safety of nuclear power plan. Polish Journal of Chemical Technology. 2016;**18**(1):15-18. DOI: 10.1515/pjct-2016-0003

[49] Deshpande PA, Polisetti S, Madras G. Analysis of oxide and vanadate supports for catalytic hydrogen combustion: Kinetic and mechanistic investigations. AICHE Journal. 2012;**58**:932-945. DOI: 10.1002/aic.12636

[50] Davis ME, Davis RJ. Fundamentals of Chemical Reaction Engineering. Courier Corporation; 2012

[51] Fogler HS. Essentials of Chemical Reaction Engineering. International ed. Pearson Education International; 2011

[52] Mutschler R, Moioli E, Züttel A. Modelling the CO2 hydrogenation reaction over Co, Ni and Ru/Al2O3. Journal of Catalysis. 2019;**375**:193-201. DOI: 10.1016/j.jcat.2019.05.023

[53] Akbari Z, Lombardo L, Züttel A. The critical role of nanoparticle geometry in turnover frequency calculation. ACS Measurement Science Au. 2025;**5**:1001-1009. DOI: 10.1021/acsmeasuresciau.5c00130

[54] Shinde VM, Madras G. Nanostructured Pd modified Ni/CeO2 catalyst for water gas shift and catalytic hydrogen combustion reaction. Applied Catalysis B: Environmental. 2013;**132–133**:28-38. DOI: 10.1016/j.apcatb.2012.11.021

[55] Sandeep KC, Bhattacharyya R, Warghat C, Bhanja K, Mohan S. Experimental investigation on the kinetics of catalytic recombination of hydrogen with oxygen in air. International Journal of Hydrogen Energy. 2014;**39**:17906-17912. DOI: 10.1016/j.ijhydene.2014.08.148

[56] Mukri BD, Hegde MS. High rates of catalytic hydrogen combustion with air over $Ti_{0.97}Pd_{0.03}O_{2-\delta}$ coated cordierite monolith. Journal of Chemical Sciences. 2017;**129**:1363-1372. DOI: 10.1007/s12039-017-1345-8

[57] Völkening S. Dual-path mechanism for catalytic oxidation of hydrogen on platinum surfaces. Physical Review Letters. 1999;**83**:2672-2675. DOI: 10.1103/PhysRevLett.83.2672

[58] Wey JP, Neely WC, Worley SD. Infrared spectroscopy at high pressure: Interaction of hydrogen and deuterium with rhodium/alumina. The Journal of Physical Chemistry. 1991;**95**:8881-8886. DOI: 10.1021/j100175a085

[59] Zhang J, Hu Z, Man T, Zhao Y, Ding M, Guo Z. A review of catalyst material for hydrogen mitigation systems in nuclear facilities. Nuclear Engineering and Design. 2024;**428**:113481. DOI: 10.1016/j.nucengdes.2024.113481

[60] Zhang W, Zou J, Sun J, Wang W, Shan L, Bao K, et al. Fabrication and sensing performance of carrier-free catalytic combustion hydrogen sensors based on electrodeposition method. International Journal of Hydrogen Energy. 2024;**61**:1356-1364. DOI: 10.1016/j.ijhydene.2024.03.005

[61] Sarioglan A, Can Korkmaz Ö, Kaytaz A, Akar E, Akgün F. A 5 kWt catalytic burner for PEM fuel cells: Effect of fuel type, fuel content and fuel loads on the capacity of the catalytic burner. International Journal of Hydrogen Energy. 2010;**35**:11855-11860. DOI: 10.1016/j.ijhydene.2010.08.068

[62] Jonin M, Haussener S. Design and energy assessment of non-premixed gas burners: Radiative and catalytic hydrogen combustion. Applied Thermal Engineering. 2025;**280**:127674. DOI: 10.1016/j.applthermaleng.2025.127674

[63] Kozhukhova AE, du Preez SP, Bessarabov DG. Catalytic hydrogen combustion for domestic and safety applications: A critical review of catalyst materials and technologies. Energies. 2021;**14**:4897. DOI: 10.3390/en14164897

[64] Fumey B, Buetler T, Vogt UF. Ultra-low NOx emissions from catalytic hydrogen combustion. Applied Energy. 2018;**213**:334-342. DOI: 10.1016/j.apenergy.2018.01.042

[65] Enga BE, Thompson DT. Catalytic combustion applied to gas turbine technology. Platinum Metals Review. 1979;**23**:134-141

[66] Alavandi SK, Etemad S, Baird BD. Low single digit NOx emissions catalytic

combustor for advanced hydrogen turbines for Clean Coal power systems. American Society of Mechanical Engineers Digital Collection. 2013:53-62. DOI: 10.1115/GT2012-68128

[67] Jiaqiang E, Cai L, Li J, Ding J, Chen J, Luo B. Effects analysis on the catalytic combustion and heat transfer performance enhancement of a non-premixed hydrogen/air micro combustor. Fuel. 2022;**309**:122125. DOI: 10.1016/j.fuel.2021.122125

Chapter 2

Hydrogen Production by Ethanol Steam Reforming

William G. Coors

Abstract

Transitioning from an energy economy based on fossil fuels to a sustainable economy based on hydrogen will require technology that is competitive with the refined petroleum fuels that hydrogen will replace. In this monograph ethanol steam reforming (ESR) in microporous membrane reactors is described that offers the necessary process intensification for distributed hydrogen production at the necessary scale. As a result, hydrogen may be produced anywhere in the world from fermentable agricultural feedstocks in much the same way that beer and wine are made. This monograph makes the case for the feasibility of producing distributed hydrogen at the scale that is competitive with petroleum-based fuels, offering a path forward for energy independence from petroleum. Part I investigates the feasibility of large-scale hydrogen production by ESR, including the amount of non-food cropland that would need to be allocated. Part II provides the experimental basis for the ESR technology.

Keywords: hydrogen, fuel ethanol, ethanol steam reforming, kaolin, clay-based microporous membrane, microporous membrane reactors, nickel nanoparticles

1. Introduction

1.1 Feasibility

Fuel ethanol is made by fermentation of carbohydrates. It is produced in large quantities around the world as a high-octane additive to gasoline. 95% ethanol (E95), as a replacement for gasoline for use in internal combustion engines (ICEs), has been gaining in popularity. It is a net-zero-carbon biofuel that is cleaner-burning and releases far less carbon dioxide into the atmosphere than conventional petroleum-based fuels. Ethanol steam reforming takes fuel development to the next level by converting fuel ethanol into hydrogen. As an energy carrier, hydrogen is preferred because it may be turned into electric power at about twice the conversion efficiency of conventional heat engines. Furthermore, the only exhaust product at the point where the hydrogen is consumed is water vapor.

Most hydrogen is produced today by steam reforming of natural gas, which releases more than 6 tons of carbon dioxide into the atmosphere for every ton of hydrogen generated. Without carbon capture and sequestration, such a method for producing hydrogen is considered unsustainable in the long run. The same technology used for methane steam reforming applies to ethanol steam reforming–the

 IntechOpen

fundamental difference being that ethanol is derived from biomass, and is thus, a net-zero-carbon energy vector. That is, the amount of CO_2 generated is equal to the amount of CO_2 extracted from the atmosphere by the biosphere. Steam reforming of biogas is also net-zero. Hydrogen can also be produced by electrolysis of water, which is a zero-carbon technology if the electric power used is zero-carbon. The feasibility of hydrogen at scale rests on the question of sustainability and the cost versus the value of sustainability. Ethanol steam reforming offers a pathway forward that is both sustainable and cost-effective. The question of hydrogen at scale from sugar beets was addressed in Coors, [1].

As a replacement for fuels derived from crude oil, the feasibility of hydrogen derived from ethanol ultimately hinges on the amount of agricultural cropland required and the infrastructure needed for sustainable distributed production of fermentable crops. The following summary provides the conversion metrics that are necessary for determining the feasibility of producing hydrogen from ethanol at the scale sufficient to impact world consumption of petroleum-based fuels. The questions of how much hydrogen, how much ethanol, and how much fermentable sugar are required is addressed. The conversion factors are based on the common metric of one kilogram of hydrogen. The conversions reflect theoretic limits and do not account for production inefficiencies. However, the values are useful for rough scalability calculations.

- How much hydrogen: 1 kg hydrogen = 3.7 liters of gasoline equivalent (lg_e).
- How much ethanol: 1 kg H_2 requires 5.9 liters of ethanol.
- How much sugar: 1 kg H_2 requires 8.6 kg of sugar.
- How much cultivated land: 1 kg H_2 requires 0.01 ha (934 kg H_2/ha)

1.2 Scale factor details

1.2.1 Energy equivalence

When considering energy equivalence for fuels, it is customary to use specific energy values. That is, the combustion enthalpy per mole or kilogram of fuel under standard conditions of temperature and pressure. These values are tabulated in thermochemical tables such as the Engineering Toolbox [2]. If the water in the exhaust is treated as liquid, the lower heating value (LHV) is used. The advantage of using specific combustion enthalpies is that the conversion efficiency of any particular engine is left out of the calculation. Conversion efficiency is a separate issue best left for consideration in the final detailed analysis. Since hydrogen is a gas at normal temperature and pressure, it is customary to use specific rather than volumetric equivalents. All the values cited in this monograph are based on the specific heat of combustion for hydrogen (LHV) of 240 kilojoules per mole (120 MJ/kg). Gasoline has a heat of combustion (LHV) of 43.4 MJ/kg (32 MJ/L gasoline). An equivalent amount of hydrogen, on a heat of combustion basis, is 32 MJ/L ÷ 120 MJ/kg H_2 = 0.27 kg H_2/liter gasoline.

Ethanol has a combustion enthalpy of 21.1 MJ/L—only 2/3 that of gasoline on a volumetric basis. Even though bioethanol is a renewable, clean-burning fuel, it constitutes a relatively poor substitute for gasoline in an internal combustion engine. On the other hand, hydrogen, based on specific weight, has much more combustion enthalpy than almost any other liquid fuel.

There are several ways to produce hydrogen at scale. Hydrogen is produced globally in million ton per day quantities by methane steam reforming of natural gas. Hydrogen produced by this process is called "gray" hydrogen because about 6 kg of CO_2 are exhausted into the atmosphere for every kg of hydrogen produced. At the other end of the carbon footprint spectrum is "green" hydrogen produced by electrolysis of water, which generates no CO_2 as long as the electricity consumed comes from non-CO_2 producing sources. A third option is to make hydrogen from biomass. This method is carbon neutral with respect to atmospheric CO_2 emissions. The CO_2 generated by this process is offset by CO_2 taken in from plants by photosynthesis. We need look no farther than our own green planet, which is green because of photosynthesis. The biosphere converts vast quantities of carbon dioxide and water into cellulose and lignin, and in some cases into fermentable fruits and vegetables like corn, rice, sugar cane and sugar beets. Much of the work has already been done by nature, and the energy required to make ethanol by fermentation is small as anyone who has made beer or hard cider at home knows. Making ethanol from fermentable feedstocks is quite easy. Fuel ethanol production, rather than ethanol for human consumption, is being practiced globally at large scale today, with about 80 million tons produced in 2021 [3] and approaching 100 million tons in 2024. Scalability of fuel ethanol is being demonstrated in many locations around the world. The technology for making fuel ethanol at large scale is quite advanced [4], but commercialization has been hampered by concerns over competition with food crops. It is imperative that cultivation of crops for fuel ethanol be carried out on land that is not suitable for food crops or for crop-rotation on marginal farmland and does not compete with food crops for water. Hydrogen from fuel ethanol is not "green" in the strict sense of zero CO_2, but it is carbon-neutral or nearly carbon-neutral. Nevertheless, the potential for dramatic reduction in greenhouse gas emissions is considerable. More importantly, dependence on imported petroleum can be reduced in countries without indigenous petroleum resources.

1.2.2 Hydrogen by ethanol steam reforming

Ethanol, as a gasoline fuel additive, is traditionally produced as pure, anhydrous alcohol by the costly and energy intensive process of distillation from fermentate that has an alcohol content of about 10%. The ideal feedstock for ESR, on the other hand is less than 40% alcohol by volume (ABV). The compactness and thermal efficiency of microporous membrane reactors means that ethanol may be steam reformed directly from fermentate with little or no distillation required. ESR requires only a fraction of the heat required for making anhydrous fuel ethanol by distillation, since the process is suited for feedstock with less than 40% ABV (e.g. vodka).

Ethanol steam reforming, ESR, requires breaking the C-C double bond, which makes the process more difficult than methane steam reforming, where there are only C-H bonds. ESR is carried out over nickel catalysts at temperatures around 800°C in various devices such as microporous membrane reactors. Reaction products typically consist of only 72% hydrogen. The other gases must be removed to yield hydrogen with the required purity.

Steam reforming is endothermic, requiring supplemental heat to sustain the reaction at a fixed temperature. The heat required for endothermic steam reforming may be obtained from some portion of the hydrogen generated—typically less than 20%.

Once the hydrogen is separated from the other reformate species, it may be burned in a zero-emission internal combustion engine or fuel cell to produce electric power.

Ethanol steam reforming, ESR, produces hydrogen and carbon dioxide by the ideal reaction,

$$C_2H_6O + 3H_2O = 2CO_2 + 6H_2 \tag{1}$$

with 75% H_2 and 25% CO_2. The amount of stoichiometric oxygen in the fuel determines how much additional oxygen is required in the form of steam. In practice, however, besides just H_2 and CO_2, the reaction generates CO, H_2O, and some low molecular weight gases such as methane. Furthermore, ESR is an endothermic reaction typically carried out at about 800°C. Of the 6 moles of hydrogen produced per mole of EtOH, approximately 1 mole of hydrogen is used to provide the heat. The net combustion enthalpy of 5 moles of hydrogen is 5 x 240 kJ/mol = 1.20 MJ/mol of EtOH. Ethanol has a combustion enthalpy of 1.34 MJ/mol [2], so on a molar basis, the difference of 137 kJ/mol of EtOH is lost as heat by ESR. The molar volume of EtOH is 17.05 mol/L, times 5 mol H_2/L EtOH, giving 85.3 moles, or 171 g of H_2 per liter of EtOH. Therefore, 5.87 liters of ethanol are required to make 1 kg of hydrogen.

1.2.3 Hydrogen purification from reformate

The reformate gas from ESR is nominally 72% H_2, which must be purified to greater than 99% if the gas is to be combusted in an internal combustion engine, or 99.999% if the gas is to be used in a fuel cell. Various methods for purifying hydrogen are practiced industrially, but these are energy- and capital-intensive, and difficult to carry out at small scale. However, palladium separation membranes and cryogenic separation offer the potential to dramatically increase the process intensification over a wide range of scalability.

1.2.4 Fermentable sugar

Sucrose from sugar beets and sugar cane as well as disaccharides from corn syrup are produced in enormous quantities around the world today, almost entirely for sweeteners. Thus, disaccharides are part of the food chain. Production of fuel ethanol from sugar fermentation must necessarily be in addition to sugar produced for food. As a source for hydrogen, the fermentation reaction is preceded by first converting sucrose into two glucose molecules by hydrolysis:

$$C_{12}H_{22}O_{11} + H_2O = 2C_6H_{12}O_6 \tag{2}$$

Two moles of glucose are then fermented to make four moles of ethanol and four moles of carbon dioxide by the reaction:

$$2C_6H_{12}O_6 = 4C_2H_6O + 4CO_2 \tag{3}$$

The molar mass of sucrose is 342.3 g/mol or 2.92 mol/kg, from which 4 moles of ethanol are produced. 11.7 moles of ethanol/kg sucrose equate to 0.68 L EtOH/kg sucrose.

By steam reforming (Eq. (1)), four moles of ethanol are converted to twenty-four moles of hydrogen. However, assuming that only five of the six moles of hydrogen are recoverable, only twenty moles of hydrogen may be generated per mole of sucrose. The weight of 20 moles of hydrogen is 40 g, so 0.117 kg of hydrogen may be produced per kg of sucrose (assuming complete conversion during fermentation).

World granulated sugar production from sugar beets and sugar cane in 2023 was 177.5 million metric tons [5]. If all this sugar were used to make hydrogen at 117 kg/ton, 20.7 million tons of hydrogen could be produced.

1.2.5 Farm acreage

According to the US Department of Agriculture (USDA) [6], the average production of refined sugar from sugar beets in 2005 was 3.5 tons per acre, or 8.6 metric tons of sugar per hectare (2.47 acres/ha). The land requirement for raising sugar beets for ethanol is substantial, particularly considering that the acreage needs to be brought into production in addition to current acreage in food production. The amount of sugar produced in the world today is enormous [6], but far below what would be required to displace petroleum as the dominant energy vector. The best croplands for raising sugar beets and sugar cane are already in sugar production for the commodity market price of about $440/ton. Any increased production will require growing crops on less productive agricultural lands or lands not suitable for food production. The American Crystal Sugar Company, the largest producer in the United States, reported 1.62 million tons of sugar from 398,000 acres along the Red River Valley of North Dakota and Minnesota, or about 4 tons/acre [7]. At 117 kg H_2 per kg sugar, this would yield 468 kg H_2/acre (1160 kg/ha).

The Spreckels Sugar Company in Brawley, California–one of the most productive sugar beet-growing regions in the United States–produced 7 tons of refined sugar per acre from 28,000 acres in the Imperial Valley [8]. Unfortunately, this plant is scheduled to close in 2025 due to declining demand for refined sugar, leaving many people out of work and forcing local farmers to find other crops to grow. The harvested beets could produce about 21 million kg of hydrogen, generating about $85 million in revenues.

Sugar beets are hardy and grow almost anywhere, albeit at reduced yields. Presumably, increased sugar production on marginal agricultural lands would not be as productive. The eastern half of the state of Colorado has about 10 million acres of such marginal cropland. At one time, Colorado was a major producer of sugar beets because beets are drought and freeze tolerant, making the arid eastern planes of the Rocky Mountains a good place to grow sugar beets. A 125-acre field irrigated by a center-pivot sprinkler could produce about 500 tons of sugar from which 50 tons of hydrogen, worth $200,000, could be harvested. Tens of thousands of such systems would be required, but 10 million acres of sugar beets could yield about 4 million tons of hydrogen, worth $16 billion annually from land with practically no other agricultural value. Once energy parity is achieved between petroleum-based fossil fuels and renewable H_2 produced by ESR from fermented sugar, world sugar production should increase rapidly beyond what is required as sweeteners. The $4/kg threshold for energy/cost parity is within reach. The prospects for making ethanol from lignocellulosic feedstocks, including solid municipal waste, are also encouraging [9].

1.3 Agricultural considerations

Most of the fuel ethanol produced in the world today comes from corn and sugar cane. Both require considerable amounts of water and are not suitable crops for marginal farmland. Sugar beets, on the other hand, produce sucrose in a tuber that grows below ground. Sugar beets thrive in almost any climate and are increasingly being considered as a feedstock for fuel ethanol. The plants have compelling properties that make them ideal as a source of fermentable sugar for fuel ethanol. They are drought resistant, requiring less water, and have a long taproot that can access water deeper than about one meter. Also, the plant tolerates brackish water at levels above what most food crops can tolerate.

But most sugar beets are processed today for making refined sugar–a food crop. The world production of sugar beets is determined by the market demand for table sugar, food sweeteners, and some alcoholic beverages. Therefore, little attention has been paid until recently to their potential as a source of fermentable sugar for fuel ethanol. For this, additional acreage is required that does not compete with acreage allotted to food crops. It is one thing to contemplate the necessary expansion of sugar production, but it is essential that the cost of H_2 energy produced match the cost of energy per MJ delivered by petroleum fuels. Everything depends on the value of hydrogen produced. The price of diesel and gasoline at the fuel pump, although wildly variable, is a common gage of energy value worldwide. If a benchmark of $4/gal is used as the threshold for energy parity, $4 per kg hydrogen is the goal. It is almost impossible to achieve this threshold with H_2 produced by electrolysis because it would require enormous quantities of electric power at $0.05 per kWh. $4 H_2 is routinely made by steam reforming natural gas. This is how more than 90% of hydrogen is produced today. But how about hydrogen by ethanol steam reforming?

At a net yield of 100 kg H_2 per ton of sugar, $4 hydrogen would mean that sugar would need to be purchased for less than about $300/metric ton (assuming $1/kg for processing). Like petroleum, the commodity price for sugar varies wildly, but has averaged about $440/metric ton over the past five years. At this price, H_2 from ESR would cost about $5/kg. $5/kg would be the cost of hydrogen derived from the sugar at the commodity price established for the food crop, the size of which, is determined by the demand for sweetener. What would happen if a huge new market for sugar as an energy crop were to emerge? Any increase in sugar beet, sugar cane, or corn production beyond the demand for sweetener would result in over-production and lower market prices, but lower sugar prices would not necessarily result in increased demand for sweetener. The downward pricing pressure on fermentable sugars intended for hydrogen production would not impact the amount of sugar or prices of premium sugar produced for sweeteners. Sugar produced beyond this demand used for H_2 production would benefit both the food sector and the emerging energy sector with new opportunities for agricultural production–that is, as long as false economies from government interference in the marketplace were not to make it more attractive to grow sugar for fuel at the expense of food, as has been done in the production of anhydrous corn ethanol. Government subsidies and carbon credits create false economies which are harmful for the consumer in the end. Furthermore, other non-food sources of fermentable sugar, such as cellulose, would proliferate as the demand for hydrogen increases.

World cropland today exceeds 1 billion hectares. The best farmland is optimized for food production. The addition of several million hectares for fuel ethanol production presents a formidable challenge, but by no means is it impossible. Developing the technology and infrastructure for fermentable sugar harvesting will require many years, but given sufficient economic incentives, fuel ethanol offers one of the few viable paths forward at the necessary scale to energy independence from crude oil.

1.4 Overview of hydrogen production technologies

Hydrogen is the logical energy vector for the future, leading the way to the long anticipated "hydrogen economy". Hydrogen can be produced in many ways. What follows is a summary of the pros and cons of the various production methods.

1.4.1 The case for PV hydrogen by water electrolysis

1. Photo-voltaic (PV) power is well-suited for making hydrogen by water electrolysis when operated at low voltage d.c. and prior to the a.c. inverter.

2. The potential is for 100 g/day/m^2 (36.5 kg/yr), so only a 20 m^2 solar array is required for 2 kg H_2/day–sufficient for a single-family unit. However, solar energy is intermittent with diurnal radiance and cloud cover, so H_2 generation is not consistent, requiring some compression and storage during peak production hours.

3. H_2 produced by PV/electrolysis is not suitable for distribution without compression and a delivery infrastructure—high-pressure pipelines, high-pressure storage tanks, delivery trucks, and distribution network.

4. The best scenarios for PV hydrogen are:

    a. Use of H_2 directly on-site for local loads–heating, AC, operating farm equipment, etc. Intermittency may be managed with low pressure storage < 100 atm.

    b. Use for arbitrage: off-peak, low-cost grid power combined with PV H_2 may be used for generating inexpensive H_2 by electrolysis and sold back onto the utility grid at higher peak rates. This scenario requires considerable on-site storage capacity.

5. PV power is better suited for large solar farms (>100 MW) that are connected to the utility grid since peak solar power typically overlaps times of peak power demand.

6. The case for H_2 generated by water electrolysis with large solar arrays is dubious because of the complication (and cost) of large-scale compression and distribution.

7. The case for making H_2 by water electrolysis using costly utility grip electricity is difficult to justify as a substitute for gasoline or diesel fuel.

8. Without suitable short-term storage H_2 generated by water electrolysis has little value in urban settings since peak generation overlaps with peak power demand. Using PV power to make H_2 intermittently only to turn it back into electric power at a later time in fuel cells or engines makes no sense since round-trip conversion efficiency necessarily means that the resulting power will be more costly.

9. Only when off-peak utility rates are lower does water electrolysis make sense. But this scenario requires compression and storage.

10. Low-cost nuclear power, which is not intermittent, could dramatically alter the economics of hydrogen produced by water electrolysis.

1.4.2 The case for fuel ethanol

1. Denatured Fuel Ethanol, E95, is a renewable, net-zero-carbon biofuel derived from fermentation of sugar or starch. Typically 95% alcohol by volume, it is an agricultural product produced from crops such as corn, sugar beets, sugar cane, rice, potatoes, etc.

2. Fuel ethanol is distinct from alcoholic beverages in that denaturant (typically methanol) is added to make it undrinkable.

3. E95 is made by distillation of fermented feedstock—usually sugars.

4. Although flammable, it is environmentally benign. It may be stored and transported in bulk until needed.

5. Even though substantial acreages of farmland are required to grow crops suitable for alcohol fermentation, farm economics are designed for seasonal harvest and farmers are experienced in handling this.

6. E95 from corn is well-established in the US, from sugar cane in Brazil, and sugar beets in Northern Europe. Crops intended for producing fuel alcohol can be optimized by methods, such as with genetic modifications that are not suitable for human consumption.

7. Large expanses of marginal and uncultivated land exist almost everywhere. Sugar beets, for example, are drought-resistant and tolerant of brackish water, making them ideal for producing fermentable sugar on lands that do not compete for resources needed for food production.

8. Fuel ethanol may be produced on-site at the farm or at a local farmer's cooperative in the same manner as with many agricultural crops.

1.4.3 The case for H_2 by ethanol steam reforming (ESR)

1. Ethanol steam reforming requires a mixture of water and ethanol not exceeding 40% ABV, or E40, requiring less distillation of fermentate than E95, but may be made from E95 by dilution with water. At only 40% ABV, E40 is non-flammable and requires less heat energy to produce.

2. Hydrogen produced on-site by ESR may be used directly in PEM fuel cells in farm equipment and to provide off-grid electric power or used in internal combustion engines and gas turbines.

3. A sugar beet farm of about 100 ha producing 30,000 kg H_2/year can operate completely off-grid without the need for gasoline or diesel fuel.

4. Larger farming operations can compress and bottle hydrogen for distribution or sell surplus E95 for cash.

5. Hydrogen produced from E40 is clean-burning and non-polluting with respect to carbon particulates and oxides of sulfur and nitrogen with operation in PEM fuel cells.

6. The combustion enthalpy of E95 in a heat engine is about 26 MJ/kg x 46,070 kg/mol = 1.2 MJ/mol. Assuming 5 moles (net) per mole of EtOH, the hydrogen combustion enthalpy is 5 x 0.24 MJ/mol (LHV) = 1.2 MJ/mol EtOH—about the same as for burning E95 in an internal combustion engine. However, energy conversion efficiency in a fuel cell is about double that of an ICE. On balance, hydrogen is a better energy vector than E95.

2. Experimental

2.1 Background

Ethanol steam reforming offers the best solution for on-demand hydrogen production, requiring little or no compression and storage. Compact microporous membrane reactors (MMRs) bring the necessary process intensification for producing hydrogen at the scale from a few grams to tons per day. Microporous membranes are made from bisque-fired kaolin-based clay having a microstructure with high open porosity and large internal surface area, enabling high gas permeabilities. Over the past decade, interest in microporous ceramic membranes using earth-abundant clay minerals has been increasing [10]. When bisque fired, the ceramic forms a network of interconnected microchannels that are suited for three application areas of interest to hydrogen production [11–13]: (1) as a catalyst support for microporous membrane reactors, MMRs, (2) gas separation—particularly for separating H_2 from CO_2 [14], (3) as porous supports for depositing palladium membranes for hydrogen diffusion [15]. The focus of what follows specifically pertains to catalytic MMRs for ethanol steam reforming (ESR).

Recent experiments with Ni-infiltrated bisque-fired kaolin have demonstrated that hydrogen can be produced from ethanol by ESR at rates that are of high commercial interest. Processing, microstructure evolution, and mechanical integrity of kaolin-based ceramics in the context of ESR and gas separation, where interfaces and surfaces are the key attributes are described.

Water and ethanol are miscible with nearly the same boiling point, making the use of a water/ethanol mixture as a feedstock for ESR convenient and practical in reactors, where the mixture is introduced and vaporized. The steam-to-carbon ratio may be adjusted by the percent alcohol by volume %ABV. **Figure 1** plots this relationship. The safe operating range for avoiding carbon formation, called “coking”, is S/C greater than 2, or about 40% ABV, which is the typical alcohol content of vodka. The S/C ratio as a function of the % ABV is plotted in **Figure 1**. 40% ABV has an intrinsic S/C ratio of 2.4. The extra 0.9 moles of water per mole of EtOH beyond what is required by stoichiometry (Eq. (1)) does not participate in the reaction but is required to prevent coke formation.

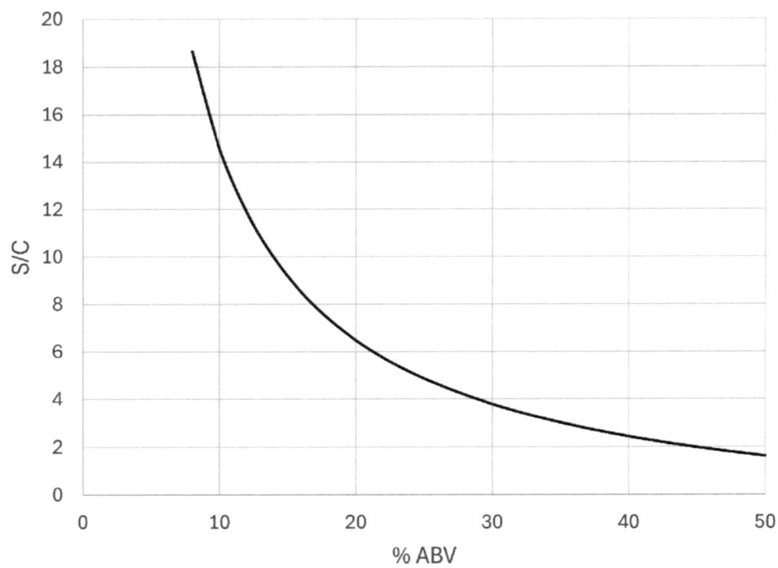

Figure 1.
Steam-to-carbon ratio (S/C) versus percent ethanol by volume (%ABV).

The reaction in a catalytic microporous membrane reactor, MMR, results in a gas mixture of about 72% hydrogen and other reaction products, CO, CO_2, CH_4 and traces of other gases due the thermodynamic limitation of the reaction. **Figure 2** shows the mole fractions of the principal gases that would exist at thermodynamic equilibrium versus temperature at one atmosphere pressure for a 40% alcohol-by-volume (ABV) reactant. It is observed that hydrogen production reaches the maximum at about 600° C, at the point where CH_4 approaches zero. This is a critical operating threshold for a membrane reactor intended for making hydrogen. The mole fractions of actual reaction products are bounded by these thermodynamic limitations. The effectiveness of the catalysts of a MMR is measured as the ratio of the hydrogen production to the upper bound imposed by thermodynamic limitation at the MMR operating temperature, which is the solid black curve in the figure. 800°C is a more typical operating temperature since the ESR reaction is endothermic and a higher temperature is warranted to keep the reactor temperature constant. The equilibrium mole fractions at 800°C and one atmosphere are H_2 = 0.707, CO = 0.169, CO_2 = 0.123, and CH_4 = 0. The sum of the carbon-containing species is 0.292, which makes the total 1.0.

Quantitative analysis of the reaction products of steam/ethanol reforming with a quadrupole mass spectrometer provides the actual mole fractions, which may be compared to the theoretical equilibrium values. Carbon monoxide may be converted to CO_2 and additional hydrogen at a lower temperature downstream by the water-gas-shift reaction to give an additional mole of hydrogen per mole of ethanol,

$$CO + H_2O = CO_2 + H_2 \tag{4}$$

The maximum theoretical hydrogen yield is given by Eq. (1) at 75%.

Measurement of the mole fractions of gases resulting from ESR is an indicator of the performance of the MMR. Hydrogen is the desired product, so the hydrogen yield compared to the theoretical, stoichiometric yield gives a measure of the efficacy of the MMR. The hydrogen yield is the mole fraction of hydrogen produced to the

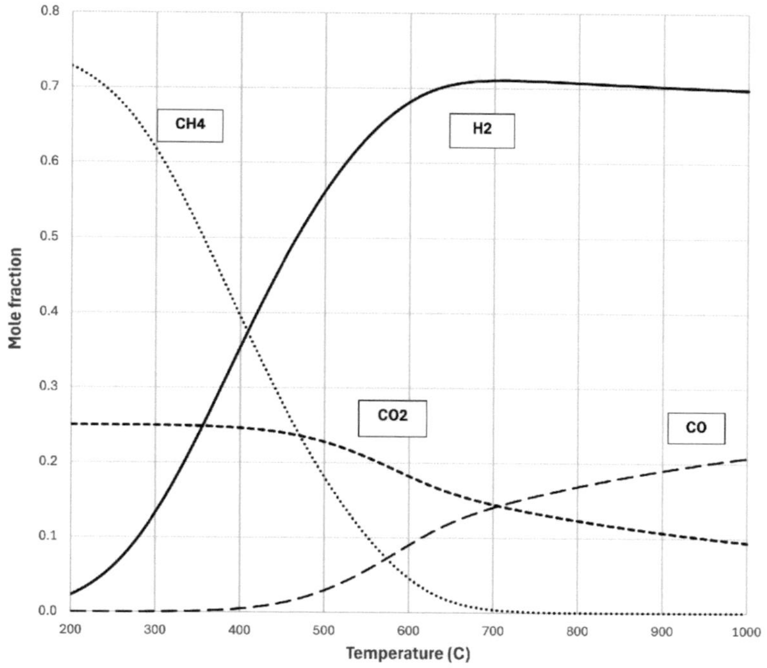

Figure 2.
Mole fractions of major steam reforming products (less H_2O) for 40% ABV at thermodynamic equilibrium vs. temperature at one atmosphere.

theoretical maximum of 75% in Eq. (1). The production of methane is mostly responsible for reduced hydrogen yield and is a key indicator of the reforming efficiency. The extent of reaction is controlled by the surface area of nickel catalyst that can be incorporated into the MMR along the gas flow path.

In a clay-based MMR the open porosity is about 1/3 of the total bulk volume. The connected pore size and distribution are critical determinants of the gas permeability. The membrane wall thickness as well as the path tortuosity through which the reacting gases permeate determines the length of time the transiting gases have for encountering a particle of nickel catalyst to undergo a heterogeneous reaction on the catalyst surface. The size and distribution of nickel particles within the flow channels of the porous clay-based support determines the catalyst surface area, and thus, the catalytic activity of the MMR.

2.2 Experimental

2.2.1 Sample preparation

From the evaluation of various candidate kaolin clays described in prior monograph [12] Edgar Plastic Kaolin (Edgar Minerals) was used in this study. Slip for casting was prepared in a *Ninja* smoothie-maker using 300 grams of EPK kaolin, 150 grams water, and 5 ml Darvan 7 deflocculant. The MMR bulbs tested in this study were fabricated from this formulation, then fired in air at 1000°C for one hour. The resulting ceramic exhibited open porosity of about 37 vol%. The pore size and distribution were determined by Hg intrusion porosimetry shown in **Figure 3**. It may be observed that the open pore size exhibits a narrow distribution below 100 nm with a

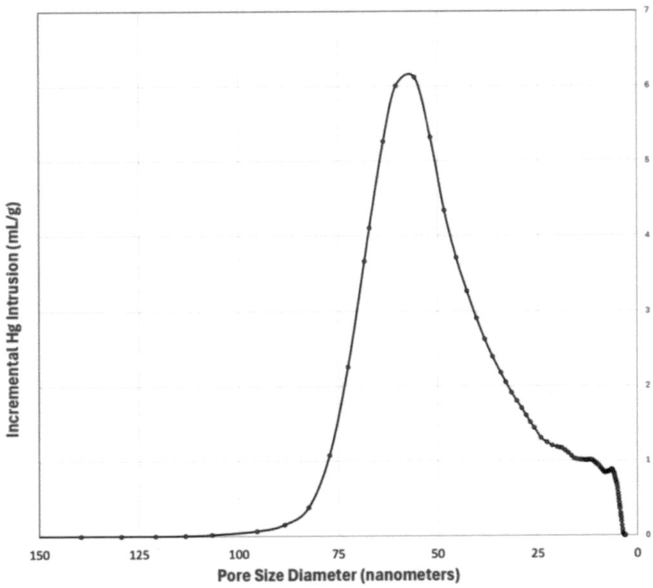

Figure 3.
Pore size and distribution of Edgar Plastic Kaolin fired for 1 hour at 1000°C.

Total intrusion volume at 59,943.07 psia:	0.2260 mL/g
Total pore area at 59,943.07 psia:	22.421 m^2/g
Median pore diameter (volume) at 3290.34 psia and 0.113 mL/g:	0.05497 μm
Median pore diameter (area) at 4246.83 psia and 11.211 m^2/g:	0.04259 μm
Average pore diameter (4 V/A):	0.04031 μm
Bulk density at 0.54 psia:	1.6549 g/mL
Apparent (skeletal) density at 59,943.07 psia:	2.6434 g/mL
Porosity:	37.3957%

Table 1.
Hg intrusion porosity summary data (Micromeretics PTA).

maximum at 58 nm. The internal surface area was determined to be 22.4 m^2/g, which is exceptional for a microporous material. The summary report (Micromeretics PTA) is given in **Table 1**.

An MMR bulb is shown in **Figure 4** (in cross-section on the left). The dark color is due to the nickel nanoparticles embedded in the ceramic. The outer bulb diameter is 1.8 cm, and the length is 7 cm, giving a total surface area of about 50 cm^2 on the shell side. The wall thickness of the hollow bulb is 1.5 mm. The bulb has a 10 mm diameter stem port at the top for making the gas-tight connection to the tube side using a graphite ferrule.

For nickel infusion, the dry weight of the bulb was recorded. Infusion of nickel catalyst was carried out using three separate infusions of the porous structure with 1.5 M nickel nitrate hexahydrate solution (Sigma Aldrich $NiNO_3.6H_2O$, SN#244074) with intermediate calcination. For each infusion step, $NiNO_3$ solution was inserted into the interior of the bulb using a syringe. The ceramic darkened on the exterior

Figure 4.
Edgar Plastic Kaolin (EPK) hollow bulb slip cast in a 2-piece plaster mold. In cross-section on the left. The dark color is due to nickel nanoparticles.

surface after about 2 minutes once the infusant penetrated the wall, indicating complete infusion. The infused bulb was then calcined for 1 hour in air at 500°C to decompose the nitrate after which the bulb was weighed to determine the net weight gain. This process resulted in about 20 mg of Ni per gram of substrate. Three subsequent infusions resulted in Ni catalyst loading of about 60 mg/g. The final step involved nucleating the metallic Ni nanoparticles by operating the bulb at 800°C on hydrogen flowing at 20 sccm for 24 hours. The 20 nanometer nickel particles may be seen decorating the plates of a kaolinite grain in **Figure 5**.

2.2.2 Testing apparatus

The ESR test rig is shown in **Figure 6** consisting of a high temperature horizontal tube furnace on the left and an HPLC liquid chromatography pump (Laboratory Alliance) for injecting the water/ethanol mixture into the test chamber on the right. The specimens under test were installed into a 1″ diameter Inconel pressure vessel using a 10 mm compression fitting and graphite ferrule. The ends of the pressure vessel, sealed standard with 1″ compression fittings, may be seen extending out of the tube furnace on each side. Steam reforming was carried out at a furnace set point of 800°C. A second type-K thermocouple was inserted into the interiosr of the bulb from the left through gas tight fittings. The steam reforming reaction is endothermic, so this second thermocouple typically registered a temperature 10 degrees lower than the

Figure 5
Kaolinite plates decorated with Ni nanoparticles in EPK microporous membrane (FESEM courtesy of Sandrine Ricote, Colorado School of Mines).

Figure 6.
ESR test apparatus. From left to right: (1) specimen thermocouple reader, (2) tube side outlet with steam trap and condensing coil, (3) MTI tube furnace, (4) shell side inlet (right), pressure gages and safety pressure release valves, (5) HPLC, (6) mass flow controllers and control valves, (7) water/ethanol reservoir.

furnace set point. This provided a qualitative indication that steam reforming was taking place.

The liquid flow, in ml/min, was set on the HPLC pump to inject a mixture of water and ethanol—typically 40%ABV—into the test chamber. The injected liquid was gasified in a coil inserted near the furnace hot zone from the right. A pressure gauge monitored the resulting pressure rise inside the reactor shell due to gasification. A pressure release valve was installed for safe operation below 20 barg. The reformate gas on the tube side of the bulb specimen exited to the outside from the left of the furnace, where it passed through a steam trap and on to a mass spectrometer for analysis.

2.2.3 Testing protocol

The testing protocol involved a preliminary pressure test with hydrogen at room temperature up to 10 barg to confirm that the seals were gas tight. This was confirmed by spraying soap solution on the connection and watching for bubbles to form. This was followed by heating the tube furnace at 10°C/min to the operating temperature of 800°C with 20 sccm hydrogen flowing through the system for a period of about 24 hours to fully reduce the nickel catalyst. The outlet gas was passed through a water trap and vented to the outside. The permeance of the specimen was measured with pure hydrogen at the operating temperature.

The atmosphere was then switched to argon and the permeance of the bulb specimen tested a second time to obtain the separation factor between Ar and H_2 at the operating temperature. With flowing Ar at 100 sccm, the outlet was fed to a mass spectrometer (MKS Cirrus 3XD with Process Eye Professional composition analysis software), calibrated beforehand with synthetic syngas, (75% H_2, 20% CO_2, 5% CO) through a mass flow controller (Alicat 500) until all the hydrogen was purged from the lines and argon was the only residual species detected. At this point, the HPLC was turned on and lines purged until the first sign of a pressure increase in the test pressure vessel—an indication that gasification was taking place. The reformate gas flowed to the mass spectrometer where the hydrogen increased as the residual argon in the supply lines was diminishing. The test was commenced once the argon signal became negligible. After an interval of approximately one hour, the catalyst was considered fully activated as the reformer reached steady state, where it operated continuously for about six hours.

3. Results and discussion

3.1 Bulb material characterization

A preliminary test was carried out on a bare bulb without any nickel catalyst. At first, some reformate species were detected, presumably resulting from homogeneous water-ethanol reactions or heterogeneous reactions with the inner walls of the Inconel vessel. But after a short time, the pores of the MMR became plugged with water and ethanol vapors and gas stopped flowing from the shell side to the tube side—confirming the need for catalyst in the flow path within the tube wall.

For the second test, a tube, infused with Ni nanoparticles by the nickel nitrate process with 60 mg/g nickel catalyst was tested. The room temperature permeance of the bulb was measured in the ESR apparatus. The permeance results for hydrogen and argon are plotted in **Figure** 7. The technique is a standard method for estimating pore size and distribution and is described in more detail in refs. [12,13]. Permeation flux (in sccm/cm^2) as measured using an Alicat 500 mass flow controller is plotted versus applied pressure in bar for pure H_2 and Ar. The curves have been adjusted to pass through the origin with zero flux at atmospheric pressure. The slopes reflect the permeance of the different gases. The units are sccm/cm^2/bar. Often, a unit for

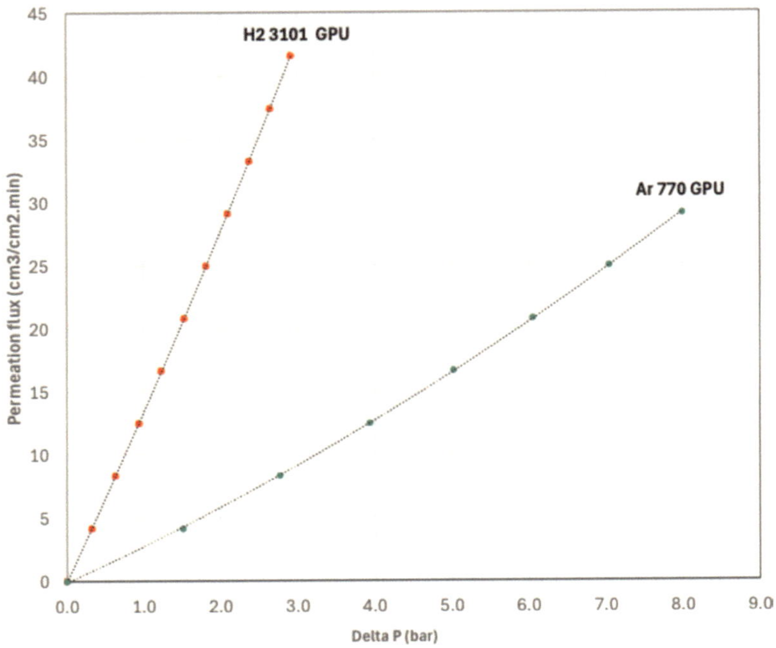

Figure 7.
Hydrogen and argon permeation at room temperature in slip cast bulb.

permeability called gas permeance unit, GPU, is used, where one GPU = cm^3(STP)/(cm^2.s.cmHg) x 10^{-6} which is proportional to sccm/cm^2/bar. The pressure-drop in bar rather than 76.0 cm Hg is used for convenience. Absolute permeance is more useful for comparing different materials, however, it is the separation factor that is more relevant for these present studies. When the same specimen is tested against two different gases, at the same gas flow rate, all that remains in the calculation is the inverse ratio of the differential pressures required to maintain the same flux. This ratio determines the gas separation factor and indicates the fraction of the permeance that occurs by Knudsen diffusion versus by ordinary plug flow. Knudsen diffusion is determined by the square root of the mass of the diffusing gas and becomes dominant for pores on the order of 50 nm and below. The theoretical separation factor for two different gases is the inverse square root of the ratio of their respective masses. For hydrogen and argon, the value is $\sqrt{(40/2)} = 4.47$.

From the plot, it may be observed that the room temperature permeance of hydrogen was 3101 GPU and that of argon was 770 GPU. The difference in the slopes confirms that the diffusion mechanism is significantly by Knudsen diffusion. The separation factor between the two gases is 4.03, which is 90% of the theoretical value based on Knudsen diffusion and is consistent with the prediction of predominantly Knudsen diffusion through the 50 nm pores measured by Hg intrusion porosimetry.

3.2 ESR performance testing

40% ABV Liquid feed was injected into the reaction vessel under pressure from the HPLC pump and gasified to a pressure of between 10 and 20 barg. At 800°C ethanol decomposes, into a mixture of H_2, CO, CO_2, CH_4, and traces of other species, such as ethane and ethylene. This gas constitutes the shell gas of the ESR reactor, passing through the walls of the membrane rector bulb, and exiting the reactor on the tube side in a tube-and-shell reactor configuration. The ESR reaction occurs as the shell gas encounters nickel catalyst embedded in the porous clay structure. **Figure 8** shows the mass spectrometer analysis of the reformate tube gas, switched to sampling shell gas after eight minutes, and switched back to reformate tube gas after 17 minutes. The effect of the MMR is demonstrated most compellingly by the increase in hydrogen mole fraction from 45% on the shell side to 72% on the tube side, meaning that the ESR reaction reached completion upon passing through the membrane. ESR is typically practiced using packed-beds and microchannel reactors, with length scales of several centimeters. **Figure 8** demonstrates that the clay-based MMRs achieves similar performance in just one millimeter of wall thickness.

Figure 9 shows the mass spectrometer analysis of the tube-side reformate of the MMR for 90 minutes of operation at a liquid feed rate of 0.5 ml/min of 40% ABV, or 0.2 ml/min of effective ethanol feed. EtOH molar density is $0.789\frac{g}{ml} \div 46.068\frac{g}{mol} = 0.0171 mol/ml$, giving 0.0034 mol EtOH per minute entering the reactor. According to the plot, about 72% leaves the reactor as H_2 Therefore, the catalytic effectiveness is 0.72/0.75 = 96%. 0.96 times 6 moles of hydrogen per mole of EtOH gives 5.76 mol H_2 per mole EtOH reformed, or 0.0196 mol H_2/min. This amounts to 1.18 mol/hr., or 2.35 grams of hydrogen per hour being produced by the MMR.

The periodic variability is the result of non-uniform gasification of liquid reactant. Liquid condensate accumulating in the vaporizer coil flashes over, causing a periodic sudden burst of gasified fuel and steam. Each time this occurs, the pressure on the shell side jumps from about 10 barg to 20 barg, then relaxes back to 10 barg. The spikes of the hydrogen concentration correspond with these sudden pressure jumps. More interestingly, the CO and CO_2 mole fractions are seen to be 180 degrees out of phase, clearly indicating that equilibrium of Eq. (1) is being shifted to the right by the

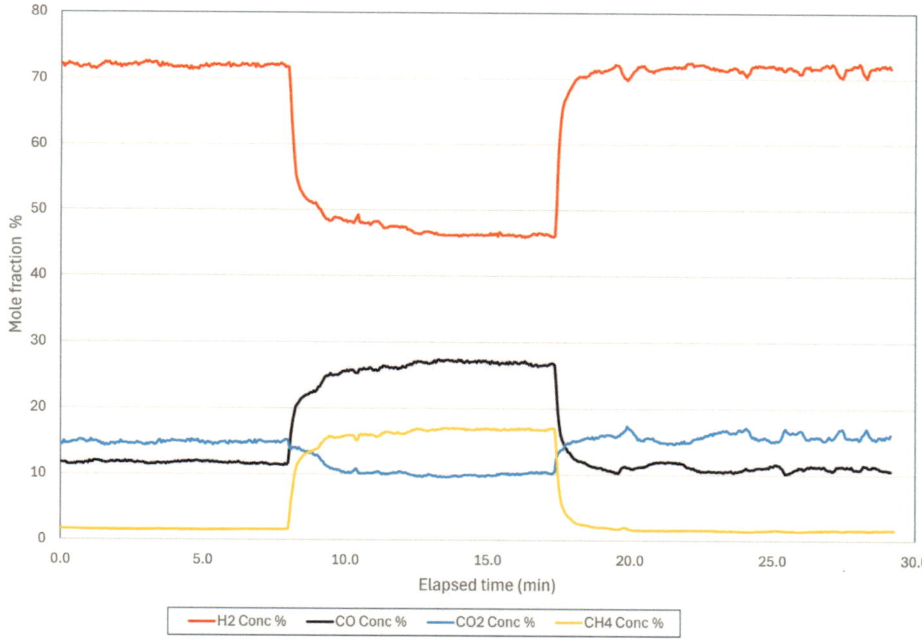

Figure 8.
Mass spectrometer analysis of ESR tube composition (0–7 min), followed by shell composition (7–17 min), and returned to tube composition.

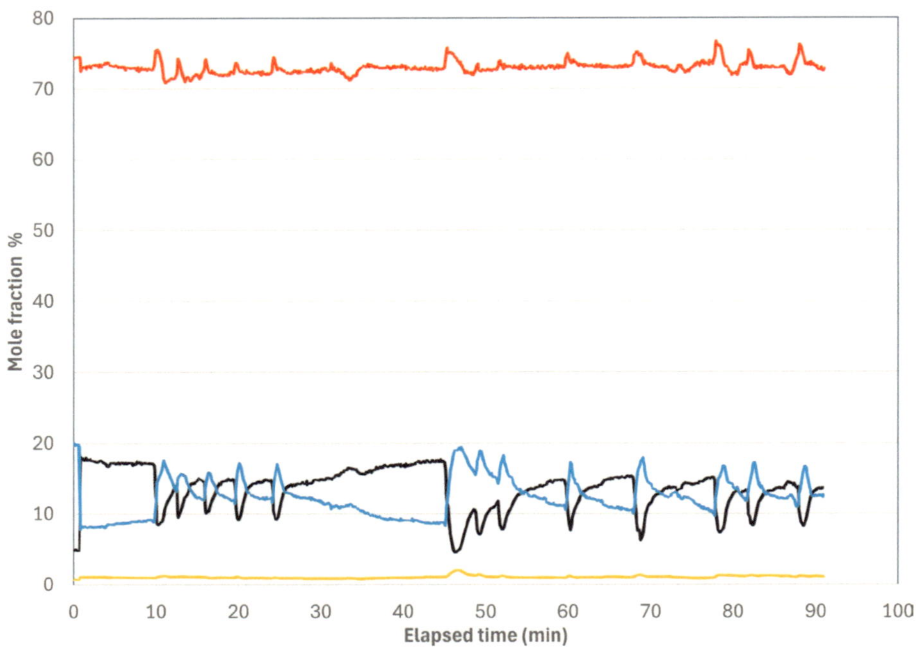

Figure 9.
ESR test run on MMR bulb with 60 mg/g infused Ni catalyst.

doubling of pressure, favoring reaction products H_2 and CO_2, and providing a beautiful confirmation of Le Chartier's Principle. Hydrogen and methane at 800°C are in the flat portion of the equilibrium diagram, so the mole fractions are less influenced by pressure changes, unlike CO and CO_2 which are more sensitive to temperature and pressure in this region as seen in **Figure 1**.

3.3 Future considerations

One of the grand challenges for implementing a compact ESR is that the reforming reaction is endothermic, and thus, requires the introduction of heat to sustain the reaction. With traditional packed-bed reformers, heat is applied from the outside, which is inefficient and difficult to carry out at small scale. To solve this problem of efficient thermal management, development is under way to extrude the clay MMRs as coils and heat the reactor from the interior. A prototype of this architecture is shown in **Figure 10**. A resistance cartridge heater was inserted into the coil, and the operation at 800°C may be observed by the orange glow between the coils. Ultimately, the MMR will be encased in insulation so that practically no heat is lost to the surroundings.

4 Conclusions

The transition from a petroleum-based energy economy to a renewable hydrogen-based one will take decades. It took 150 years for the world to become addicted to crude oil. Two world wars have been fought over it and dark storm clouds for a third have been gathering for years. Withdrawal from petroleum addiction will also likely not be painless once oil interests are threatened. However, although the societal advantages of hydrogen as a replacement for fossil fuels has been appreciated for a

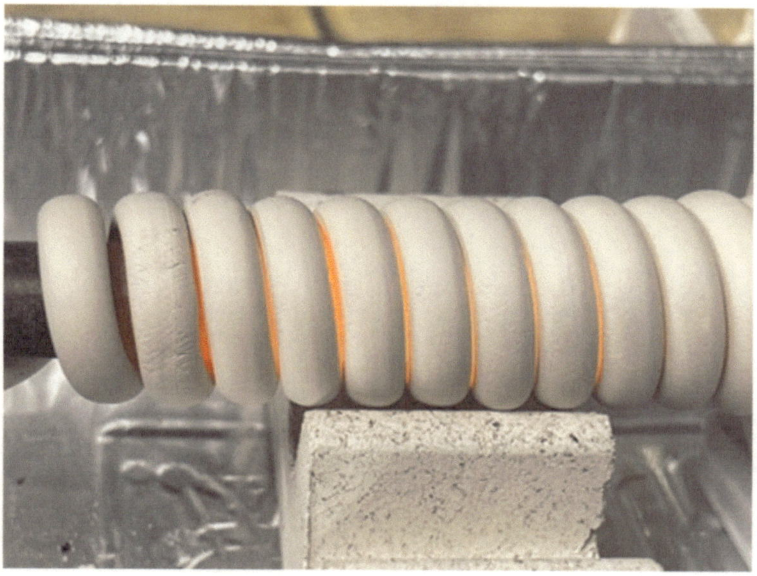

Figure 10.
Extruded coil with internal heating.

long time, hydrogen has failed to gain a foothold due to its high cost. Hydrogen made from fermented sugar has the potential to break through this $4/kg cost barrier to provide the abundant and inexpensive energy people have come to expect without the need for carbon taxes and government subsidies. Unlike crude oil, which is controlled by a small cadre of oil-producing nations, hydrogen made by ethanol steam reforming offers a path forward to energy independence for anyone on the planet. Many technical challenges remain–the biggest being scaling production to levels necessary to trigger demand. This will not happen until hydrogen achieves price/energy parity with fossil fuels. Barring some sort of hydrogen energy bonanza, such as tapping into geologic hydrogen trapped in the earth's mantle, steam reforming offers a cleaner, more accessible, and more secure energy future for all.

Ethanol steam reforming (ESR) in a microporous membrane reactor, MMR, with nickel catalyst has been described in this monograph. The mole fractions of reaction products were determined using a mass spectrometer with software providing quantitative gas analysis. MMR bulbs were prepared from clay-based kaolin by traditional slip casting. Ni catalyst nanoparticles where prepared by infusion with nickel nitrate into the porous fired ceramic. The catalytic effectiveness of the MMR was demonstrated for conversion of ethanol to hydrogen at nearly the maximum possible yield–a remarkable achievement for a flow channel of only about 1 millimeter in length. Two orders of magnitude in process intensification were, thus, demonstrated with clay-based microporous membrane reactors. Hydrogen production of about 2.4 grams per hour was demonstrated in a compact form-factor that is suitable for scaling up for commercial production.

Acknowledgements

Much appreciation to Dr. Sandrine Ricote at the Colorado School of Mines for the FESEM image of **Figure 5** and for assistance with manuscript editing. Thanks to Dr. Robert Kee and Huayang Zhu at the Colorado School of Mines for generating the data for **Figure 2**. Thanks to Dr. Ivar Reimanis at the Colorado School of Mines for encouragement and moral support. Thanks to Dan Pratte for the use of his laboratory and mass spectrometer.

Author details

William G. Coors
Hydrogen Helix, LLC, Golden, Colorado, USA

*Address all correspondence to: grover@h2helix.com

References

[1] Grover Coors W. Hydrogen at scale by ethanol steam reforming. ResearchGate. 2024. 10.13140/RG.2.2.30014.66880

[2] The engineering toolbox. Available from: https://www.engineeringtoolbox.com/fuels-higher-calorific-values-d_169.html

[3] Global ethanol production by country or region. U.S. DOE, Alternative Fuels DataCenter. 2022. Available from: https://afdc.energy.gov/data/10331

[4] Dziugan P et al. Evaluation of the fermentation of high gravity thick sugar beet juice worts for efficient bioethanol production. Biotechnology for Biofuels. 2013;**6**:158. Available from: http://www.biotechnologyforbiofuels.com/content/6/1/158

[5] Amrouk E, Palmeri F. World sugar market at a glance. Append: openknowledge.fao.com

[6] USDA. The economic feasibility of ethanol production from sugar in the United States. 2006. Available from: https://www.fsa.usda.gov/Internet/FSA_File/ethanol_fromsugar_july06.pdf

[7] The American Crystal Sugar Company. Append: https://www.crystalsugar.com

[8] Spreckels sugar company. Available from: https://www.spreckelssugar.com

[9] Robak K, Balcerek M. Current state-of-the-art in ethanol production from lignocellulosic feedstocks. Microbial Research. 2020;**240**:126534

[10] Harabi A et al. Development and characterization of tubular composite ceramic membranes using natural alumino-silcate for microfiltration applications. Materials Characterization. 2015;**103**:18-27

[11] Grover Coors W. Clay-based catalyst support for microchannel reactors. ResearchGate. 2023. 10.13140/RG.2.2.24557.05602

[12] Grover Coors W. Fabrication and characterization of clay-based microporous membranes. 2025. 10.13140/RG.2.2.33887.16802

[13] Grover Coors W. Fabrication and testing of ceramic microporous membrane reactors. 2024. 10.13140/RG.2.2.31351.60322

[14] Grover Coors W. Gas diffusion in porous supports for hydrogen separation membranes. 2023. 10.13140/RG.2.2.14849.43364

[15] Grover Coors W. Overview of Pd membrane hydrogen flux measurements and concepts. 2023. 10.13140/RG.2.2.33576.14089

Chapter 3

Low-Carbon Hydrogen Technologies: Methane Pyrolysis for Production of Low-Carbon Hydrogen and Solid Carbons

Ashton Swartbooi

Abstract

Methane pyrolysis offers a viable route for producing low-carbon hydrogen, providing an eco-friendly substitute for traditional techniques such as steam methane reforming. This method thermally breaks down methane (CH_4) into hydrogen gas (H_2) and solid carbon, preventing direct CO_2 emissions during production. Commonly known as "turquoise hydrogen," this approach generates a valuable solid carbon by-product, which can be captured or used in industrial processes, improving economic feasibility. The technology utilises current natural gas distribution systems, allowing for decentralised hydrogen generation while incorporating renewable energy sources to drive the high-temperature reaction. This chapter thoroughly examines methane pyrolysis, addressing essential chemical mechanisms, catalytic and non-catalytic routes, reactor configurations, and operational issues such as catalyst deactivation and carbon handling. It explores the importance of operating conditions, kinetic modelling, and new commercial initiatives by firms such as Monolith Materials and Hazer Group, showcasing scalability and market prospects. The advantages for the environment and the economy depend on aspects such as methane leaks, the carbon intensity of energy sources, and the final application of solid carbon products, including carbon black or graphite. Life cycle assessments show that methane pyrolysis, when using clean energy, can attain a carbon footprint much lower than the conventional methods. Methane pyrolysis connects fossil fuel infrastructure to a low-carbon future, positioning it as a key technology for the hydrogen economy, with continuous advancements in catalysts and reactor designs set to improve its scalability and influence.

Keywords: methane decomposition, turquoise hydrogen, low-emission hydrogen, solid carbon, carbon reduction

1. Introduction

The current global energy system is in the midst of transformation, largely driven by climate goals, technological advances, and changing markets. To push the needs for a decarbonized future, hydrogen has been identified as a leading energy carrier and thus plays a central role. Its versatility is key, because it can act as a clean fuel for industries like shipping and aviation, a clean feedstock for chemical and petrochemicals, and more areas requiring decarbonisation [1–4].

However, a large percentage of hydrogen produced today is derived from fossil fuels through processes like steam methane reforming, which releases large amounts of CO_2. This is why developing low-carbon and green hydrogen technologies has become a global priority. In 2023, global hydrogen demand was estimated at approximately 97 Mt., a figure expected to increase to over 100 Mt. in 2024. This demand in the market represents a huge opportunity for cleaner alternatives [5–7].

A promising low-carbon alternative is by utilising methane pyrolysis, also called methane decomposition. Hydrogen produced in this manner is often referred to as "turquoise hydrogen" [8]. During this process, a molecule of methane (CH_4) is broken down using heat into hydrogen gas (H_2) and solid carbon, thus avoiding direct CO_2 emissions at the point of production. Unlike "blue" hydrogen (which captures CO_2) or "green" hydrogen (from renewable electricity), pyrolysis produces a solid carbon co-product. This carbon can be sequestered, sold, or even re-used in circular industrial processes, creating an additional revenue stream that can improve the project's economics [6–8].

Interest in methane pyrolysis has grown due to a combination of maturing technology, market pressures, and policy incentives. The core chemistry—splitting methane into hydrogen and carbon—is a simple concept, though achieving high efficiency and continuous operation is complex due to the stability of the methane molecule as well as catalytic deactivation. While there are still many discussions around an agreed carbon pricing mechanism, pathways with low lifecycle emissions are becoming more attractive. A wave of companies like Monolith Materials, Ekona Power, and Hazer Group are now advancing pilot and demonstration plants, testing different reactor designs and technologies and bringing the technology closer to commercial reality [9, 10].

To understand the full process, the promise of methane pyrolysis must be viewed holistically. Its environmental benefit depends on where the methane comes from, how much leaks during extraction and transport, and the carbon footprint of the energy used to power the process. The final treatment of the solid carbon—whether it is used in long-lasting products or sequestered—also determines the net emissions [8].

Economically, pyrolysis must compete with established methods like steam methane reforming with carbon capture. Its competitiveness hinges on natural gas prices (the feedstock of methane), energy costs, reactor design, the value of the carbon co-product, and the presence or absence of carbon taxes or subsidies. Under the right conditions, it can be cost-competitive, especially if the produced carbon is of a higher value and can be sold for a high price [6, 11–13].

This book chapter provides an integrated and holistic look at methane pyrolysis, touching on all fundamentals related to the process, including fundamental chemical reactions and catalysts to reactor designs, commercial projects currently in operation, and economic models. Starting with fundamental atomic level chemistry, and linking that with process level design and economics, this chapter hopes to provide a practical view at evaluating methane pyrolysis' role in the energy transition [7, 8].

2. Fundamental methane decomposition mechanisms

Methane pyrolysis requires the breaking or cleaving of the strong carbon-hydrogen (C–H) bonds in the methane molecule. To achieve this requires large energy input and is normally done at high temperatures. Methane is then converted into hydrogen and solid carbon without producing CO_2 [8, 14, 15].

The standard reaction is given in Eq. (1) below:

$$CH_4 \rightarrow 2H_2 + C_{(s)} \qquad \Delta H^0_{298} = +74.9\ kJ/mol \tag{1}$$

Methane pyrolysis can typically be achieved without a catalyst at very high temperatures (usually over 1000°C), but this can be lowered through the use of a catalyst. The actual mechanism has been studied by various authors and is usually very complex, since it involves radical chain reactions and the formation of intermediate hydrocarbons before solid carbon finally deposits out [8, 14, 15].

2.1 The radical reaction pathway (non-catalytic)

For non-catalytic decomposition and thus pure gas-phase reactions, the reaction proceeds through a series of steps initiated by radical formation. The process is initiated when a methane molecule splits into a methyl radical (CH_3^*) and a hydrogen atom (H^*) [8, 14, 15].

- Initiation:

$$CH_4 \rightarrow CH_3^* + H^*$$

This step is usually considered the slowest and requires the most energy.

The methyl radical and hydrogen atom then trigger chain reactions, with larger hydrocarbons being formed which eventually fully decompose to carbon and hydrogen:

- Propagation:

$$CH_3^* + CH_4 \rightarrow C_2H_6 + H^*$$
$$C_2H_6 \rightarrow C_2H_4 + H_2$$
$$C_2H_4 \rightarrow 2C_{(s)} + 2H_2$$

- Termination:

$$CH_3^* + CH_3^* \rightarrow C_2H_6$$

The chain reaction ends when radicals re-combine and thus do not take part in any further reactions.

It has been shown that the activation energy for this non-catalytic path is around 355–420 $kJ.mol^{-1}$. This high activation energy shows the stability of the methane molecule and necessitates the needs for a catalyst to decrease the activation barrier [8].

2.2 Catalytic pathways

By using a heterogeneous catalyst, the activation energy is dramatically reduced, resulting in lower operating temperatures. Transition metal catalysts like nickel (Ni) or iron (Fe) have been shown to provide active sites that allow methane to decompose at much lower temperatures (600–900°C instead of >1000°C) [8, 16].

The reaction on a metal catalyst surface follows a similar series of steps without the building of longer hydrocarbon chains, however. The stepwise decomposition can be described by:

1. Adsorption: A methane molecule is bound to the catalyst surface.

2. Stepwise dehydrogenation: The methane loses its hydrogen atoms one by one, leaving a carbon atom bound to the surface.

3. Hydrogen recombination: The hydrogen atoms on the surface pair up and desorb as H_2 gas.

4. Carbon nucleation: The carbon atoms on the surface cluster together to form solid carbon.

As with the non-catalytic route, the first C—H bond break is usually the limiting step and on a nickel catalyst, the activation energy can be approximately 87 $kJ.mol^{-1}$, which offers a massive reduction from the pure gas-phase reaction (**Figure 1**) [8, 16].

Different catalysts can also play a key role when being used in the process and can also lead to different carbon products being formed. Nickel is highly active resulting in high conversions but can be quickly deactivated as carbon coats its surface due to the fast kinetics. Similarly, iron catalysts might not be as active as nickel, and they tend to form carbides that promote the growth of filamentous carbon (like carbon nanotubes). This can lead to the requirement of different

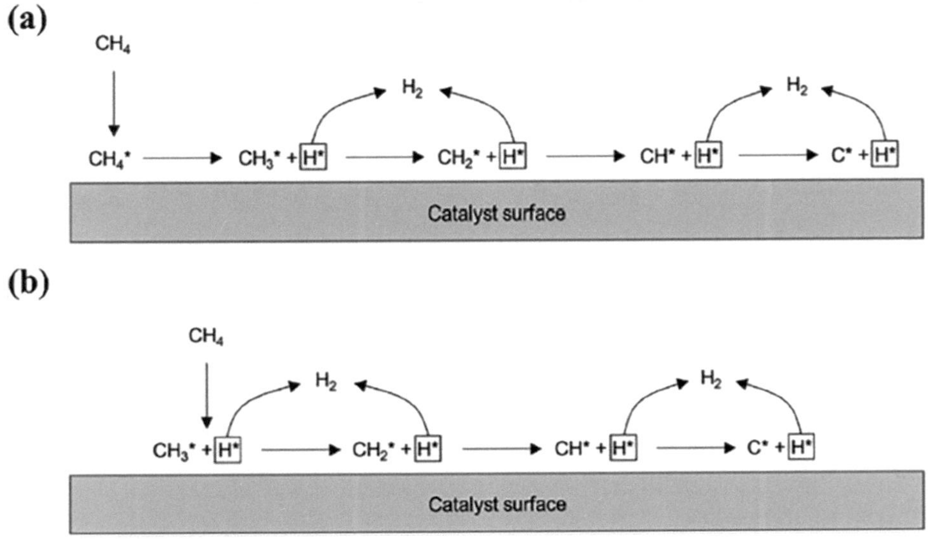

Figure 1.
Mechanisms for the adsorption of methane molecules onto a catalyst surface through (a) molecular adsorption mechanism and (b) dissociative adsorption mechanism. Taken from Song and Park, 2024 [17].

catalytic packing or reactor set-ups depending on which parameter was the focus of optimization, for example nickel would favour high hydrogen yields and low-grade carbons, while iron favours high-value carbon formation with lower hydrogen yields. Some researchers have also looked at reactors in series to maximise both products [18, 19].

2.3 The role of operating conditions

The rate of the reaction and the type of carbon produced are highly sensitive to temperature, pressure and residence time since all of these affect the equilibrium of the reactions. From Eq. (1) and following the Le Chatelier principle, higher temperatures and lower pressures generally favour the forward reaction and formation of hydrogen and carbons, which then also favours deactivation on certain catalysts. High temperatures, however, also favour the formation of more ordered, graphitic carbon [8].

2.4 Kinetic modelling

To design efficient reactor models, kinetic models need to be developed to help predict how fast the reaction will proceed under various conditions. A proper kinetic model not only helps with unit (reactor) design but also plays a key role in the process design, which subsequently feeds into the economics and operational cost calculations. Common kinetic models are usually based on experimental results and include:

- Power-law models for non-catalytic systems.

- Langmuir-Hinshelwood models for catalytic systems. This model accounts for the fact that the reaction can slow down at high methane pressures as the catalyst active sites become saturated. A dual Langmuir-Hinshelwood model can also be developed when hydrogen is expected to be competing for the same active sites.

- Nucleation-growth models that specifically describe how carbon builds up on the catalyst and leads to its deactivation over time [20].

3. Reactor technologies

Turning the chemistry of methane pyrolysis into a practical, large-scale process is the duty of the reactor. The reactor's design not only controls methane conversion, but the flow and catalyst used inside the reactor also determines the type of carbon produced as well as the overall economics. The main challenge for methane pyrolysis in general is how to supply the heat efficiently while managing the solid carbon that forms, while at the same time preventing it from clogging the system and shutting the operation down [7, 8].

Researchers have explored a wide range of reactor designs, which can be grouped into a few key categories, and presented in **Table 1** below:

Technology	Pros	Cons	Other comments
Fixed-bed reactors	Simple to build and operate, excellent for fundamental studies.	Catalyst deactivates quickly as carbon coats it. Poor heat distribution creates hot spots. Requires stopping production to regenerate catalyst, making continuous operation difficult.	Often used in labs for testing catalysts. Temperature ranges from 600 to 1000°C, with methane conversion of 40–90% depending on conditions [21].
Fluidised-bed reactors	Great heat and mass transfer, scalable, suitable for continuous operation.	Catalyst particles can be worn down by attrition. Fine carbon dust can be carried away with the gas, requiring filters.	Constant motion of catalyst particles creates excellent heat distribution and allows continuous addition of fresh catalyst and removal of carbon-coated particles [22].
Molten media reactors	Natural prevention of fouling, uniform temperature, potential for very high efficiency.	Higher initial capital costs due to complex design.	Uses a bath of molten metal (e.g., tin) or molten salt. Carbon floats to the top or disperses, preventing fouling. [9, 23–25].
Plasma reactors	Fast reaction (milliseconds), compact size.	High electricity consumption.	Uses high-voltage electrical discharge or microwaves to create super-hot plasma that decomposes methane [26].
Pulsed combustion reactors	High energy efficiency, pulsation helps distribute carbon and prevent clogging.	Complexity in operation, especially pulsing.	Uses controlled, tiny combustion pulses to create intense heat waves to crack methane.
Solar-thermal (emerging concept)	Truly renewable-driven process using concentrated sunlight for heat.	Challenges in scaling and cost.	Innovative concept using giant mirrors to generate extreme heat needed for pyrolysis.
Membrane reactors (emerging concept)	Pushes chemistry to produce more hydrogen at lower temperatures.	Complexity in integrating membranes.	Integrates hydrogen-selective membranes to continuously extract hydrogen from the reaction zone.

Table 1.
Different advantages and disadvantages for different reactors used for methane pyrolysis.

4. Catalysts and catalytic processes

For non-catalytic methane pyrolysis, high temperatures are required, and hence, a catalyst is required to reduce this high heat requirement. By lowering the activation energy barrier required to break the C-H bond, it ensures that the reaction can be operated at more manageable and practical temperatures (600–900°C). A good catalyst must be active, but also resistant to being deactivated by the very carbon it helps create [8, 16, 18, 19].

Catalysts provide a surface where methane molecules can bond and dissociate step-by-step, lowering the activation energy significantly. They also play a huge role in determining what form the solid carbon takes—whether it is disordered carbon black or valuable carbon nanotubes [8, 20]. The carbon property is determined by the elemental make-up of the catalyst, since each element can provide a different type of carbon.

4.1 Types of catalysts

Transition metals: Nickel (Ni), iron (Fe), and cobalt (Co) are the most studied because they are highly active and readily available. Each species has its own advantages and disadvantages:

Ni: Highly active but can be deactivated quickly when coated with carbon.

Fe: Tends to form carbides that promote the growth of filamentous carbon nanotubes.

Bimetallic catalysts: Combining metals can create synergistic effects. For example, adding a little copper to nickel can help stabilise it and slow down deactivation.

Carbon-based catalysts: Ironically, carbon itself can act as a catalyst. Materials like activated carbon or carbon nanotubes, especially when doped with nitrogen or other elements, can catalyse the reaction. They are low-cost and avoid the issue of metal deactivation [27].

4.2 Catalyst activity: Deactivation versus regeneration

While carbon is a desired product from the pyrolysis reaction, it is also the primary enemy for many metal catalysts due to coking, where the carbon builds up and encapsulates the active metal sites. To combat this, reactors often have regeneration cycles where the carbon is carefully burned off with air or treated with hydrogen. This, however, is not desirable where the type of carbon formed is of a high value and needs recovery. The latest research is focused on designing "carbon-tolerant" catalysts, like certain iron-based ones, where the carbon grows as filaments away from the active site, allowing the catalyst to remain active for much longer [8, 18].

5. Commercial plants and industrial applications

Over the past decade, methane pyrolysis plants have moved from lab-scale to pilot and early commercialisation scale demonstrations. The fast pace of these developments can be linked to the biggest driver namely climate policies but also market interest in low-carbon hydrogen as companies move towards global net-zero targets. Companies across the globe are now proving the technology can work at a meaningful scale [9, 10], with the main push coming from:

- Decarbonisation: Industries need a clean alternative to CO_2-heavy steam methane reforming, especially in hard-to-abate sectors.

- Economics: The solid carbon co-product—whether carbon black, graphite, or nanotubes—creates a valuable second revenue stream.

Producing hydrogen *via* pyrolysis is reported to have a carbon footprint as low as 1–3 kg of CO_2 equivalent per kg of H_2 (depending on the energy source and process methane leaks), a massive reduction compared to conventional steam methane reforming methods. Several economic analyses suggest that hydrogen can be produced competitively, especially when the carbon is sold [6, 11–13].

5.1 Key projects leading the way

- Monolith materials (USA): Their Olive Creek 1 plant in Nebraska uses a plasma-driven molten media reactor to decompose natural gas into hydrogen and high-purity carbon black. They sell both, with the carbon revenue significantly offsetting the cost of hydrogen production [9, 10].

- Hazer group (Australia): Their approach uses a low-cost iron ore catalyst in a fluidised-bed reactor. The process simultaneously produces hydrogen and high-purity graphite, a material in high demand for battery anodes. Their pilot plant has demonstrated continuous operation, a key step towards larger scales.

- Ekona power (Canada): Ekona uses a pulsed combustion reactor that creates intense heat waves to crack methane. This design efficiently manages the carbon, ejecting it with each pulse to prevent clogging. Their pilot unit highlights a different, innovative path to continuous operation.

- BASF (Germany): The chemical giant is developing a molten metal bubble-column reactor, using a tin bath. They plan to create a robust, continuous process that could be integrated into their own industrial sites for hydrogen and carbon production [25].

These diverse projects prove that there is not just one right way to do methane pyrolysis and this all depends on how the carbon by-product will be handled or formed. However, focussing on different reactor technologies and demonstrations shows both technical feasibility and a growing market interest. The future of these projects will depend on improving reactor durability, securing stable markets for the carbon, and supportive policy frameworks [7–9].

6. Integration into the hydrogen economy

For methane pyrolysis to make a real impact, it must fit seamlessly into the emerging low-carbon energy system yet at the same time compete with blue or green hydrogen technologies. Methane pyrolysis' unique advantage is that it acts as a bridge, connecting the existing fossil fuel infrastructure of today to the renewable-powered hydrogen economy of tomorrow [8]. This places turquoise hydrogen in a strategic middle ground, offering immediate low-carbon hydrogen benefits without waiting for a fully built-out renewable grid [2–4].

6.1 Leveraging existing infrastructure

A major benefit of pyrolysis is its compatibility with systems already in place today:

- Natural gas grids: The same pipelines that deliver natural gas for heating and power can feed distributed pyrolysis units. This allows for decentralised hydrogen production right where it is needed, avoiding the huge cost of building new hydrogen transport networks.

- Industrial clusters: The high-purity hydrogen can be fed directly into existing industrial processes, such as ammonia plants or refineries, for immediate decarbonisation [13].

When using natural gas as a feedstock, and thus still fossil-based, renewable energy sources can be used to provide the energy for the high-temperature reaction including:

- Using renewable electricity for electric or plasma heating.
- Using concentrated solar thermal energy to provide the heat directly.
- Using waste heat from other industrial processes, like steel or cement manufacturing.

This creates a hybrid system that reduces the overall carbon footprint and provides a practical transition pathway [15].

7. Carbon product benefits and applications

Unlike other hydrogen production methods that treat carbon as a waste gas (through the formation of CO_2) resulting in significant cleaning or (regulatory) penalty costs, pyrolysis ensures that the carbon can be sold. This transforms the economics and sustainability of the entire process [8].

7.1 The full spectrum of carbon products

The type of carbon produced can be tailored by adjusting the reactor conditions and formulation of the catalyst. The resulting carbon formed ranges from common industrial materials to advanced nanomaterials:

- Carbon black: A fine powder used as a reinforcing agent in tyres (about 70% of the market) and as a pigment. Replacing conventional carbon black production (which emits CO_2) with sustainable pyrolysis derived carbon black also ensures the environmental benefit.
- Graphite: High-purity synthetic graphite is essential for lithium-ion battery anodes. Demand is soaring with the electric vehicle revolution, making this a high-value target.
- Carbon nanotubes (CNTs): These are tiny, incredibly strong tubes of carbon with the exceptional electrical and thermal properties. They are used in advanced composites, electronics, and next-generation batteries. Even producing a small amount can dramatically improve a plant's revenue.
- Amorphous carbon/activated carbon: This porous, high-surface-area material is used for water purification, air filters, and gas absorption [20].

It can be seen that Fe-based catalyst mostly produces graphitic and filamentous carbons as discussed earlier. If we take the data from **Figure 2** and convert it to

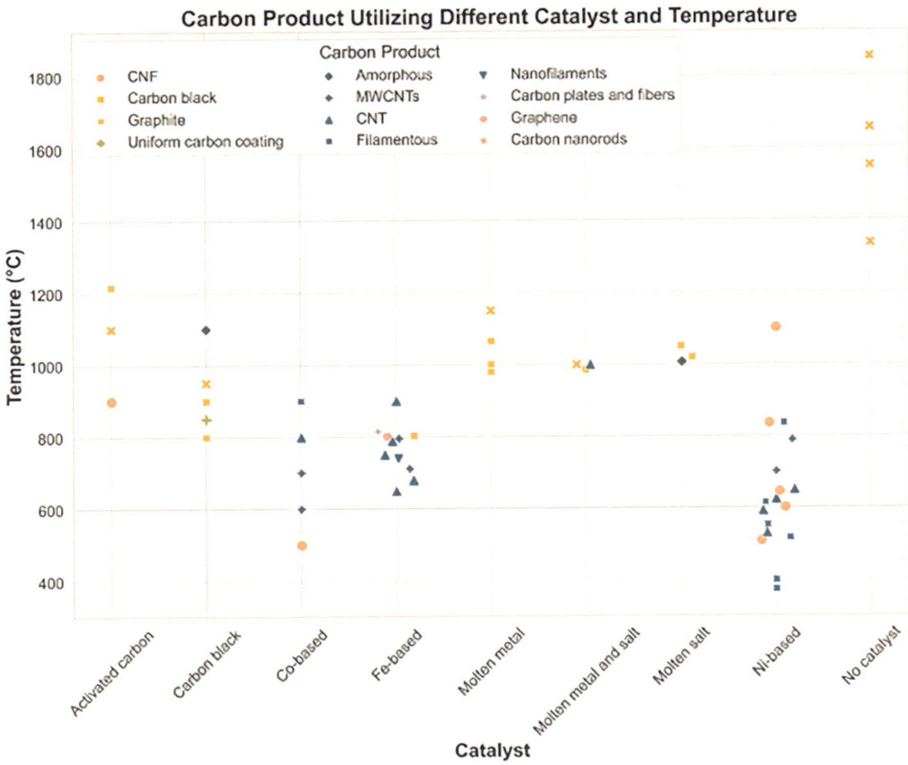

Figure 2.
Different carbon species formed when using different catalyst and reactor configurations. Taken from Alhamed et al. [28].

understand how the reactor configurations play a role, the resultant figure is given below.

Data from **Figure 3** is also given in **Table 2** below for ease of reference.

7.2 Environmental and economic impact

The economic benefit of the carbon by-product cannot be ignored or overstated:

Economic lever: Selling 3 tons of carbon black for every ton of hydrogen can generate enough revenue to cover a significant portion of the hydrogen's production cost.

Climate benefit: Every ton of solid carbon produced is a ton that is not released into the atmosphere as CO_2. If this carbon is used in long-lived products like construction materials or graphite electrodes, the process can even be considered carbon-negative, as it effectively sequesters carbon [8].

7.3 Policy and positioning

Governments are starting to recognise the benefits of low-carbon (turquoise) hydrogen. Policies in the EU and the US (like the 45 V tax credit in the Inflation Reduction Act) are creating financial incentives for low-carbon hydrogen production, which pyrolysis can qualify for if it meets specific emissions thresholds [1].

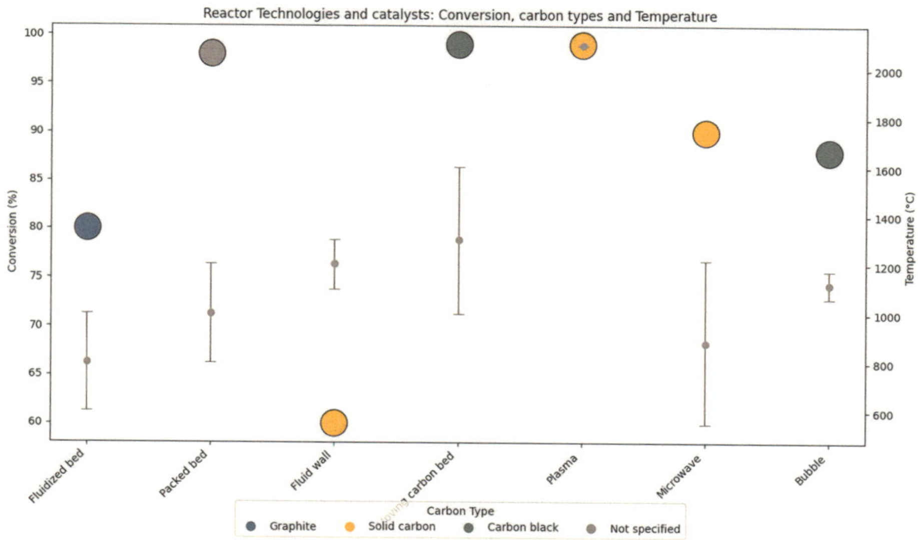

Figure 3.
Different reactor configurations resulting in different carbons. Bubbles are read on the left axis (conversion); bubble colour relates to the type of carbon formed; points and error bars are read on the secondary axis (operating temperature).

Reactor	Catalyst	Conversion	Carbon	Temperature	Reference
Fluidized bed	Iron oxide	80	Graphite	900	[7, 29]
Fluidized bed	Fe/Al_2O_3	80	Solid carbon	600–1000	[7, 29]
Packed bed	Carbon-encapsulated Fe nanoparticles	95.7	Not specified	800–1000	[7, 30]
Packed bed	Activated carbon	98	Not specified	1200–1400	[7, 21]
Fluid wall	Not specified	60	Solid carbon	1100–1300	[7, 31]
Moving carbon bed	Carbon granules	99	Carbon black	1000–1600	[7, 32]
Plasma	Plasma (non-catalytic)	>99	Solid carbon	2100	[7, 33]
Microwave	Coal char catalyst	90	Solid carbon	550–1216	[7, 34]
Bubble	Molten tin (Sn)	78	CNT	1175	[7]
Bubble	Molten Ni-Bi alloy	95	Carbon black	1065	[7]
Bubble	Molten Gallium (Ga)	91	Carbon black (amorphous)	1119	[7]
Bubble	Molten Ni-Sn	90	Graphite, carbon black	1000	[7]

Table 2.
Overview of reactor technologies and carbon formed.

8. Techno-economic considerations

For methane pyrolysis to become a mainstream source of hydrogen, it must make economic sense. The unique twist is that its business case depends on both products formed during the cracking, that is hydrogen and solid carbon. This dual-revenue stream is its greatest advantage, but it also makes the economics more complex [6, 11–13].

8.1 What drives the cost?

The Levelised Cost of Hydrogen (LCOH) is the all-important metric. For pyrolysis, it is calculated as:

$$(Annual\ Capital\ Cost + Annual\ Operating\ Cost - Carbon\ Revenue)/Annual\ Hydrogen\ Production$$

The Carbon Revenue term is what benefits the economics as this can be seen to be reducing the LCOH from the above equation. The key cost drivers however can be broken down into the following components:

- Feedstock: The price of natural gas is the single biggest factor, making up 40–60% of operating costs. Investigations into other sources of methane (bio-methane) are investigated but sometimes have problems with scaling.

- Energy: The cost of the electricity or heat needed to drive the reaction.

- Capital (CAPEX): The upfront cost to build the plant. Reactors like molten metal systems have higher initial costs but are built for long, stable operation.

- Carbon revenue: Selling carbon black for $1000 per ton creates a much stronger business case than selling lower-value amorphous carbon [11–13].

8.2 The power of policy (again)

Government policies are powerful levers that can tip the scales:

- Carbon pricing: A tax on CO_2 emissions makes grey hydrogen more expensive, narrowing the cost gap with turquoise hydrogen. However, because carbon taxing varies across different countries, this shows that methane pyrolysis would benefit in countries where the tax is high.

- Subsidies: Government incentives for clean hydrogen or products can dramatically reduce the effective cost of hydrogen from pyrolysis, potentially making it cheaper than fossil-based alternatives [1].

9. Environmental, life cycle assessment and circularity

A full cradle-to-grave life cycle assessment (LCA) should be used to properly measure the true environmental impact of methane pyrolysis. Such a life cycle would include the sourcing of the gas as well as the use of the final carbon product [8].

9.1 The full picture

A full LCA would examine:

- Methane leaks: If methane (a potent greenhouse gas) leaks during extraction and transport, it can significantly undermine the climate benefits. Controlling these leaks is critical.

- Energy source: Because of the high temperature requirements, the reactor heat can come from burning natural gas, which would push the emissions higher. If, however it is powered by renewable electricity or solar thermal, the carbon footprint is significantly reduced.

- Carbon's end-of-life: The ultimate climate benefit also depends on what happens to the solid carbon from the process. The benefit is limited when it is used in tyres that have a shorter lifespan, but larger improves when it is locked away in the long-lived graphite electrodes or building materials [8].

When all factors are accounted for, methane pyrolysis powered by clean energy and with controlled methane process leaks can have a lifecycle carbon footprint of just 1–3 kg of CO_2 equivalent per kg of H_2. This is a 70–95% reduction compared to conventional grey hydrogen. Furthermore, if the process uses methane from biogenic sources (like landfills or manure), the overall cycle can be considered to be carbon negative.

9.2 Broader benefits

Over and above just the climate benefits, methane pyrolysis also benefits by:

- Reduced water use: Pyrolysis uses over 95% less water than electrolysis or SMR, a major advantage in water-stressed regions.

- Circular economy: It turns a greenhouse gas into a valuable material, creating a circular carbon pathway where carbon is used and reused rather than released [8].

9.3 Biogas as a circular approach

Biogas, which is derived through anaerobic digestion of organic waste, can serve as a renewable feedstock for methane pyrolysis. Many countries limit the landfilling of organic matter due to the formation of methane, which is a more potent greenhouse gas. By rather capturing the methane from agricultural, municipal, or industrial waste, hydrogen can be produced which would limit the reliance on fossil-based fuels. The solid carbon which is also formed in the process has use cases in industries such as tyre manufacturing or as a soil enhancer. Through this process, a full circular and closed material loop is created.

Biogas does, however, present some challenges, especially related to the cleaning and upgrading to achieve high purity bio-methane, which feeds into the techno-economics of the process. A case-study for a South African facility based on biogas feed [35] has shown that hydrogen produced from biogas can in fact be competitive with green and even blue hydrogen production facilities. The same study also showed that the price of the carbon sales was one of the largest sensitivities in the economic analysis. This is also supported by many other similar studies [36, 37].

The scalability of biogas supply is however a limiting factor, which rather points to the fact that such a process favours decentralised hydrogen production facilities rather than setting up one large, centralised facility as would be the case for green hydrogen production. By valorising waste-derived biogas and its by-products, this approach aligns well with circular economy principles.

10. Future outlook and research directions

Methane pyrolysis is fast progressing and stands at a very good threshold—it has moved from a theoretical concept to a proven technology with operating pilot plants. The path going forward for the technology would focus on refinement, scaling and integration [7, 8].

The global hydrogen demand is steadily increasing with many countries agreeing to meet the net zero targets as outlined in the Paris Agreement. This hydrogen demand can best be demonstrated by **Figure 4** below which shows the projected demand for hydrogen for the top 35 countries, taken from Ref. [38].

It can be seen that a large increase in electrolytic hydrogen production is planned, with the biggest demand coming from the transport, ammonia and methanol sectors. Latest market studies, however, have adjusted the demand that can be realised with green hydrogen alone, pointing to the current high cost of green hydrogen and the premium that it demands.

Methane pyrolysis, by using existing infrastructure and reactor technologies, can fill in the short-term demand for hydrogen where industries need to decarbonize, thus, providing a bridging technology up until the point where large green hydrogen plants can provide feasible hydrogen costs.

From the discussions above, the key areas for innovations in the space of methane pyrolysis needs to focus on the following topics which would allow methane pyrolysis to be highly competitive with SMR, but at the same time addressing the low emissions required to reach net zero targets:

- Smarter reactors: The next generation of reactors should focus on modular, container-sized units for distributed production, improved materials to withstand extreme temperatures, and designs that make continuous carbon removal even more efficient.

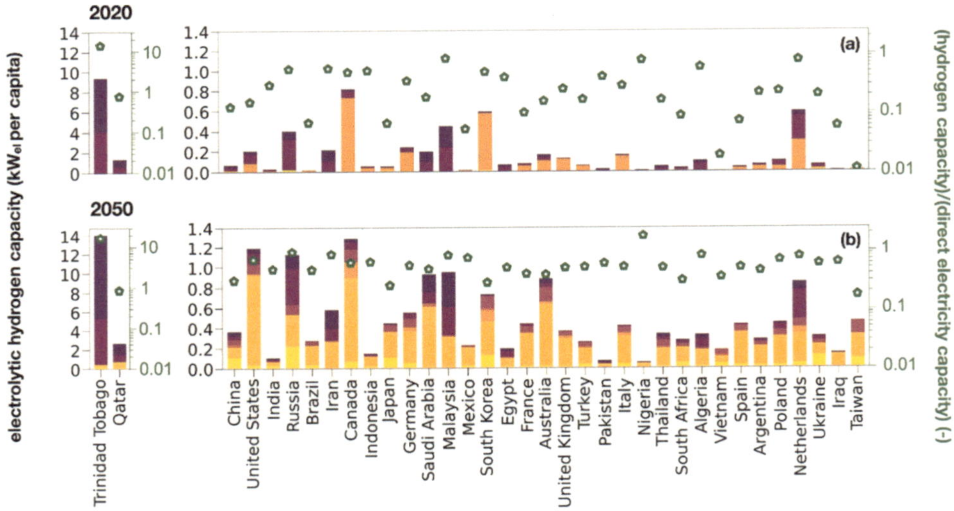

Figure 4.
Global hydrogen outlook for the top 35 countries until 2050. Taken from [38].

- Next-generation catalysts: Research is pushing for catalysts that work at even lower temperatures, last longer before needing regeneration, and give precise control over the type of carbon produced. AI and machine learning are now being used to discover new, high-performance catalyst materials at an unprecedented speed.

- The circular carbon loop: The carbon loop needs to be closed, and this could involve using "renewable methane" made by combining green hydrogen with captured CO_2 (a process called methanation) and then feeding that synthetic methane into the pyrolysis reactor. This creates a cycle where carbon is constantly recycled rather than extracted from the ground [8].

11. Conclusion

In conclusion, methane pyrolysis represents a powerful and reasonable bridge to a sustainable hydrogen future. By leveraging existing infrastructure, producing no direct CO_2 emissions, and creating valuable carbon materials, it offers a compelling combination of environmental and economic benefits.

While challenges around carbon handling and catalyst longevity remain, the pace of innovation is fast and rapidly increasing with many demonstration plants at various scales coming on-line. It is thus not merely a transitional technology, but it has the potential to be a long-term player in a decarbonised world, by turning the problem of fossil carbon into a portfolio of clean energy and advanced materials [7, 8].

Author details

Ashton Swartbooi
HySA & CCU Group, Chemicals Cluster, Council for Scientific and Industrial Research (CSIR), Pretoria, South Africa

*Address all correspondence to: aswartbooi@csir.co.za

References

[1] International Energy Agency. Net Zero by 2050: A Roadmap for the Global Energy Sector. OECD Publishing; 2021. DOI: 10.1787/c8328405-en

[2] Saha P, Akash FA, Shovon SM, et al. Grey, blue, and green hydrogen: A comprehensive review of production methods and prospects for zero-emission energy. International Journal of Green Energy. 2024;**21**:1383-1397

[3] Aydin MI, Dincer I. An assessment study on various clean hydrogen production methods. Energy. 2022;**245**: 123090

[4] Dawood F, Anda M, Shafiullah G. Hydrogen production for energy: An overview. International Journal of Hydrogen Energy. 2020;**45**:3847-3869

[5] Satyapal S. Hydrogen and Fuel Cell Technologies Office: Annual Merit Review Plenary [Presentation Slides]. U. S. Department of Energy, Hydrogen and Fuel Cell Technologies Office; Jun 2021. Available from: https://www.energy.gov/sites/default/files/2021-06/hfto-amr-plenary-satyapal-2021.pdf

[6] Shokrollahi M, Teymouri N, Ashrafi O, et al. Methane pyrolysis as a potential game changer for hydrogen economy: Techno-economic assessment and GHG emissions. International Journal of Hydrogen Energy. 2024;**66**: 337-353

[7] Patlolla SR, Katsu K, Sharafian A, et al. A review of methane pyrolysis technologies for hydrogen production. Renewable and Sustainable Energy Reviews. 2023;**181**:113323

[8] Sánchez-Bastardo N, Schlögl R, Ruland H. Methane pyrolysis for zero-emission hydrogen production: A potential bridge technology from fossil fuels to a renewable and sustainable hydrogen economy. Industrial and Engineering Chemistry Research. 2021; **60**:11855-11881

[9] Mesrobian C. Taking Methane Pyrolysis from Concept to Industrial Plant [Presentation Slides]. ARPA-E; Jan 2021. Available from: https://arpa-e.energy.gov/sites/default/files/migrated/2021-01/08%20OK%20-Monolith_ARPAE_MethanePyrolysis2021_v3.pdf

[10] Hardman NJ, Taylor RW, Hoermann AF, Johnson PL, Cardinal CJ-P, Hanson RJ. Carbon Black from Natural Gas (WO2017048621A1). WIPO; 2017. Available from: https://patents.google.com/patent/WO2017048621A1/en

[11] Angikath F, Abdulrahman F, Yousry A, et al. Technoeconomic assessment of hydrogen production from natural gas pyrolysis in molten bubble column reactors. International Journal of Hydrogen Energy. 2024;**49**:246-262

[12] Keipi T, Tolvanen H, Konttinen J. Economic analysis of hydrogen production by methane thermal decomposition: Comparison to competing technologies. Energy Conversion and Management. 2018;**159**:264-273

[13] Li Y, Yu H, Jiang X, et al. Techno-economic analysis for hydrogen-burning powerplant with onsite hydrogen production unit based on methane catalytic decomposition. Energy Conversion and Management. 2023;**277**: 116674

[14] Skinner GB, Ruehrwein RA. Shock tube studies on the pyrolysis and oxidation of methane. Journal of Physical Chemistry. 1959;**63**:1736-1742

[15] Palmer HB, Lahaye J, Hou KC. Kinetics and mechanism of the thermal decomposition of methane in a flow system. Journal of Physical Chemistry. 1968;**72**:348-353

[16] Saraswat SK, Sinha B, Pant K, et al. Kinetic study and modeling of homogeneous thermocatalytic decomposition of methane over a Ni–Cu–Zn/Al_2O_3 catalyst for the production of hydrogen and bamboo-shaped carbon nanotubes. Industrial and Engineering Chemistry Research. 2016;**55**: 11672-11680

[17] Song J, Park S. Review of methane pyrolysis for clean turquoise hydrogen production. Journal of Analytical and Applied Pyrolysis. 2024;**183**:106727

[18] Lua AC, Wang HY. Decomposition of methane over unsupported porous nickel and alloy catalyst. Applied Catalysis B: Environment and Energy. 2013;**132**:469-478

[19] Qian JX, Liu DB, Basset JM, et al. Methane decomposition to produce hydrogen and carbon nanomaterials over costless, iron-containing catalysts. Journal of Cleaner Production. 2021;**320**:128879

[20] Baker R, Barber M, Harris P, et al. Nucleation and growth of carbon deposits from the nickel catalyzed decomposition of acetylene. Journal of Catalysis. 1972;**26**:51-62

[21] Mokashi M, Shirsath AB, Çelik A, Lott P, Müller H, Tischer S, Maier L, et al. Methane pyrolysis in packed bed reactors: Kinetic modeling, numerical simulations, and experimental insights. Chemical Engineering Journal. 2024;**485**: 149684. DOI: 10.1016/j.cej.2024.149684

[22] Pathak S, McFarland E. Iron catalyzed methane pyrolysis in a stratified fluidized bed reactor. Energy & Fuels. 2024;**38**:12576-12585

[23] Upham DC, Agarwal V, Khechfe A, et al. Catalytic molten metals for the direct conversion of methane to hydrogen and separable carbon. Science. 1979;**2017**(358):917-921

[24] Zaghloul N, Kodama S, Sekiguchi H. Hydrogen production by methane pyrolysis in a molten-metal bubble column. Chemical Engineering Technology. 2021;**44**:1986-1993

[25] Bode A. Methane Pyrolysis – A Potential New Process for Hydrogen Production Without CO_2 Emission [Conference Presentation Slides]. Hannover, Germany: Niedersächsische Energietage; 6 Nov 2019. Available from: https://www.efzn.de/fileadmin/Sites/EFZN/Images/Ereignisse/2019-11-NET12/Vortraege/NET2019_FF1_04_Bode_Rev1.pdf

[26] Li T, Rehmet C, Cheng Y, et al. Experimental comparison of methane pyrolysis in thermal plasma. Plasma Chemistry and Plasma Processing. 2017; **37**:1033-1049

[27] Al-Hassani AA, Abbas HF, Daud WW. Production of COx-free hydrogen by the thermal decomposition of methane over activated carbon: Catalyst deactivation. International Journal of Hydrogen Energy. 2014;**39**: 14783-14791

[28] Alhamed H, Behar O, Saxena S, et al. From methane to hydrogen: A comprehensive review to assess the efficiency and potential of turquoise hydrogen technologies. International Journal of Hydrogen Energy. 2024;**68**: 635-662

[29] Qian JX, Enakonda LR, Wang WJ, et al. Optimization of a fluidized bed reactor for methane decomposition over Fe/Al_2O_3 catalysts: Activity and regeneration studies. International

Journal of Hydrogen Energy. 2019;**44**: 31700-31711

[30] Kundu R, Ramasubramanian V, Neeli ST, et al. Catalytic pyrolysis of methane to hydrogen over carbon (from cellulose biochar encapsulated iron nanoparticles). Energy and Fuels. 2021; **35**:13523-13533

[31] Schultz I, Agar DW. Decarbonisation of fossil energy via methane pyrolysis using two reactor concepts: Fluid wall flow reactor and molten metal capillary reactor. International Journal of Hydrogen Energy. 2015;**40**:11422-11427

[32] Çelik A, Ben Othman I, Müller H, et al. CO_2-free production of hydrogen via pyrolysis of natural gas: Influence of non-methane hydrocarbons on product composition, methane conversion, hydrogen yield, and carbon capture. Discover Chemical Engineering. 2024;**4**: 29. DOI: 10.1007/s43938-024-00067-4

[33] An H, Cheng Y, Li T, et al. Numerical analysis of methane pyrolysis in thermal plasma for selective synthesis of acetylene. Fuel Processing Technology. 2018;**172**:195-199

[34] Moghaddam AL, Hejazi S, Fattahi M, et al. Methane pyrolysis for hydrogen production: Navigating the path to a net zero future. Energy Environmental Science. 2025;**18**:2747-2790

[35] Swartbooi A, Kapanji-Kakoma KK, Musyoka NM. From biogas to hydrogen: A techno-economic study on the production of turquoise hydrogen and solid carbons. Sustainability. 2022;**14** (17):11050. DOI: 10.3390/su141711050

[36] Nahar G, Mote D, Dupont V. Hydrogen production from reforming of biogas: Review of technological advances and an Indian perspective. Renewable and Sustainable Energy Reviews. 2017; **76**:1032-1052

[37] Pinilla JL, de Llobet S, Moliner R, et al. H2-rich gases production from catalytic decomposition of biogas: Viability of the process associated to the co-production of carbon nanofibers. International Journal of Hydrogen Energy. 2017;**42**:23484-23493

[38] Tonelli D, Rosa L, Gabrielli P, et al. Global land and water limits to electrolytic hydrogen production using wind and solar resources. Nature Communications. 2023;**14**:5532. DOI: 10.1038/s41467-023-41107-x

Chapter 4

Perspective Chapter: Hydrogen Production Via Proton Exchange Membrane (PEM) Water Electrolysis

Shawn Gouws and Claudelle Sybilline Anensong Djadock

Abstract

This chapter provides an in-depth exploration of the various factors around green hydrogen production from water electrolysis in response to the global drive to mitigate carbon emissions. Eleven colors of hydrogen are used to illustrate the various processes in industries where hydrogen is produced as a primary source or as a by-product, for example, in green hydrogen produced from solar or wind energy. Several industries today depend heavily on black, brown, or gray hydrogen for fuel production in the saturation of molecules in reforming or cracking processes in a petrochemical process. Carbon emissions in the steel or cement industries must be mitigated to decarbonize these processes by introducing green hydrogen from water electrolysis. The different water electrolysis methodologies that are currently available include proton exchange membrane water electrolysis (PEMWE), alkaline water electrolysis, anion exchange water electrolysis, and solid oxide water electrolysis. Their advantages and disadvantages are explored to demonstrate why PEMWE is more favorable than other water electrolyzers. The different analytical techniques characterize the material science of the catalysts by X-ray diffraction, X-ray fluorescence, scanning electron microscopy, and transmission electron microscopy analysis, and the electrochemical characterization by cyclic voltammetry, linear scan voltammetry, and chronopotentiometry for stability studies. The last section details the requirements for hydrogen production, transport, and storage, emphasizing the importance of preventing catastrophic effects of hydrogen explosions or hazardous release of hydrogen into the atmosphere that could create cloud explosions.

Keywords: hydrogen production, proton exchange membrane water electrolysis, electrocatalysts, electrochemical characterization, hydrogen safety

1. Introduction

Conventional hydrogen is produced from steam methane reforming, where methane, a natural gas, is the feedstock and emits carbon dioxide as a byproduct.

One of the key drivers globally is to mitigate carbon dioxide to achieve a net-zero carbon emission process by generating hydrogen from other sources, such as water splitting, as a by-product from nuclear processes, or fracking of hydrogen from the earth's crust. Hydrogen production could be the answer to the effects of global warming and depletion of the ozone layer.

Green hydrogen is derived from water electrolysis connected to a renewable energy source, such as solar or wind energy. The process is called "green" because it does not generate carbon emissions. It is a clean and sustainable alternative to conventional fossil fuel hydrogen production, which is classified as black or gray hydrogen.

The following discussion on hydrogen's importance in the industry and chemical sectors outlines the various methods of producing hydrogen for industrial applications and uses [1]. In its molecular form, hydrogen is a dihydrogen molecule that is a stable, colorless, odorless, and tasteless gas with a very low melting point and boiling point. A key discussion point in this chapter will be the production of green hydrogen via proton exchange membrane (PEM) water electrolyzer, which involved electrical current generated from renewable energy.

2. The importance of hydrogen and hydrogen production

Carbon dioxide is the most important anthropogenic greenhouse gas (GHG), with over 35 billion tons emitted annually from burning fossil fuels, leading to an atmospheric concentration of approximately 380 ppm. This value is expected to rise soon [1]. Only half of the CO_2 released into the atmosphere is naturally recycled through photosynthesis, while the remaining CO_2 contributes to global warming [2, 3]. Global warming negatively impacts our environment in several ways, such as increasing global average air and ocean temperatures and causing the loss of ecosystems [2–4]. Carbon dioxide emissions are a global concern; 25 years ago, under the auspices of the United Nations Environment Program (UNEP), the World Meteorological Organization (WMO), and the Intergovernmental Panel on Climate Change (IPCC), governments globally established a framework to provide scientific information on climate change, its environmental impacts, and to specify potential solutions to mitigate CO_2 in the atmosphere [5]. In response, countries and regions have been making significant efforts to promote the transition of energy structures from fossil fuels to low-carbon or carbon-free options, aiming to keep the global average temperature increase below 2°C by the end of the twenty-first century, in terms of the Paris Agreement. To reduce carbon emissions and achieve low-carbon or carbon-free energy, it is imperative to replace traditional fossil fuels with clean, sustainable fuels derived from renewable energy sources such as wind and solar power, geothermal, and biomass energy.

In this respect, hydrogen fuel is a promising carbon-free energy source with a zero-carbon footprint; it is clean, nonpolluting, and has advanced globally as a substitute for oil and fossil fuels [6]. The attractive properties of hydrogen as an energy carrier include its high energy density of 140 MJ kg^{-1}, which surpasses that of typical fuels (50 MJ kg^{-1}), its low volumetric density of approximately 0.08 kg m^{-3}, and its distinction as the most environmentally friendly energy vector [1, 7, 8], especially when produced *via* clean processes. These qualities have consistently attracted researchers worldwide [9]. Hydrogen can be stored in various forms—gas, liquid, or solid—depending on the requirements and target applications and can later be re-electrified when energy demand arises. Gaseous hydrogen is stored under high pressure, up to 800 bar, while cryogenic cooling at 21 K is utilized for liquid hydrogen storage. A promising approach for solid hydrogen storage is the chemisorption or physisorption process [10]. The value of the hydrogen energy storage market is significant. In 2024, the market reached US$16.64 billion, with projections suggesting it could hit US$20 billion by 2028 [11]. This growth has been fueled by increased research and development in hydrogen energy. More than 30 nations have issued hydrogen road maps; the industry has noted over 200 hydrogen projects and significant investment commitments, and governments globally have pledged more than US$70 billion in public support. Should all the projects materialize, overall investment in hydrogen would exceed US$300 billion by 2030, representing 1.4% of global energy expenditure [12, 13]. By 2050, the annual demand for hydrogen could increase to 560 million tons—ten times that of 2015—fulfilling 18% of global energy demand [14].

Hydrogen is primarily used in industrial applications such as fertilizers, refining, steel, ammonia and methanol production, petrochemicals, fuel cells, and the chemical industries [1, 8, 9]. Green hydrogen is gaining the most attention owing to its zero-carbon emissions and diligent alignment with global sustainability goals. The production capacity of green hydrogen is expanding rapidly, with companies like Green Hydrogen International leading the way with a project pipeline of 87.4 million tons of capacity [15, 16]. The numerous benefits of hydrogen production have received significant attention in recent years. The various widespread approaches to hydrogen production, along with their disadvantages, advantages, efficiency, and cost, are provided in **Table 1** [17–20].

For the last half-century, hydrogen production has undergone intensive reviews and has been skillfully developed through research. Generally papers address the fundamentals of proton exchange membrane water electrolysis, including membrane

Hydrogen production methods	Advantages	Drawbacks	Efficiency (%)	Cost (kg/ USD)
Steam Reforming	Developed technology and existing infrastructure	Produces CO, CO_2; unstable supply		74–85 2.27
Partial Oxidation	Established technology	H_2 production; produces heavy oils and petroleum coke	60–75	1.48
Autothermal Reforming	Well-established technology and existing infrastructure	Produces CO_2; uses fossil fuels	60–75	1.48
Thermolysis	Clean and sustainable	O_2 byproduct; copious feedstock; high capital costs; toxicity; corrosion problems	20–45	7.98–8.40
Photolysis	O_2 as a by-product, abundant feedstock, no emissions	Low efficiency; ineffective photocatalytic material; requires sunlight	0.06	8–10
Electrolysis	Established technology; zero-emission; existing infrastructure; O_2 as a byproduct	Storage and transportation problems	60–80	10.30

Table 1.
Advantages, disadvantages, efficiency, and costs of various hydrogen production methods [17–19].

electrode assembly (MEA), electrocatalysts, gas diffusion layers (GDLs), bipolar plates (BPs), and their degradation mechanisms. Several other papers report prevailing challenges, recent advances, and possible mitigation strategies for realizing PEM water electrolysis aimed at green and low-cost hydrogen production [18, 19, 21–23].

Yet other authors focus on hydrogen production from coke oven gas and coal gasification [24, 25], steam methane reforming, autothermal reforming, syngas chemical looping, and chemical looping reforming [26].

The production of blue hydrogen [27–29], turquoise, and green [30] is reported. However, a purely economic assessment of different production methods by Bartels et al. [31] found that gray hydrogen was the most economical method at the time of the analysis. In their analysis of 14 different H_2 production methods, Nikolaidis and Poullikkas [32] found that thermochemical pyrolysis and gasification have the greatest potential to become competitive. Steam methane reforming is still regarded as the most cost-effective production method, as reported by Yukesh et al. [33], which provides a techno-economic assessment of various production technologies. Ozturk and Dincer [34] compared different hydrogen colors and found that the current PEM water electrolysis has a limited environmental impact, the lowest normalized efficiency when powered by solar energy, and may become a cheaper technology. This was further elucidated by reports made by Dincer [35], Acar and Dincer [36], and Longden et al. [37].

3. Hydrogen color

3.1 Definitions of hydrogen color

Several approaches have been commonly agreed upon to produce H_2, such as steam reforming hydrocarbon, biomass, coal gasification, biological sources, hydride metal hydrolysis, electrolysis, and photoelectrochemical (PEC) water splitting (see **Table 1** [20]). Numerous sources for producing hydrogen can be used, such as coal, natural gas, water, and biomass, making it possible to assign different colors to each common method, depending on the source of the hydrogen produced, the direct emissions from hydrogen molecules, and the environmental impact. The methods for producing hydrogen are frequently classified based on different colors to make it easier to identify the alternative being considered for hydrogen production, without needing to go into the details of the process. Literature reports, papers, press releases, and national plans mention ten colors of hydrogen: black, brown, gray, blue, turquoise, green, orange, purple, pink, yellow, and red, based on sources of energy, the process for hydrogen production, and the CO_2 emissions. Here the focus is on hydrogen colors of gray, blue, turquoise, green, purple, and yellow (**Figure 1**).

Hydrogen classification differs from one study to another. Some authors consider all hydrogen production methods that rely on fossil fuel inputs without carbon capture, utilization, and storage (CCUS) as gray hydrogen, including brown or black hydrogen [7]. Others refer to gray hydrogen as hydrogen produced from steam methane reforming without CCUS; green hydrogen, such as hydrogen produced from water electrolysis powered by renewable energy sources (RES), and blue hydrogen, are typically produced from steam methane reforming, including CCUS [40]. Some authors assign purple [41], pink [42], or red [43] to nuclear-energy-powered processes to produce hydrogen. Red and pink have been further divided according to the process employed to produce hydrogen, notably thermolysis/thermochemical and

Color of Hydrogen	Technology	Process
Black	Gasification	This is hydrogen produce from the gasification of coal.
Brown		The traditional process of producing hydrogen from coal emits high volumes of carbon dioxide into the atmosphere.
Gray	Reforming	Hydrogen from natural gas in the reforming unit produces fewer carbon emissions than brown or black hydrogen.
Turquoise	Pyrolysis	Methane is split into hydrogen and solid carbon, a critical feedstock, without carbon dioxide emissions.
Blue	Reforming CCS	Carbon dioxide capture technology is combined with gray hydrogen to minimize greenhouse gas emissions.
Red	Electrolysis	Produced by the high-temperature catalytic splitting of water using heat from nuclear reactors.
Purple		Produced via electrolysis using nuclear power and heat.
Pink		Hydrogen is produced from water splitting in an electrolyzer connected to nuclear power
Orange	Capture of concentrated CO_2	Sometimes used to describe hydrogen productiDadrockCO2 sequestration.
Yellow	Electrolysis	Hydrogen is produced from water splitting in an electrolyzer connected via the national grid.
Green	Electrolysis	Hydrogen is produced from water splitting in an electrolyzer connected via renewable energies such as wind and solar.
White	Natural mining	Layers of the continental crust, or volcanic gases, geysers, and hydrothermal systems. geological regions and rock formations

Figure 1.
Colors of hydrogen [7, 38, 39].

electrolysis. White hydrogen is assigned to a product naturally occurring in a chemical reaction [44] or a resultant by-product of chemical processes [45], which is not relevant in a hydrogen economy future.

3.1.1 Gray hydrogen

According to Howarth and Jacobson [28], gray hydrogen is the hydrogen generated from fossil fuel (96%), especially from steam methane reforming (SMR) and coal gasification processes. The SMR process is responsible for approximately 75% of all hydrogen production worldwide. Hydrogen and carbon dioxide are produced by heating water and high-pressure natural gas. Steam then splits the methane in the reformer into syngas, having CO and H_2 as the main components. The waste product CO_2, which is estimated to be around 830 Mt. CO_2 per year, is released directly into the atmosphere. Ten tons of carbon dioxide are produced for each ton of hydrogen extracted [46], making gray hydrogen detrimental to the climate [45]. Other authors ascribe the hydrogen produced via the SMR process as a by-product and designate it white hydrogen [44, 45, 47].

Coal gasification, another production process, results in gray hydrogen, which some authors classify as brown or black hydrogen [42, 45]. This production method is frequently used since coal reserves are the largest fossil energy source worldwide. With its large coal reserves, and in response to high natural gas prices, China produces copious amounts of hydrogen from coal gasification [42]. Coal gasification and SMR are completely mature technologies; both processes produce hydrogen with an efficiency of up to 85% [48].

The cheapest form of hydrogen is gray hydrogen, ranging from 0.7 to 2.3 US$/kg$H_2$, followed by black and brown hydrogen (1.3–2.5 US $ /kg H_2 [49–51]); however, the cost could increase with an increase in CO_2 taxes [50–52]. Gray hydrogen is used in the petrochemical industry and to produce ammonia [33].

3.1.2 Blue hydrogen

Unlike gray hydrogen, blue hydrogen production utilizes natural gas in steam methane reforming, followed by carbon capture storage (CCS) [1, 41, 42, 45]. During this process, natural gas is split into hydrogen and CO_2. Carbon dioxide emitted in the steam methane reforming (SMR) process is stored or processed industrially through CCS or carbon capture utilization (CCU) technologies, even if a certain amount of CO_2 capture has not been defined [1, 45]. This suggests that blue hydrogen does not produce any CO_2 emissions. However, the long-term impacts of storage are uncertain, and leakage can still negatively affect the environment and contribute to climate change.

Blue hydrogen may have different carbon intensity values, depending on the equipment and the percentage of CCUS applied. For 95% CCUS, carbon intensity varies from 0.8 to 4.8 kg CO_2 eq./kg [51]. For example, for 5000 km transport distance and 90% CCUS, 1–3.9 kg CO_2 eq./kg H_2 is reported, while for 1700 km and 98% CCUS, it can be as low as 0.2–0.4 kg CO_2 eq./kg H_2 [53, 54]. In many projects, the captured carbon was intended to be used in oil refining processes or for urea production and eventually released back into the atmosphere. Whether blue hydrogen helps mitigate CO_2 or contributes is unclear.

The cost of blue hydrogen ranges from 1.4 to 3.2 US$ /kg H_2 [50, 51], increasing since the addition of CCUS increases the cost of production. The EU sponsored €587 million carbon capture and sequestration demonstration projects to be added to fossil power plants; however, none have been implemented to date [54].

3.1.3 Turquoise hydrogen

Turquoise hydrogen is produced *via* a process called methane pyrolysis in which natural gas is broken down by methane pyrolysis into hydrogen and solid carbon [42, 45]. Unlike the production of gray and blue hydrogen, the by-product of turquoise hydrogen is solid carbon, filamentous carbon, or carbon nanotubes [55]. Carbon remains permanently bound and is not combusted during further processing; this process is also CO_2-neutral. However, in the production of turquoise hydrogen, extracting the raw material, such as natural gas, often produces emissions. As a result, turquoise hydrogen is usually not completely climate-neutral, given the entire production process and the downstream carbon processing as a by-product [45]. The solid carbon can be found in various forms, and its price depends on its type: €150–400/t for coke.

€500–1000/t for carbon black, €1500–1800/t for activated carbon, and up to €1 Mio/t for carbon filaments [56]. Although pyrolysis has not yet been industrially applied in hydrogen production, methane pyrolysis for turquoise hydrogen production is currently one of the most discussed technologies. Apart from green and gray hydrogen, its technology has received significant investment, owing to the solid carbon that eliminates CCUS; a 1.9–4.8 kg CO_2 eq./kg H_2 impact has been reported [49]. The cost of turquoise hydrogen is estimated at 1.6–3.4 US$/kg H_2, slightly higher than blue hydrogen [49].

3.1.4 Other colors of hydrogen

Some colors are assigned based on the method of hydrogen production. Nuclear power drives the electrolyzers for purple, pink, and red hydrogen. Yellow hydrogen is hydrogen produced from a mixture of renewable and fossil fuels. Red and pink hydrogen have no direct carbon emissions, but they have a serious environmental spinoff from radiation and the risk of accidents. Low associated CO_2 emissions of 0.3–0.6 and 0.1 kgCO_2 eq./kg H2, respectively, have been reported [49]. Pink and red hydrogen could be an economically viable, cost-effective, and cost-competitive technology of 3.8–7 US$/kg H_2 and 2.2–2.6 US$/kg H_2, respectively [49, 51]. However, the cost of pink hydrogen depends primarily on the cost of the nuclear power plant.

3.1.5 Green hydrogen

Green hydrogen is produced on a CO_2-neutral basis through water electrolysis using electricity from renewable energy sources like solar, wind, and hydropower. In electrolysis, water is split into its oxygen and hydrogen constituent elements. Since neither the production of green hydrogen nor the end products, hydrogen and oxygen, are harmful to the environment or the climate, green hydrogen is climate-neutral [41, 42, 45]. Among the hydrogen colors mentioned above, green hydrogen is the only one produced with zero harmful greenhouse gas emissions. The resulting reaction produces only hydrogen and oxygen, emitting zero carbon dioxide [42]. For large-scale green hydrogen, larger electricity generation capacities from renewable energies are required. Electricity prices must also be affordable if green hydrogen is to be economically competitive. To produce 1 kg of green hydrogen (theoretically from stoichiometry) requires approximately 9 kg of water. This also applies to orange and pink hydrogen, while gray hydrogen, utilizing SMR, needs only 7 kg of water. However, a more realistic assessment of water consumption can be conducted in a future study [57, 58].

Of all the costs of the various colors, green hydrogen costs vary the most, from US$1.9 to 8.2/kg H_2 [49, 51], because of the different RES technologies that can be used to power the electrolyzers. Some countries, such as Chile, and some North African countries produce low quantities of green hydrogen using solar PV, whereas northern European

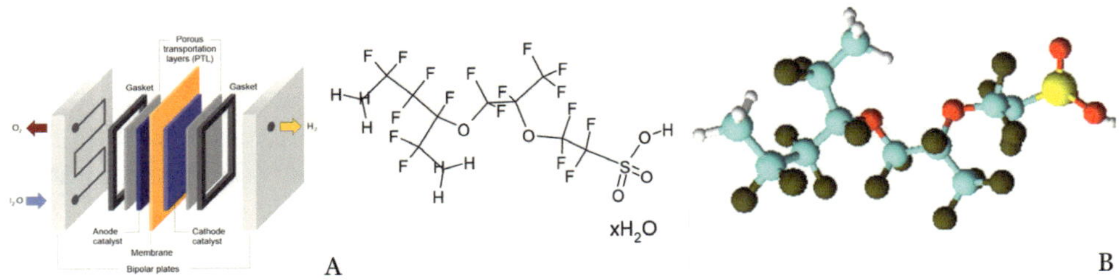

Figure 2.
(a) PEM water electrolysis technology, (b) PFSA is a linear fluorocarbon polymer composed of tetrafluoroethylene and perfluoroalkyl vinyl ether monomers.

countries will be at the high end of the range [59]. Green wind energy costs depend on whether the wind is onshore or offshore, centralized or decentralized. The cost of electricity depends on the capital cost of RES as the demand for green hydrogen grows, and policies provide incentives for constructing more green hydrogen plants; green hydrogen is expected to become cost-competitive by 2030 [12, 13].

Although green hydrogen remains puzzling in some cases, it has recently become the main focus of study and interest, as shown in **Figure 2**. Scientists and organizations have tried to normalize the definition and to reduce the number of terms used to avoid confusion. Instead of color, the International Energy Agency (IEA) uses the term "renewable hydrogen" or "low-emission" or "low-carbon hydrogen" as the hydrogen produced *via* electrolysis is powered by renewable electricity or *via* biomass-based routes with less—or even zero—CO_2 emissions with CCUS [51]. In such a case, green hydrogen can be renewable hydrogen. The latest IEA report [59] suggests that discussions around the different colors of hydrogen should be limited to carbon emissions as a way to avoid the confusion caused by using hydrogen colors. The EU refers to green hydrogen as "clean energy" if more than 90% of electricity used to run the electrolyzers comes from RES without adding extra renewable power capacity to the grid [60]. However, hydrogen production does not occur in just a few hours, and renewable electricity may vary by location. In this situation, the short supply of renewable electricity and its associated costs will limit the hours available for hydrogen production [60]. The EU has clarified that the green hydrogen label will not be used for hydrogen produced from nuclear-powered electrolysis. This recent plan has mentioned only green and low-carbon hydrogen. A 14.5 kg CO_2/kgH_2 threshold was adopted for "low-carbon" hydrogen (corresponding to a 50% reduction relative to the CO_2 emissions of hydrogen from coal gasification). The term "clean and renewable" hydrogen has also been defined when the emissions are lower than 4.9 kg CO_2/kgH_2 (corresponding to a 65% reduction relative to the "low-carbon" hydrogen threshold) [61]. According to the IEA report in 2021, 0.04% of green hydrogen was produced globally from 94 Mt. of hydrogen. Practically all the hydrogen still comes from fossil fuels (83%), mostly from gray hydrogen (62%), followed by 19% from a mixture of black and brown hydrogen, and 0.7% from blue hydrogen. The rest was a by-product in the chemical industry [51].

4. Types of electrolyzer for hydrogen production

The benefit of hydrogen and its growing global demand have stimulated researchers to rapidly develop efficient and sustainable hydrogen production technologies, and water electrolysis technologies (WET) have recently become a major interest. WET utilizes intermittent electricity from energy sources such as solar or wind power to produce large-scale hydrogen and oxygen. Deiman and Troostwijk were the first authors, back in 1789, to demonstrate the concept of water electrolysis by using gold electrodes in water with an electrostatic generator to provide an electrostatic discharge [62]. Later, in 1888, Lachinov built an industrial synthesis of hydrogen and oxygen via electrolysis [63]. Since the twentieth century, industrial electrolyzers have been developed using various electrolytes. The EWS technologies, which depend on different types of electrolytes (acidic, neutral, and basic), operating temperatures, and ionic transport (OH- and H+), can be categorized into several broad groups (**Table 2**): Alkaline Water Electrolyzer, Anion Exchange Membrane Water Electrolyzer, Polymer Electrolyte Membrane Water Electrolysis, Solid Oxide Water Electrolysis, and Proton Exchange Membrane Water Electrolyzer, typically abbreviated as AWES, AEMWES, SOWES, and PEMWES, respectively [17]. The characterization of these types of water electrolyzers are summarized in **Table 2**.

Parameters	SOWES	AWES	AEMWES	PEMES
Operating temperature (°C)	700–850°C	70–90	40–60	50–80
Operating pressure (bar)	1	1–30	< 35	< 70
Electrolyte	Yttria-stabilized zirconia (YSZ)	Potassium hydroxide (KOH) 5–7 molL-1	DVB polymer support with KOH/NaHCO3 1 mol L^{-1}	PFSA membranes
Separator	Solid electrolyte (above)	ZrO_2 stabilized with PPS mesh	Solid electrolyte (above)	Solid electrolyte (above)
Electrode/ catalyst (oxygen side)	Perovskite-type (e.g., LSCF, LSM)	Nickel-coated perforated stainless steel	High surface area nickel or NiFeCo alloys	Iridium oxide
Electrode/ catalyst (hydrogen side)	Ni/YSZ	Nickel-coated perforated stainless steel	High surface area nickel	Platinum nanoparticles on carbon black
Porous transport layer anode	Coarse nickel mesh or foam	Nickel mesh (not always present)	Nickel foam or carbon cloth	Platinum-coated sintered porous titanium
Porous transport layer cathode	None	Nickel mesh	Nickel foam or carbon cloth	Sintered porous titanium or carbon cloth
Bipolar plate anode	None	Nickel-coated stainless steel	Nickel-coated stainless steel	Platinum-coated titanium
Bipolar plate cathode	Cobalt-coated stainless steel	Nickel-coated	Nickel-coated stainless steel	Stainless steel; gold-coated titanium
Frames and sealing	Ceramic glass	PSU, PTFE, EPDM	PTFE, silicon	PTFE, PSU, ETFE

Parameters	SOWES	AWES	AEMWES	PEMES
Current density (Acm-2)	0.5–1	< 0.6	0.2–2	0–4 (up to 10)
Voltage range (V)		1.4–3	1.4–2.0	1.4–2.5
H_2 purity	N/A	99.8%	99.99%	99.99999%
Charge carrier	O_2-	OH-	OH-	H+
Efficiency	89% (laboratory)	59–70%	52–67%	65–82%
Capacity [Nm3h-1]	N/A	5	0.25–1	0.265–30
Stack lifetime [h]	N/A	100,000	>5000	50,000–80,000
Advantages	Low cell voltage electrolysis; Efficiency close to 100% at current densities of 1A/cm2)	Well established; Low cost	PGM-free electrocatalysts; Low alkaline corrosion	Immediate response; High current densities; High-purity hydrogen
	Long start-stop time; Rapid degradation; High temperature; Interdiffusion of the cell components; Corrosion products.	Low current densities; Alkaline corrosion; Expensive maintenance costs; Low hydrogen purity; High gas crossover.	Low OH conductivity; Short lifetime; Membrane degradation; High catalyst loading.	Acidic corrosion; Noble metal catalysts; High-cost components.
Development status	R&D	Large-scale application	R&D	Small-scale application

Table 2.
Summarizes the different types of water electrolyzsers.

4.1 Solid oxide electrolyzer (SOWES)

The solid oxide electrolyzer (SOWES) was first introduced by Donitz and Erdle in the 1980s [64] who found that, by conducting long-term tests in a single cell with electrolysis tubes of series-connected cells, an enrichment of hydrogen in the vapor stream of up to 85% was achieved at a current density of 0.3 A/cm^2 and 100% Faraday efficiency at a voltage of only 1.07 V. The greater efficiency, along with the ultrapure flexibility of hydrogen, highlights the appeal of the solid oxide electrolyzer [65]. The SOWES system operates at high temperatures up to 850°C using an ionic conductivity yttrium-stabilized zirconia electrolyte (YSZ). The SOWES process utilizes O_2 ion conductors (~0.05 S/cm at 800°C) derived from nickel/yttrium-stabilized zirconia [65] to produce oxygen at the anode side, while the water supplied to the cathode is then transferred to the anode through the solid electrolyte to generate hydrogen. Current research focuses on improving electrolyte conductivity, minimizing reactant crossover, and optimizing electrode stability. Several ceramic proton-conducting materials have been studied; among them, scandium-stabilized zirconia (ScSZ) stands out as an alternative to YSZ owing to its superior oxygen ion conductivity of ~0.128 S/cm at 800°C. Additionally, doped ceria with intermediate conductivity and temperatures of 0.079 Scm − 1 at 800°C and 0.012 S/cm at 600°C, as well as perovskite LaGaO3 doped with Sr. and Mg, namely, $La_{0.9}Sr_{0.1}Ga_{0.8}Mg_{0.2}O_3 - \delta$ (LSGM) (0.12 S/cm at 800°C) with high conductivity, have also been reported [66, 67]. The primary drawback of this electrolyzer is the high operating temperature, which results in degradation and a lack of stability, along with the elevated capital costs associated with the solid oxide electrolyzer (SOE) system that must be addressed before large-scale commercialization [67].

The state of the art for SOWES for hydrogen production has reported achieving a current density of 0.3–1 mA/cm^2 at operating temperatures of 700–850°C and 1 bar, with an efficiency of 40–50 kWh_{el}/kgH_2, and a lifetime exceeding 20 kWh in a cell size of 200 cm^2 [68]. It has also reported a current density of 0.3–1 mA/cm^2 at operating temperatures of 700–850°C, 1 bar, with an efficiency of 40–50 kWh_{el}/kgH_2 and a lifetime of more than 20 kWh in a cell size of 200 cm^2. The production capacity is expected to be 2 mA/cm^2 at temperatures below 600°C, at pressures greater than 20 bar, with an efficiency of <40 kWh_{el}/kgH_2 and a lifetime of 80 kWh in a 1000 cm^2 cell size projected for 2050 [68].

4.2 Alkaline electrolyzer (AWES)

AWES is a well-established electrolyzer type for hydrogen production. In AWES, technology anodic and cathodic electrodes are made of nickel (Ni)-coated, stainless steel-based electrocatalysts, immersed in a KOH (5 M) solution as the electrolyte at lower temperatures of 70–90°C [69]. An asbestos/ZrO_2-based diaphragm separates the two electrodes, producing hydrogen and oxygen gases separately. The low cost of using non-noble electrodes and robust stack durability has garnered significant attention as a state-of-the-art method for large-scale hydrogen production. However, using a liquid KOH electrolyte and a diaphragm as a separator for the anodic and cathodic electrodes can lead to low efficiency, low purity of hydrogen, and potential safety concerns if critical concentrations of hydrogen and oxygen are reached. Furthermore, the 0.252 mm-thick diaphragms result in high ohmic resistances between the two electrodes, significantly hindering the transport of OH- and drastically reducing current density at a given voltage [68]. Today, the investment cost of AWES is about US$ 500–1000/kW, and the system has a lifetime of 90,000 h [69, 70]. The drawback associated with AWES is low current densities of 0.1–0.5 A/cm^2 owing to moderate OH^- mobility and the usage of corrosive KOH electrolytes. To address this, several research institutes and organizations are actively working on developing new electrode materials and separators [69, 71–74]. A recent report outlines a cobalt-sulfo-boride compound as a bifunctional electrocatalyst for high-current-density in alkaline water electrolysis [75]. Akash et al. [75] reveal that the developed electrocatalyst was validated through a 50 h chronoamperometry stability test and a recyclability test involving 10,000 cycles of linear sweep voltammetry. Furthermore, Co–S–B was tested in an alkaline zero-gap water electrolyzer, achieving 1 A/cm^2 at 2.06 V and 60°C. Despite the drawbacks observed, AWES is a mature technology for industrial hydrogen production up to the multimegawatt range in commercial applications across the globe. It is known for reliability, scalability, and cost-effectiveness. Pricing varies based on capacity, custom features, and installation requirements. For smaller applications, procurement professionals expect prices in the range of US$500–US$1500 per kW, with total project costs influenced by system size, auxiliary equipment (e.g., compressors, gas separators), and installation. Finally, a new approach in the alkaline electrolysis, anion-exchange membranes (AEM) is under the development to increase conductivity. This innovative technology appears to be a promising development in alkaline water electrolysis.

4.3 Anion exchange membrane water electrolyzer (AEMWES)

AEMWES is an innovative version of AWES technology. The system replaces asbestos diaphragms with an anion exchange membrane (quaternary ammonium ion exchange membranes) for alkaline water electrolysis. Transition metal catalysts and a distilled water/low concentrated alkaline solution (1 M KOH) are used instead of noble metal catalysts, and a high concentration (5 M KOH solution) is avoided as the electrolyte [76]. Here, quaternary ammonium ion exchange membranes (AEM) are separators, transporting OH- from cathode to anode sites [76]. At this stage, water is reduced to H_2 and OH^- by receiving two electrons from the anode ($H_2O + 2\ e \rightarrow H_2 + 2\ OH^-$). At the anode, OH- recombines to form O_2 and H_2O, producing two electrons ($4\ OH^- \rightarrow 2H_2O + O_2 + 2e^-$) [76]. This technology can achieve a higher current density of 2 A cm^{-2} at 2.0 V, compared to 0–0.6 A cm^{-2} in AWES. The key challenges are limited stability (currently 35,000 h with a targeted stability of 100 1.00.000 h) and a high hydrogen production cost. The current hydrogen production cost is US$1279 kW/$H_2$ (2020), but a targeted cost reduction to US$ $\leq$ 300 kW/H_2 is expected by 2050 [77]. To address these challenges, several research studies are developing AEM water electrolyzers [78, 79]. A recent AEMWES performance report has been based on a poly(fluorenyl-co-aryl piperidinium) (PFAP)-based anhydrous cathode. Chen et al. [80] developed poly(fluorenyl-co-aryl piperidinium) (PFAP)-based anion exchange materials, including membrane electrolyte and binder, with high ionic conductivity and durability under alkaline media. The developed electrolyte with platinum group metal (PGM) catalysts (IrO_2 and Pt/C) achieved a new record cell performance of 7.68 A/cm^2 two at 2 V with 1 M KOH, and PGM-free materials (Ni- Fe) displayed an outstanding current density of 1.62 A cm^{-2} at 2.0 V. Additionally, PGM and PGM-free materials demonstrated excellent durability of more than 1000 h of operation at a constant current density of 0.5 A cm^{-2} at 60°C without any voltage degradation. Extensive research on AWE and AEMWE continues, focusing on developing membranes and electrocatalysts to create stable AEMs and high-efficiency and low-cost catalysts. The significant drawbacks of the previously mentioned water electrolysis technologies have directed unparalleled attention to PEMWES for green hydrogen production.

4.4 PEM electrolyzer systems (PEMWES)

Proton exchange membrane (PEM) technology was developed in the mid-1960s [81] and was primarily used for life support in space and submarines, focusing on oxygen generation rather than hydrogen generation. Russell et al. [82] first reported on PEM water electrolysis for hydrogen generation in 1973. The authors demonstrated the remarkable performance of PEM water electrolysis at the General Electric Company, achieving a current density of 1 to 2 A/cm^2 at a voltage range of 1.88–2.24 V, with stability exceeding 15,000 hours. Compared to AWES and AEMWES, PEM water electrolysis technology (**Figure 2a**), also called solid polymer electrolyte (SPE) electrolyzers, utilizes an ultrathin solid polysulfonated membrane (PFSA: Nafion, fumapem) as a proton conductor [83, 84]. PFSA is a linear fluorocarbon polymer composed of tetrafluoroethylene and perfluoroalkyl vinyl ether monomers, with terminal side chains containing sulfonic acid groups that facilitate proton transport within the membrane when suitably hydrated (**Figure 2b**) [62–66, 81, 85–88].

The solid polymer electrolyte is preferred owing to its high proton conductivity (0.07–0.08 S/cm at room temperature) and lower thickness (20–300 μm) compared to SOWES (~0.05 S/cm at 800°C) and AEMWES (0.252 mm) [10]. The polymer must also possess exceptional chemical and thermal stability, high electrochemical performance, and hydrophilicity to retain moisture within the catalyst layer [10]. The PEMWES operates at low temperatures (20–80°C), and the gas permeability of the proton exchange membrane enhances proton transport while isolating cathode and anode electrodes, resulting in large current densities (>2 A/cm^2), excellent conversion efficiency (80–90%), and high hydrogen purity (>99.99%). On the other hand, catalyst properties such as porosity and electrochemical surface area (ECSA)—as well as cell operating conditions, for example, current density and voltage—are other essential factors to consider when determining the efficiency of hydrogen production. The electrochemical water splitting process involves anodic oxygen evolution reaction (OER) and cathodic hydrogen evolution reaction (HER) in a water electrolysis PEM cell. While the OER process proceeds *via* four-electron transfer to produce oxygen with a large overpotential, the HER is a simple two-step process involving two electrons and occurs at a very low overpotential. From this point of view, OER processes are kinetically sluggish compared to HER. The basic reaction of water electrolysis is described in Eq. 1.

$$2H_2O + \text{Electricity}\left(237\,\text{kJ.mol}^{-1}\right) \rightarrow O_2\,(g) + 2H_2\,(g) + \text{Heat}\left(48.6\,\text{kJmol}^{-1}\right) \quad (1)$$

The key parameters (energy efficiency, operational temperature, current density, and long-term durability) quantitative comparison of PEMWEs versus other technologies are reported in **Table 3** [76–82].

In summary, while both PEMWEs and AEMWEs operate using PGM-free catalysts, PEMWEs are recognized for their high performance related to high efficiency, high current densities, and high membrane chemical and thermal stability at low temperature. AEMWEs offer potential cost advantages. Besides, AWEs provide long-term durability and are a well-established technology, but the lower efficiency and current density compared to PEMWEs are outlined. Understanding the trade-offs between efficiency, current density, and durability is essential for optimizing these technologies for different applications.

Thus, highly efficient catalysts are essential to enhance the reaction kinetics of OER and improve the overall efficiency of PEMWE cells. Noble metals and their oxides are the most efficient electrocatalysts for the HER and OER reactions under all pH conditions. Platinum serves as the benchmark catalyst for the HER, while ruthenium oxide/iridium oxide are the most effective for the OER [10, 89].

Keys parameters	AWEs	AEMWEs	PEMWEs
Efficiency (%)	65	>90	80–85
Operational temperature (°C)	60–90°C	wider temperature range	50–80
Current density (A.cm^{-2})	0.2–0.4	0.2–2.0	1–2
Long-term durability(h)	60.000 to 90.000	1000 at 60°C	20.000–80.000

Table 3.
Key parameters comparison of PEMWEs versus AEMWEs and AWEs.

4.4.1 Oxygen evolution reaction (OER)-electrocatalysts for hydrogen production

Because of the pivotal role of oxygen evolution reaction (OER) in water electrolysis, developing robust, highly active, and stable electrocatalysts for OER-PEM electrolyzers is important. Until now, electrocatalysts for OER-PEM have predominantly centered on Ir- and Ru-based materials [17, 90–92], their oxides [93, 94], and alloys [95, 96]. Various nanostructures of Ir- and Ru-based electrocatalysts have been studied to enhance OER activity and stability in acidic media [97–100]. Titanium-based materials, such as TiO_2-supported Ir at IrO(OH)x core–shell nanoparticle catalyst [101], Ti-PTL/TiO_2 - x/Ir catalyst [102], TiC material [103], noble metal catalysts, and protective coatings, are needed not only to provide long-term stability to cell components but also to ensure optimal electron conductivity and cell efficiency. Besides Ir/Ru-oxides and their material-based counterparts, researchers have also developed alternative cost-effective, high-performance, and durable proton exchange membranes and electrocatalyst materials for OER-PEM in acidic electrolytes [17, 104–107]. Recently, in 2025, Jiaqi et al. [104] reported that a PEM electrolyzer with Raney-IrOx anodes achieved 3 A. cm^{-2} at 1.8 V with a low geometric Ir loading of approximately 0.3 mg Ir/cm^2, meeting the technically important power-specific Ir utilization target of 0.05 gIr/kW.

In 2023, Siracusano et al. [105] reported the high activity of Ag/Ti-suboxides as a non-PGM anode electrocatalyst deposited on a 212 NAFION® membrane for PEM water electrolysis in an acidic environment, showcasing the promising performance of 0.6 A cm^{-2} at 2 V/cell at 80°C with excellent stability (degradation rate < 14 μV/h during a 1000-hour test). The catalyst loadings for the anode and cathode were 12 and 1.25 mg cm^{-2}, respectively, for Ag/TiO_{2n-1} and 40% Pt/C.

Spinel Co_3O_4 has also gained significant attention owing to its cost-effectiveness and impressive electrocatalytic properties among various cobalt oxides. Its lattice structure features Co^{3+} ions in octahedral sites and Co^{2+} ions occupying tetrahedral sites. Notably, the Co^{2+} sites, residing in tetrahedral positions, act as the active centers for OER [108]. However, its application in acidic OER environments is hindered by the undesirable formation of unstable CoO_2, which further decomposes to yield CoO and release oxygen [107]. To overcome this challenge, researchers have explored diverse approaches to enhance the stability of spinel Co_3O_4 or other non-noble transition metals in acidic environments. Peiyu et al. [109] synthesized Na-doped Co_3O_4, prepared *in situ* by programmed electrostatic spraying. A Nafion N117 proton exchange membrane separated the anodic and cathodic compartments. A PEMWE was constructed with Na-Co_3O_4 catalyst-coated CP as the anode and commercial 20 wt% Pt-deposited Ti felt as the cathode. A PEMWE device assembled using Na-Co_3O_4/CP as the anode achieved a current density of 100 mA/cm^2 at a voltage of only 1.62 V (using a Nafion N117 membrane). Using the same method, Qisheng et al. [110] reported a Cr-doped Co_3O_4 catalyst, achieving a current density of 1.5 A/cm^2 at only 2.17 V that remained stable at 500 mA/cm^2 for 100 hours in PEMWE. The introduction of chromium into cobalt spinel (Co_3O_4) catalyst regulates the electronic structure of cobalt, resulting in a high oxidation state and increased Co–O covalency within a stable structure. Moreover, Fanfan et al. [111] introduced P atoms into the lattice of spinel Co_3O_4 (P-Co_3O_4) to replace octahedrally coordinated Co^{3+} *via* a hydrothermal process followed by thermal treatment. The obtained P-Co_3O_4 demonstrated satisfying stability over 30 hours in a PEM electrolysis device with an overpotential of 400 mV@10 mA/cm^2 in 0.1 M HClO4.

4.4.2 HER-PEM electrocatalyst for hydrogen production

On account of their high performance and stability, Pt and Pt-based materials are documented as the pioneer electrocatalysts for the cathode side in PEM electrolyzers for hydrogen production [17, 21, 90, 105, 107, 112, 113]. Many strategies (such as low loading, structure control, and even transition metal substitution) have been developed to reduce the usage of noble metals and, simultaneously, improve their catalytic performance. Ayers et al. [114] reported the synthesis of Pt/Ru core–shell catalysts, where the Pt is placed on the surface of Ru particles. The Nafion N117 membrane with 5 cm × 5 cm active area was used as electrolyte, the Pt/C cathode was loaded with 0.1–0.3 mg of Pt/Ru, and the IrO_x anode had an Ir loading of 0.1 mg/cm^2. The electrolyzer stack was initially allowed to operate at 1.8 A/cm^2 for up to 500 h at 50°C or 80°C and 435 psi differential pressure. The cell's performance showed the steady state (potential cell of 2.3 V during 600 h) performance for a cell operated at 2 A/cm^2, 50°C temperature. So, to date, the core–shell catalysts have been limited to the gas diffusion layers configuration in electrolyzers owing to current manufacturing capability.

Martin et al. [115] used the electro-spraying deposition technique to prepare electrodes with a Pt loading of 0.1 mg Pt cm^2 and an active area of 4 cm^2. Polytetrafluoroethylene (PTFE) was added to prepare electrodes by incorporating the corresponding amount of PTFE to achieve catalytic deposits with 5–30 wt% PTFE (in solids). The maximum performance of MEAs (Pt/C-PTFE) operates at 0.2 A cm^{-2} and undergoes a long transient of about 300 h until a steady-state voltage of 15 μV h^{-1} is reached. Despite the low Pt loading, the degradation rate of the 30 wt% PTFE MEA aligns well with the values found in the literature for the same operating temperature (180°C). Similarly, Kannan et al. [116] reported a voltage decay of 12.5 μV h^{-1} at 180°C with a cathode Pt loading of 0.72 mg Pt cm^{-2} and an operating cell at 200 mA cm^{-2}, demonstrating long-term durability of more than 5500 h. A comparable low cathode Pt loading of 0.72 mg Pt cm^{-2} and 0.35 mg Pt cm^{-2} at a high temperature of 180°C was reported by Gabialti et al. [117] and Pinar et al. [118], respectively. Transition metal substitution electrocatalysts for HER-PEM have also been reported; Pd/NCNTs [119], MoS [120], FeP Nanocatalyst [121], and Boron-Doped Carbon nanoparticles supported palladium [122].

4.4.3 Membrane electrode assembly (MEA)

The core of a PEMWES is the membrane electrode assembly (MEA), where the crucial electrochemical reactions occur. The MEA is prepared with proton exchange membrane, ionomer solution, and catalysts [83, 84]. Proton polymer membranes mainly focus on Nafion membranes (Nafion 115, 117, and 212). The fabrication of the MEA plays an important role in determining the electrolyzer's performance, production costs, and longevity. The methods often used for water electrolysis MEA are the catalyst-coated membrane (CCM) and catalyst-coating substrate (CCS) method [123]. In the CCM approach, the selected catalysts are mixed with ionomer in an appropriate solvent and dispersed under sonication [69, 70, 123]. The resulting ink is coated on both sides of the membrane, at the anode and cathode, forming the so-called CCM, which is then fabricated with anode and cathode gas diffusion layers to form the MEA. This CCM-MEA route provides good contact between the membrane and the obtained MEA; it usually has small contact resistance and thus improves proton transfer conductivity.

In the CCS fabrication process, the catalysts are first coated onto the substrate and form the anode and cathode sides, which are then used to fabricate CCS-MEA. This CCS-MEA production does not suffer from the swelling of the membrane but is limited by the insufficient cohesion of the catalyst layers with the membranes. Reasonably, the CCS-MEA has large contact resistance and low proton-transfer conductivity [90, 123]. Besides enhancing the performance, these procedures optimize the noble metal loadings by increasing catalyst utilization.

4.5 Technology readiness level

The AWES (which mostly relies on orange hydrogen) and PEMWES (which relies on green hydrogen) are both electrolyzers that are commercially available and have a technology readiness level (TRL) of 9 [51]. Although AWES for hydrogen production has been on the market longer than PEMWES, recently PEMWES has received significantly increased attention, leading to a reduction in the price of the electrolyzers. Several manufacturers worldwide have lately developed large-scale PEMWES for hydrogen production in industrial applications [1]. It is expected that the stability of PEM water electrolysis could reach up to 100.000 h with negligible loss of performance, satisfying the requirements of industrial utilization. There is wide agreement that PEM water electrolysis could be the most environmentally friendly and efficient technology for producing high-purity green hydrogen from water, coupled with intermittent renewable energy sources. However, the large-scale practical application of commercial PEMWES is prevented by the high costs of its materials, including specific catalysts and construction materials: (a) under the highly corrosive acidic environment of PEMWES, only a few precious metal-based electrocatalysts, usually Pt and Ir/Ru-O_2 for cathode and anode, respectively, are available for practical activity and stability; (b) titanium materials are widely used to construct the stack components of PEMWES, which increases electrolysis cost; and (c) expensive Nafion-type membranes are used as electrolytes in PEMWES.

4.6 Future research on PEM technology for hydrogen production

One of the important key advantages of PEMWEs is the production and compression of hydrogen at the same time. For that, some components of PEMWEs, such as MEA optimization, fluctuating conditions of renewable energy sources integrated with PEMWEs, as well as future challenges, need to be addressed. As mentioned in paragraph 4.4.3, MEA is the core of PEMWEs. Optimized MEA means improved proton exchange perfluoro sulfonic acid (PFSA) membrane, ionomer solution, catalysts, and the assembly process associated with two bipolar plates [83, 84]. Various reports on the effect of anode catalyst loading, temperature, membrane thickness, flow field design, and anode porous transport layer have been mentioned [124]. Lirong et al. [125] reported that increasing the Ir catalyst from 1 to 3.5 mg at the anode site increases the ECSA as well as the performance of PEMWEs. Additionally, Liso et al. [126] show that the use of a Nafion-117 membrane at a high temperature of 80°C decreases the efficiency of the cell due to the dissipation [127] of heat. The influence of MEA compression was further investigated by Martin et al. [128] revealed that the higher compression enhances cell performance while increasing mass transport. The effect of PTL thickness on liquid water was also reported by Chen et al., and they found that PTL thickness significantly influences liquid water access to the catalyst layer, which eventually affects the cell's performance and the degradation of MEA. So, future research trends to optimize MEA's for water electrolyzers, particularly for scale-up and integrating with renewable energy sources such as photovoltaic or wind fluctuations, revolve around several interrelated themes across material science that involve the development of advanced ionomer development [115] that will enhance proton conductivity while reducing gas crossover while improving mechanical strength. Durable catalyst layers that could lower PGM content with non-precious metals with high activity and stability, together with nanostructured materials such as carbon or conductive ceramics to improve catalyst dispersion [105, 108–110, 121]. Through architectural design, the gradient gas layer can be enhanced by optimizing (i) the water-gas distribution. (ii) Improving ionic conductivity while minimizing thickness to reduce the ohmic losses lead to higher power output efficiency and improving overall performance of MEAs and (iii) Study chemical and mechanical degradation under real conditions, including catalyst dissolution and delamination.

The expected demand of hydrogen scales up electrolyzer is 4–5 TW capacity by 2050 [129]. However, diverse scale-up challenges include maintaining consistency in the manufacturing process to produce uniform catalyst layers for MEA production, and rigorous quality controls must be in place. Another potential problem due to stacking is maintaining temperature consistency and hydration across all the MEA; this can be built into the design to incorporate smart flow field and humidification control systems needed for scale-up. This will include substantial resources innovation, in terms of materials, supplies, and the manufacturing methods used to produce electrolyzers at Large scale application. For example, PEM electrolyzers still rely on Ir-catalyst at the anode site, perhaps around 7–8 tons of Ir/year is extracted globally with the cost of R 2508198, 74 R (Ir/kg). Therefore, stable supply chains of materials for all electrolyzer technologies and a reduction in cost by the introduction of non-precious metals without losing performance are urgently required.

PEMWE is well-suited for integrating renewable sources (solar, wind) due to its adaptation to varying power input. Still, this integration presents some challenges related to system stability, efficiency, and long-term durability of the electrolyzer [130]. This requires real-time monitoring and control systems to adapt MEA operation to the profiles of PV and wind output. The integration of battery backup power to buffer intermittent power as a hybrid system and storage allows for buffering intermittent power sources like solar or wind while improving reliability and enabling a more consistent energy supply.

5. PEM electrocatalyst characterization tools

5.1 Analytical characteristics

Structure-activity property relationships and reaction mechanisms have been a significant challenge for acidic OER-PEM catalysts. During electrochemical reaction, most electrocatalysts undergo reversible, irreversible, or reconstruction structural changes. Furthermore, the synthesis approach (such as hydrothermal/solvothermal, precipitation, sol-gel, impregnation, solid-state synthesis, thermal decomposition, electrodeposition, etc.) of the electrocatalyst affects its shape, phase composition, size, and efficiency in catalyzing OER [17]. Therefore, to elucidate the electrocatalysis structure, shape, size, and so on, the clear mechanism, after and before electrolysis, several operando techniques, include scanning microscopy (SEM-EDS, HR-TEM), spectroscopy analysis (FT-IR/Raman, XPS, XAS), XRD, and DFT have been established.

Microscopy techniques such as scanning electron microscopy (SEM-EDS) and transmission electron microscopy (HRTEM) can provide insights into the elemental composition and distribution, surface morphology, and particle size of the catalyst from the nanometer to micrometer scale. The morphology, or physical shape and structure of the electrocatalyst are key determinants of its performance [131]. Different morphologies, including nanosheets, octahedral, nanorods, nanoparticles, nanotubes, and others [18, 132, 133], significantly impact the electrochemical surface area (ECSA), electron conductivity, and mass transport between reactants and products. The surface area of an electrocatalyst exposed to an electrolyte directly affects the number of active OER sites available. A higher ECSA generally translates to improved OER activity. Certain morphologies, such as nanostructures and nanosheets, offer a large ECSA owing to their significant surface-to-volume ratio [17]. Fourier Transform Infrared (FT-IR) spectroscopy is frequently used to determine the functional groups, their binding mode, and the variations of key intermediates in chemical reactions [113]. X-ray photoelectron spectroscopy (XPS) is a useful technique for surface analysis that operates on the principle of the photoelectric effect proposed by Albert Einstein in around 1905. This technique was used as a standard surface characterization to detect and quantify the chemical and oxidation state of elements present in material at a nanometer range of 1–10 nm in catalytic materials [134, 135]. X-ray absorption spectroscopy (XAS) is a very sensitive technique used to illustrate the electron transfer and probe the structure and composition of heterogeneous catalysts, revealing the nature of the active sites and establishing links between structural motifs in a catalyst, local electronic structure, and catalytic properties [113, 136]. X-ray diffraction (XRD) is the most important method for determining structures. This method describes structures such as bond lengths, bond angles, and the relative positions of ions and molecules in a unit cell, which allows one to predict the structure of compounds and explain the trends in many properties. Recently, theoretical tools have been intensively employed to complement experimental studies. Computational approaches use computer simulations or modeling based on theoretical chemistry to study the properties of compounds. Density Functional Theory (DFT) is one of the promising computational modeling approaches used in the field of inorganic chemistry [137–139] owing to the limited storage, high cost, and time involved in real-time laboratory experiments. The DFT method not only is used to study the electronic properties (HOMO, LUMO, MESP, DOS, etc.), or to investigate the chemical reactivity (μ, η, S, *etc*) of compounds, but also has progressed over the years in accessing deeper and greater insights into surface chemistry [140]. Using this technique, the real active sites for the electrocatalyst during the HER and OER can be identified.

A better understanding of the degradation mechanisms of MEAs designed for polymer electrolyte membrane (PEM) water electrolysis is based on a novel short-side chain (SSC) perfluorosulfonic acid (PFSA) membrane, (e.g. Aquivion®), with various supported Pt/C cathode and nanosized $Ir_{0.7}Ru_{0.3}O_x$ solid anode noble metal loadings that were also reported by S. Siracusano et al. [141], who referred to the *ex situ* characteristics such as XRD, SEM/TEM, XPS, and so forth. The XRD pattern reveals a rutile structure of the IrO.RuO. Solid solution and the cubic structure of Pt after a 1000 h durability test at 1 A·cm^{-2} for the low catalyst loading MEA.

Additionally, a slight change in crystallite size and a larger relative particle size increase were recorded for the cathode. This could probably represent a significant electrochemical active surface area (ECSA) loss when considering the inverse relationship between particle size and surface area and the smaller initial crystallite size. The TEM images of the anode layer scraped from the MEA show that the IrRuO NPS are embedded into the Aquivion ionomer before and after the 1000 h durability test (low-loaded MEA). However, after the durability test, the anode layer appears to be less agglomerated than at the beginning of the durability test. Moreover, recent trends have outlined the use of operando characterization tools to investigate low cost, low toxicity, high abundance, interesting redox properties, and various valence states of non-precious metal catalysts (Mn, Ti, Co, Ni, and Cu) as an alternative to Pt, Ir, and Ru precious metal [105, 106, 110, 111]. For example, the incorporation of Na^+ into Co_3O_4 results in an increase in particle size compared to pure Co_3O_4, as revealed by SEM analysis. XRD and SAED patterns confirm the polycrystalline structure of Na-Co_3O_4, with a high dispersion of Na+ ions in the crystal lattices. Large Na+ ions (~1.12 Å) introduced into the Co_3O_4 lattice (0.63 to 0.65 Å) tend to occupy the octahedral sites of Co^{3+}. This selective occupation leads to the distortion of the reticular network, hence the formation of oxygen vacancies (detected by high-resolution O1s XPS spectra) that allow rapid oxygen evolution and durability of the electrocatalyst, Na-Co_3O_4 at the electrode under acidic conditions (0.5 M H_2SO_2). The oxidation state of Na and Co $2p_{3/2}$ and Co $2p_{1/2}$ was also confirmed by XPS survey.

The post operando characterization PEMWE through SEM and XPS reveals that the morphology, composition, and electronic states of the spent Na-Co_3O_4 (AA) electrode remain nearly unchanged compared to their pristine states.

At the same time, the operando characterizations were also used to investigate the catalysts used in or bifunctional electrode. This type of electrode, which is part of the MEA, can catalyze both the OER in electrolysis mode and the Oxygen Reduction reaction (ORR) in fuel cell mode in PEMWEs. However, the catalyst selection for a bifunctional electrode is one of the formidable challenging problems. Recently, Padilla et al. [142] have reported the inkjet low-loading mixture of Pt, IrOx as a bifunctional oxygen electrode for a unitized regenerative fuel cell (URFC) and characterized it through SEM and CV. The characterization of Ir/TiOx, IrO_2 on Pt was also reported.

In situ visualization analysis, such as high-speed optical imaging, neutron imaging, and synchrotron radiography approaches, has also been established to disclose the behavior of the bubble inside the electrolyzer cells to prevent electrolyzer efficiency and MEA degradation [113]. These nondestructive, complementary imaging techniques use more penetrating X-rays and neutrons for visualizing and quantifying many processes that occur inside electrolyzers at internal interfaces between layers in the membrane electrode assembly (MEA) or elsewhere inside the cell [28, 128], which is not possible with conventional imaging techniques. Neutrons can infiltrate metallic materials and are highly sensitive to water and H_2; the two-phase flow in a PEMWE was then imaged using *in operando* neutron imaging. K. Lee et al. [143] used *in operando* neutron imaging at operating current densities of 9 $A{\cdot}cm^{-2}$ as a platform to analyze the role of reactant flow rates on the performance indicator for gas-evolving electrochemical energy conversion and to elucidate the underlying mass transport behavior in the PEM electrolyzer. These findings unequivocally demonstrate the importance of analytical characteristics of PEM-electrocatalysts for hydrogen production.

5.2 Electrochemical characteristics

Electrochemical characterization approaches such as polarization curves, electrochemical impedance spectroscopy (EIS), and cyclic voltammetry (CV) are suitable for assessing integrated systems [144]. The linear scanning voltammetry (LSV) and CV are the most widely used voltammetry techniques for studying redox reactions of organic and inorganic species because they are unmatched in their ability to provide information on the steps involved in electrochemical processes with only a modest expenditure of time and effort in acquiring and interpreting data. Overpotential (η) is the most important parameter to define the activity of an electrocatalyst, which is commonly determined from the Linear Scanning Voltammetry (LSV) in both positive and negative zones. Tafel potential or Tafel slope is one of the primary studies in evaluating the kinetic aspect of the studied electrocatalyst, derived from the well-known Butler-Volmer equation, which predicts the kinetics of the electrocatalytic reaction at the interface of the electrode-electrolyte.

The Tafel slopes could be calculated as follows: the ohmic loss region of the current (I) versus potential (E) data. Calculate the current density (j) by dividing the current density (I) by the electrode GCE surface area A/cm^2. The current density is converted to a logarithmic scale and plotted against the applied potential, E. The Tafel slope region is generally identified within the range where the kinetics are controlled by electron transfer, not diffusion. The ohmic drop correction was done by subtracting the current (I) and the solution resistance (R_s) given by IR_s from the measured potential. Given in Eq. 2.

$$Tafel\ slope,\ b = \frac{\Delta E}{\Delta \log(j)} \tag{2}$$

The resulting Tafel slope provides insight into the electron transfer process and can indicate whether the reaction is under activation control or mass transport control.

Electrochemical impedance spectroscopy (EIS) is an analytical technique used to analyze interfacial properties related to bio-recognition events at the electrode surface. The EIS could be used to explore mass transfer, charge transfer, and diffusion processes. Accordingly, the EIS can study intrinsic material properties or specific processes that could influence an electrochemical system's conductance, resistance, or capacitance. Chronopotentiometry (CP), in its most basic form, consists of applying a single voltage step at time t_0 and then measuring the current that results from the applied potential. The simplicity of CP makes it an ideal technique for analyzing a basic voltammetry operation and evaluating the stability of the material at the electrode surface during electrochemical reactions.

6. Hydrogen production PEM electrolyzer system safety

The growing challenge of rapidly scaling up sustainable hydrogen production requires careful attention to the safety of production, storage, transport, and equipment. Policymakers must consider the safety option. Three factors illustrate a conventional way to avoid combustion: fuel (H_2), oxidizer (O_2), and ignition. The strategy is to keep the hydrogen away from the oxidizer, avoiding accidental mixing.

6.1 PEM electrolyzer system safety strategies

What happens if H_2 and O_2 combine during hydrogen production? Water electrolysis is a complex system involving several physical phenomena. Hydrogen and oxygen both have distinct properties. The two miscible gases can be amplified if a fault allows them to mix within an electrolyzer system. Hydrogen is highly flammable, with one of the widest flammability ranges of any fuel and will ignite at concentrations ranging from 4 to 75% in air [145]. Thus, producing a combustible mixture of hydrogen is much simpler than other fuels like propane, which will only ignite at 210% concentration in air. Hydrogen requires as little as 0.02 mJ of energy to ignite in air. If H_2 and O_2 mix accidentally, the results can be catastrophic. Hydrogen ignites easily in pure oxygen and will combust with significantly more energy. If the oxygen concentration increases, the explosion will be more severe. A typical example is a hydrogen storage tank explosion at an electrolysis unit in Gangneung, *South Korea, which led to two deaths, six injuries,* and over $30 million in damage [145, 146]. The preliminary results from the legal investigation indicate that the explosion was caused by a static spark in the hydrogen buffer tank, combined with oxygen in the hydrogen stream. This oxygen contamination occurred due to a malfunctioning separation membrane in the electrolyzer, which allowed oxygen to diffuse into the hydrogen stream. The most effective mitigation strategy is to utilize the H_2 product immediately, which helps mitigate the risk of combustion.

What happens if O_2 is accumulated in the catalyst layer during the PEM electrolyzer process? PEM electrolysis starts by breaking down water molecules at the anode side, causing the production of O_2 at the anode and H^+ ions that traverse the membrane to form H_2 at the cathode side. Unintentional accumulation of O_2 bubbles in the PTL anode side, where the liquid water is transported to the catalyst layer, can hinder the liquid water supply and create localized hotspots, leading to increased mass transfer losses and reducing the effective surface area available for the electrochemical reaction. This causes degradation of the catalyst layer and the MEA, negatively impacting performance and durability and restricting the large-scale commercialization of PEMWEs [147, 148]. There is a clear need for more extensive studies to investigate and better understand the impact of bubbles on the performance of PEM electrolysis systems and safety issues. What happens if O_2 is accumulated in the air during the PEM electrolyzer process? Numerous electrolysis systems purge their oxygen into the atmosphere. According to the stoichiometry equation of PEM electrolysis, O_2 in volume is almost half of the H_2 produced. Regarding different molar weights, oxygen is produced eight times more than that of hydrogen, at ambient pressure. For example, it is estimated that if 5 GW of electrolyzer capacity is generated in Sweden, it will produce about ~750.000 tons of H_2/year for 45 weeks of production, and 6.0 million tons of O_2/year will be produced as a by-product. The supply of O_2 in the atmosphere enhances oxygen-enriched atmospheres, significantly increasing the risk of ignition and fire due to the reactivity of oxygen. In such a case, compounds that may not burn under normal conditions may burn energetically in an oxygen-rich milieu and propagate faster [149]. Research also reveals that the chemistry of Earth's atmosphere in prehistoric times altered dramatically with increased oxygen levels and the replacement of methane (some methane is still present today). It is assumed that the accumulation of oxygen in the atmosphere led to one of the first ice ages on Earth. Methane is a greenhouse gas because it traps the sun's heat and warms the planet. Replacing methane with oxygen caused global temperatures to cool enough to form ice sheets stretching from the poles to the tropics [150]. Though the oxygen can be harmlessly emitted, it is wise to explore the prospect of oxygen utilization for electrolyzer capacities. As the second most widely used gas after nitrogen, oxygen can be used in various industries, such as medical, industrial, and food processing industries [149]. Using O_2 from PEMWES concerns could supply medical oxygen to a capacity of 2500 hospital beds, requiring approximately 12.1 million Nm^3 of oxygen per year. A 20 MW electrolyzer powered by an offshore wind farm could provide purified oxygen to be distributed *via* cylinders at 200 bar [150].

Mitigation strategies: Other plausible strategies to avoid the risk during H_2 production *via* PEM electrolysis include:

i. Ventilation: Install mechanical ventilation fans to sweep the electrolyzer enclosure and reduce hydrogen accumulation.

ii. Venting and disposal: Evaluate concentrations for all operational modes and create procedures for venting O_2 and H_2 during system start-up, operation, and shutdown.

iii. Purging: Purge H_2 systems with nitrogen or other inert gases before and after use (optional).

iv. Detection: Use personal monitors, leak sniffers, imaging cameras, and fixed H_2. Incorporate detection into procedures and automated safety controls. Industry relies on hydrogen gas detectors to detect leaks. Industrial gas detection (IGD) has two technologies suitable for detecting hydrogen: pellitory and electrochemical sensors.

v. Safety planning: Regularly train all personnel working around hydrogen or electrolysis systems. Develop clear plans for maintenance, operation, and emergencies [146].

6.2 Hydrogen storage and transport safety

Hydrogen storage and transportation are the main obstacles to developing a fully hydrogen-powered economy. It is currently more cost-effective to relocate large hydrogen consumers to producers' sites than ship hydrogen to them long distances. Ambitious policies and investments are needed now to support the hydrogen storage, transportation, and conversion infrastructure that will be essential for the future hydrogen market to function properly. Hydrogen as an energy carrier occupies an enormous volume in normal conditions (1 kg of hydrogen at ambient temperature and atmospheric pressure takes a volume of 11 m^3), making it difficult to store [151]. Generally, hydrogen can be stored in pure form as compressed gas or liquid, or in materials (solid form). Four type of cylinders are commonly used for compressed hydrogen storage: Type I: all-metal cylinders (45 MPa), Type II: all-metal hoop-wrapped composite cylinders (20–30 MPa), Type III: fully wrapped composite cylinders with metallic liners (30–45 MPa), and Type IV: fully wrapped composite cylinders (>45 MPa) [49, 152]. Thus, different safety measurements must be taken according to the hydrogen delivery model (if in gaseous or liquid form). Gaseous hydrogen is stored under high pressure (between 35 and 70 MPa) and low energy density (40 kg/m^3@70 MPa), making it difficult to meet the requirements for a long cruising range and the related risks regarding leakage and explosion. The key to gaseous hydrogen storage is to improve hydrogen storage density while ensuring safety [152]. The main exploitation of hydrogen gas lies in its electrical energy consumption during the filling process, that is, 20 to 40% of the total cost. Safety and regulatory compliance issues must be considered, given the potential for leaks in exploiting hydrogen gas at very high pressures. Since the transport distance increases with the length of the tubes, which induces an increase in the price of H_2/kg, there is a gradual increase in safety issues. A recent report has stated that for transport distances of 50 to 500 km, prices vary from approximately 0.6–7 US$ of H_2/kg. To reduce the cost of hydrogen gas while improving the safety angle of storage and transport, three aspects must be considered: (i) developing new high-performance materials to reduce the weight and price of containers while improving safety, (ii) optimizing compressor efficiency to reduce energy consumption, and (iii) adopting advanced monitoring and control technologies to improve storage system security while reducing security costs.

Liquid hydrogen storage is mainly under cryogenic cooling, below -250°C, and the energy density can reach 70 kg/m^3. The high energy density of liquid hydrogen storage reduces the hydrogen storage space. The evaporation latent heat of liquid hydrogen is very large, which has advantages in thermal management and safety protection. Companies such as BMW and Toyota have all launched liquid hydrogen fuel cell vehicle prototypes with a cruising range of 500 km. However, the high energy consumption of 12 kWh/kg and the evaporation loss of the liquid hydrogen tank are also problems [152]. Safety and regulatory costs for liquid hydrogen storage account for about 10–20% of the total cost, as it is stored at very low temperatures. In addition, the transportation cost of liquid H_2 is about 2–3.1 USD/kg over a transportation distance of 50–500 km. Cost optimization can be achieved by (i) improving the efficiency of hydrogen liquefaction through technological innovation to reduce energy consumption and safety; (ii) developing more efficient insulation materials and tank designs to minimize hydrogen evaporation losses during storage, and ultimately reduce the risk of explosion; and (iii) developing specialized low-temperature safety technologies and equipment to improve overall system safety [49]. Such low temperatures and high pressures pose colossal methodical queries, the problems related to refrigeration costs, safety, manipulation, and losses by evaporation, and require extensive hands-on experience. Distributing hydrogen over long distances is problematic and dangerous. Pipelines will be required for long distances; there is a high risk of leakage and a greater possibility of deflagration and detonation. Additionally, the large-scale hydrogen storage in gas and liquid form is still a technical challenge and it will take time to develop technology. Effective storage can rely on underground and underwater reservoirs. Geologic hydrogen storage includes salt caverns, saline aquifers, depleted oil and gas reservoirs, or engineered hard rock reservoirs. It is the best option for large-scale and long-term storage rather than tanks, which are more suitable for short-term and small-scale storage. However, the accumulation and the reaction mechanism of hydrogen with minerals in underground or geologic reservoirs can lead to several issues, including geochemical reactions that alter reservoir and caprock properties, potential microbial activity that can consume hydrogen and produce impurities, and changes in reservoir pressure that can impact wellbore stability and induce seismicity. Furthermore, hydrogen's high penetrability and low solubility in water can pose challenges for long-term containment in some geological formations [153].

Solid hydrogen storage, which involves both physisorption and chemisorption, is a viable option. Hydrogen storage in chemical storage materials includes ammonia, methanol, methane, methylcyclohexane, sodium borohydride, and so on. These materials can store significant volumes of hydrogen under standard conditions, which will help promote the adoption of a hydrogen economy because they can be internationally traded and easily used in many industrial processes [1]. Production and transport of ammonia are already commercialized worldwide. However, the technology and commercial readiness of methanol directly from hydrogen rather than natural gas is still incomplete and must improve before being fully commercialized. It is essential to emphasize that converting into methanol, ammonia, and so on requires a huge energy input for compression and decompression processes. Physical state solid storage focuses on MOFs and COFs as materials for hydrogen storage. Solid hydrogen storage offers advantages such as low operating pressure between 0.1–5 MPa, a high energy density of 100–130 kg/m^3; it does not require gas cylinders or compressors, nor ultralow-temperature insulation devices, greatly reducing the complexity of the system and providing economical and safer storage. Hydrogen storage materials can exist stably at normal pressure and temperature, so that if there is damage, there will not be a large amount of hydrogen leakage, so avoiding the hazards caused by the high pressures and extreme temperatures required for gas and liquid storage, and facilitating easier shipping and handling [152, 154]. To mitigate the risks associated with compressed hydrogen storage, solid-state hydrogen storage materials are increasingly used in hydrogen refueling stations and as backup power supplies and are integrated into portable devices and vehicles [155]. Certain vehicle manufacturers, such as Honda, Toyota, and GM have integrated cylindrical hydrogen storage devices (~10–100 L) to optimize the utilization of constrained vehicle volumes [155]. These innovative storage and transport techniques, along with the crucial collaboration between researchers, industry stakeholders, and policymakers, could pave the way for enhanced safety, efficiency, and more practical hydrogen technologies in future clean energy solutions.

7. Conclusions

The chapter deals with the various factors around green hydrogen production from water electrolysis. The authors took you, as the reader, on a journey in search of an answer to the question: Why is green hydrogen an energy source so important to mitigate carbon emissions in a net-zero mission? The colors of hydrogen were described to illustrate the various processes in industries where hydrogen is produced as a primary source or as a by-product, for example, in red or pink hydrogen produced from nuclear energy. Several industries today heavily depend on black, brown, or gray hydrogen for fuel production in the saturation of molecules in reforming or cracking processes in a petrochemical process. Carbon emissions in the steel or cement industries must be mitigated to decarbonize these processes by introducing green hydrogen from water electrolysis. The different water electrolysis methodologies that are currently available include proton exchange membrane water electrolysis (PEMWE), alkaline water electrolysis, anion exchange water electrolysis, and solid oxide water electrolysis. Their advantages and disadvantages were discussed to demonstrate why PEMWE is more favorable than other water electrolyzers. The different analytical techniques need to characterize the material science of the catalysts by XRD, XRF, SEM, and TEM analysis and the electrochemical characterization by cyclic voltammetry, linear scan voltammetry, and chronopotentiometry for stability studies. The last section briefly discusses the requirements for hydrogen production and storage, emphasizing the importance of preventing catastrophic effects of hydrogen explosions or hazardous release of hydrogen into the atmosphere that could create cloud explosions.

Acknowledgements

The authors thank Nelson Mandela University for its support of the project and financial support to Dr. Claudelle Anensong for conducting a postdoctoral fellowship in the green hydrogen project research group. Financial support also comes from NRF (CSUR) and the Technology Innovation Agency (TIA) for seed fund support for this project and all the students involved in completing their MSc studies.

Conflict of interest

No conflict of interest to be declared.

Author details

Shawn Gouws* and Claudelle Sybilline Anensong Djadock
Department of Chemistry, Nelson Mandela University, Gqeberha, South Africa

*Address all correspondence to: shawn.gouws@mandela.ac.za

References

[1] IEA. Global Hydrogen Review: Hydrogen Production. 2024. pp. 1-298. Available from: https://www.iea.org/reports/global-hydrogen-review-2024/hydrogen-production

[2] IPCC. 2007: Climate Change 2007: Synthesis Report. Contribution of Working Groups I, II and III to the Fourth Assessment Report of the Intergovernmental Panel on Climate Change. Geneva, Switzerland: Core Writing Team, R.K. Pachauri, and A. Reisinger, IPCC; 2007. pp. 1-104

[3] Arrhenius S. On the influence of carbonic acid in the air upon the temperature of the ground. Philosophical Magazine and Journal of Science. 1896;**41**(5):237-276

[4] Dawson TP, Jackson ST, House JI, Prentice IC, Mace GM. Beyond predictions: Biodiversity conservation in a changing climate. Science. 2011;**332**(6025):53-58

[5] Pires JCM, Martins FG, Alvim-Ferraz MCM, Simões M. Recent developments on carbon capture and storage: An overview. Chemical Engineering Research and Design. 2011;**89**(9):1446-1460

[6] Kazim A, Veziroglu TN. Utilization of solar–hydrogen energy in the UAE to maintain its share in the world energy market for the 21st century. Renewable Energy. 2001;**24**(2):259-274

[7] Ajanovic A, Sayer M, Haas R. The economics and the environmental benignity of different colours of hydrogen. International Journal of Hydrogen Energy. 2022;**47**:24136-24154

[8] Abdin Z, Zafaranloo A, Rafie A, Merida W, Lipinski W, Khalilpour KR. Hydrogen as an energy vector. Renewable and Sustainable Energy Reviews. 2020;**120**:109620

[9] Jun C, Yu H. Water electrolysis based on renewable energy for hydrogen production. Chinese Journal of Catalysis. 2018;**39**(3):390-394

[10] Nurettin S, Sertac B, Ugur F, Antonio S. A comprehensive review of the state-of-the-art of proton exchange membrane water electrolysis. Materials Science for Energy Technologies. 2025;**8**:44-65

[11] Berezin A, Gorodnova N, Sergi BS, Handoko CT, Permana CT. Prospects for energy transition to hydrogen fuel: Analysis of world experience and Russian practice. International Journal of Energy Economics and Policy. 2023;**13**(4):641-653

[12] Veza I, Idris M, Fattah IMR. Circular economy, energy transition, and role of hydrogen. Mechanical Engineering for Society and Industry. 2022;**2**(2):54-56

[13] Grand View Research. Hydrogen Energy Storage Market Size, Share & Trends Analysis Report by Technology (Compression, Liquefaction), by Physical State (Solid, Liquid, Gas), by Application (Residential, Commercial), by Region, and Segment Forecasts, 2024-2030. 2023. pp. 1-90

[14] Hydrogen Council. Significant growth for Hydrogen Council, publishes report on cost competitiveness. Fuel Cells Bulletin. 2020:13-14

[15] GlobalData. Net Capacity for Green Hydrogen in the Pipeline Worldwide in Q1 2024 by Leading Company (In 1,000 Tons Per Annum) [Graph],

Statista. 2024. Available from: https://www.statista.com/statistics/1374638/globalnet-capacity-for-green-hydrogenby-leadingcompany/ [Accessed: October 16, 2024]

[16] Konstantin G, Permana CT, Handoko CT. The growing demand for hydrogen: Current trends, sectoral analysis, and future projections. Unconventional Resources. 2025;**6**:100176

[17] Tamilarasi B, Jithul KP, Jay P. Non-noble metal-based electro-catalyst for the oxygen evolution reaction (OER): Towards an active & stable electro-catalyst for PEM water electrolysis. International Journal of Hydrogen Energy. 2024;**58**:556-582

[18] Mohd NS, Teuku H, Jonathan G, Abu BS. High-pressure PEM water electrolyser: A review on challenges and mitigation strategies towards green and low-cost hydrogen production. Energy Conversion and Management. 2022;**268**:115985

[19] Shirvanian P, Van BF. Novel components in proton exchange membrane (PEM) water electrolyzers (PEMWE): Status, challenges and future needs. A mini review. Electrochemistry Communications. 2020;**114**:106704

[20] Ali M, Pervaiz E, Noor T, Rabi O, Zahra R, Yang M. Recent advancements in MOF-based catalysts for applications in electrochemical and photoelectrochemical water splitting: A review. International Journal Energy Research. 2021;**45**(2):1190-1226

[21] Shiva, Kumar S, Himabindu V. Hydrogen production by PEM water electrolysis – A review. Materials Science for Energy Technologies. 2019;**2**:442-454

[22] Siracusano S, Van DN, Backhouse R, Merlo L, Baglio V, Arico AS. Degradation issues of PEM electrolysis MEAs. Renewable Energy. 2018;**123**:52-57

[23] Feng Q, Yuan XZ, Liu G, Wei B, Zhang Z, Li H, et al. A review of proton exchange membrane water electrolysis on degradation mechanisms and mitigation strategies. Journal of Power Sources. 2017;**366**:33-55

[24] Adnan M, Kucuk H, Topal ME, Akbulut U, Dincer I. A comprehensive review on hydrogen production from coal gasification: Challenges and opportunities. International Journal of Hydrogen Energy. 2021;**46**(50):25385-25412

[25] Li J, Cheng W. Comparative life cycle energy consumption, carbon emissions and cost for hydrogen production from coke oven gas and coal gasification. International Journal of Hydrogen Energy. 2020;**45**(51):27979-27993

[26] Salkuyeh YK, Saville BA, MacLean HL. Techno-economic analysis and life cycle assessment of hydrogen production from natural gas using current and emerging technologies. International Journal of Hydrogen Energy. 2017;**42**(30):18894-18909

[27] Bauer C, Treyer K, Antonini C, Bergerson J, Gazzani M, Gencer E, et al. On the climate impacts of blue hydrogen production. Sustainable Energy & Fuels. 2022;**6**:66-75

[28] Howarth RW, Jacobson MZ. How green is blue hydrogen? Energy Science and Engineering. 2021;**9**(10):1676-1687

[29] Khan MA, Daiyan R, Neal P, Haque N, MacGill I, Amal R. A framework for assessing the economics of blue hydrogen production from steam methane reforming using carbon capture, storage & utilisation. International Journal of Hydrogen Energy. 2021;**46**(44):22685-22706

[30] Gerloff N. Comparative life-cycle-assessment analysis of three major water electrolysis technologies while applying various energy scenarios for a greener hydrogen production. Journal of Energy Storage. 2021;**43**:102759

[31] Bartels JR, Pate MB, Olson NK. An economic survey of hydrogen production from conventional and alternative energy source. International Journal of Hydrogen Energy. 2010;**35**(6):8371-8384

[32] Nikolaidis P, Poullikkas A. A comparative overview of hydrogen production processes. Renewable Sustainable Energy Review. 2017;**67**:597-611

[33] Yukesh KR, Kavitha S, Preethi OP, Karthikeyan G, Kumar NV, Dai-Viet JRB. Techno-economic assessment of various hydrogen production methods–A review. Bioresources Technology. 2021;**319**:124175

[34] Ozturk M, Dincer I. A comprehensive review on power-to-gas with hydrogen options for cleaner applications. Journal of Hydrogen Energy. 2021;**46**(62):31511-31522

[35] Dincer I. Green methods for hydrogen production. International Journal of Hydrogen Energy. 2012;**37**(2):1954-1971, 2021

[36] Acar C, Dincer I. Review and evaluation of hydrogen production options for a better environment. Journal of Cleaner Production. 2019;**218**:835-849

[37] Longden T, Beck FJ, Jotzo F, Andrews R, Prasad M. 'Clean' hydrogen? – Comparing the emissions and costs of fossil fuel versus renewable electricity based hydrogen. Applied Energy. 2022;**306**:118145

[38] Droege T. What are the Colors of Hydrogen? Williams Companies; 2021. Available from: https://www.williams.com/2021/04/23/what-are-the-colors-of-hydrogen/

[39] Dodgshun J. Hydrogen: Clearing up the Colours. 2021. Available from: https://www.enapter.com/newsroom/hydrogen-clearing-up-the-colours

[40] Incer-Valverde J, Korayem A, Tsatsaronis G, Morosuk T. "Colors" of hydrogen: Definitions and carbon intensity. Energy Conversion and Management. 2023;**291**:117294

[41] Chandra S. What is the Hydrogen Rainbow? 2021. Available from: https://www.linkedin.com/pulse/whathydrogen-rainbow-sandeep-chandra/ [Accessed: May 08, 2025]

[42] Global Power Technology Leader. What is the Hydrogen Rainbow? Cummins Inc., Global Power Technology Leader; 2021. Available from: https://www.cummins.com/news/2021/11/16/what-hydrogen-rainbow [Accessed: May 08, 2025]

[43] Arcos JMM, Santos DMF. The hydrogen color spectrum: Techno economic analysis of the available technologies for hydrogen production. Gases. 2023;**3**(1):25-46

[44] Aimikhe VJ, Oghenegare EE. Recent advances in white hydrogen exploration and production: A mini review. Journal of Energy Research and Reviews. 2023;**13**(4):64-79

[45] Samson OJ, Tawanda M, Japheth OO, Mxolisi BS. Advancing the hydrogen production economy: A comprehensive review of technologies, sustainability, and future prospects. International Journal of Hydrogen Energy. 2024;**78**:642-661

[46] Lebrouhi BE, Djoupo JJ, Lamrani B, Benabdelaziz K, Kousksou T. Global

hydrogen development - A technological and geopolitical overview. International Journal of Hydrogen Energy. 2022;**47**(11):7016-7048

[47] Roeth M. The Many Colors of Hydrogen. North America: J FleetOwner; 2021. Available from: https://www.fleetowner.com/perspectives/ideaxchange/article/21151562/the-many-colours-of-hydrogen

[48] Dawood F, Anda M, Shafiullah GM. Hydrogen production for energy: An overview. International Journal of Hydrogen Energy. 2020;**45**(7):3847-3869

[49] Monjur MHB, Zahed S. Hydrogen as an alternative fuel: A comprehensive review of challenges and opportunities in production, storage, and transportation. International Journal of Hydrogen Energy. 2025;**102**:1026-1044

[50] WEC. Working Paper: Regional Insights into Low-Carbon Hydrogen Scale Up|World Energy Insights. United Kingdom: World Energy Council; 2022. pp. 1-46

[51] IEA. Global Hydrogen Review. Paris: IEA; 2022. pp. 1-294

[52] Mulligan J. The Failed Europeanisation of Carbon Capture and Storage (CCS): A Comparative Analysis of the Impact of Domestic Factors on Six CCS Demonstration Projects. Berlin: Freie Universität; 2020. pp. 1-23

[53] Energiewende A, Guidehouse. Making Renewable Hydrogen Cost-Competitive: Policy Instruments for Supporting Green H_2. Berlin: Agora Energiewende; 2021. pp. 1-96

[54] Newborough M, Cooley G. Developments in the global hydrogen market: The spectrum of hydrogen colours. Fuel Cells Bulletin. 2020;**11**:16-22

[55] Amin AM, Croiset E, Epling W. Review of methane catalytic cracking for hydrogen production. International Journal of Hydrogen Energy. 2011;**36**(4):2904-2935

[56] Leal Pérez BJ, Medrano Jiménez JA, Bhardwaj R, Goetheer E, van Sint AM, Gallucci F. Methane pyrolysis in a molten gallium bubble column reactor for sustainable hydrogen production: Proof of concept & techno-economic assessment. International Journal of Hydrogen Energy. 2021;**46**(7):4917-4935

[57] Schneider S, Bajohr S, Graf F, Kolb T. State of the art of hydrogen production via pyrolysisof natural gas. Chemical Biology Engineering Reviews. 2020;7(5):150-158

[58] Altgelt F, Micheli M, Sailer K, Crone K. Water Consumption of Powerfuels Demand, Supply, and Policy Recommendations to Foster Environmental Sustainability. Berlin: Deutsche EnergieAgentur GmbH (dena); 2021. pp. 1-14

[59] IEA. Towards Hydrogen Definitions Based on their Emissions Intensity. Paris: International Energy Agency; 2023. pp. 1-89

[60] Kadri S. Commission Sets out Rules for Renewable Hydrogen. Bruxelle: European Commission; 2023. pp. 1-2. Available from: https://ec.europa.eu/commission/presscorner/detail/en/ip_23_594 [Accessed: May 08, 2025]

[61] Gregor E, Sara S. EU Rules for Renewable Hydrogen. BRIEFING Towards Climate Neutrality. EPRS| European Parliamentary Research Service; 2023. pp. 1-10. Available from: https://www.europarl.europa.eu/RegData/etudes/BRIE/2023/747085/EPRS_BRI(2023)747085_EN.pdf

[62] IRENA. Green Hydrogen Cost Reduction: Scaling up Electrolysers to Meet the 1.5 C Climate Goal. Abu Dhabi: International Renewable Energy Agency; 2020. pp. 1-103

[63] Chisholm G, Zhao T, Cronin L. Hydrogen from water electrolysis. Storing Energy. 2022:559-591

[64] Dönitz W, Erdle E. High-temperature electrolysis of water vapor—Status of development and perspectives for application. International Journal of Hydrogen Energy. 1985;**10**:291-295

[65] Brisse A, Schefold J, Zahid M. High temperature water electrolysis in solid oxide cells. International Journal of Hydrogen Energy. 2008;**33**(20):5375-5382

[66] He S, Zou Y, Chen K, Jiang SP. A critical review of key materials and issues in solid oxide cells. Interdisciplinary Materials. 2023;**2**(1):111-136

[67] Sapountzi MF, Gracia MJ, Kees-Jan Weststrate CJ, Hans Fredriksson OA, Niemantsverdriet JW. Electrocatalysts for the generation of hydrogen, oxygen and synthesis gas. Progress in Energy Combustion Science. 2017;**58**:1-35

[68] Sayed-A H, Toldya ÁI, Lappalainenb M, Himanena O, Bajamundi C, Santasalo AA. Strategies and challenges for reducing green hydrogen cost: Operation mode and revenue streams. Renewable and Sustainable Energy Reviews. 2025;**223**:116065

[69] Shiva Kumar S, Lim H. An overview of water electrolysis technologies for green hydrogen production. Energy Reports. 2022;**8**:13793-13813

[70] Schmidt O, Gambhir I, Hawkes A, Nelson J, Few S. Future cost and performance of water electrolysis: An expert elicitation study. International Journal of Hydrogen Energy. 2017;**42**(52):30470-30492

[71] Liu S, Li B, Mohite SV, Devaraji P, Mao L, Xing R. Ultrathin MoS_2 nanosheets *in situ* grown on rich defective $Ni_{0.96}S$ as heterojunction bifunctional electrocatalysts for alkaline water electrolysis. International Journal of Hydrogen Energy. 2020;**45**(55):29929-29937

[72] Qazi UY, Javaid R, Zahid M, Tahir N, Afzal A, Lin X-M. Bimetallic NiCo–$NiCoO_2$ nano-heterostructures embedded on copper foam as a self-supported bifunctional electrode for water oxidation and hydrogen production in alkaline media. International Journal of Hydrogen Energy. 2021;**46**(36):18936-18948

[73] Xu Z, Yeh C-L, Chen J-L, Lin TJ, Ho K-C, Lin RY-Y. Metal-organic frameworks derived 2D NiCoP nanoflakes from layered double hydroxide nanosheets for efficient electrocatalytic water splitting at high current densities. ACS Sustainable Chemistry and Engineering. 2022;**10**(35):11577-11586

[74] Wu X, Scott K. A Li-doped Co_3O_4 oxygen evolution catalyst for non-precious metal alkaline anion exchange membrane water electrolysers. International Journal of Hydrogen Energy. 2013;**38**(8):3123-3129

[75] Akash S, Riya ABJ, Aniruddha B, Suraj G, Matjaž S, Rupali P, et al. Designing bifunctional electrocatalysts based on complex cobalt-sulfo-boride compound for high-current-density alkaline water electrolysis. Energy & Fuels. 2024;**38**(19):18965-18975

[76] Liu YP, Liang X, Chen H, Gao RQ, Shi L, Yang L, et al. Iridium-containing water-oxidation catalysts in acidic

electrolytes. Chinese Journal of Catalysis. 2021;**42**(7):1054-1077

[77] Omid T. White Paper: Hydrogen Production Cost by AEM Water Electrolysis: Document ID: FM-7024-B. Canada: Ionomer Innovations Inc.; 2020. pp. 1-12

[78] Thangavel P, Kim G, Kim KS. Electrochemical integration of amorphous NiFe (oxy)hydroxides on surface-activated carbon fibers for high-efficiency oxygen evolution in alkaline anion exchange membrane water electrolysis. Journal of Materials Chemistry A. 2021;**9**:14043-14051

[79] Guo W, Kim J, Kim H, Ahn HS. Cu–Co–P electrodeposited on carbon paper as an efficient electrocatalyst for hydrogen evolution reaction in anion exchange membrane water electrolyzers. International Journal of Hydrogen Energy. 2021;**46**(38):19789-19801

[80] Chen N, Paek SY, Lee JY, Park JH, Lee SY, Lee YM. High-performance anion exchange membrane water electrolyzers with a current density of 7.68 A cm^{-2} and a durability of 1000 hours. Energy & Environmental Science. 2021;**14**:6338-6348

[81] Smolinka T, Bergmann H, Garche J, Kusnezoff M. The history of water electrolysis from its beginnings to the present. In: Systems and Applications. 2022. pp. 83-164

[82] Russell JH, Nuttall LJ, Fickett AP. Hydrogen generation by solid polymer electrolyte water electrolysis. American Chemical Society. 1973;**18**(3):24-40

[83] Xu W, Scott K. The effects of ionomer content on PEM water electrolyser membrane electrode assembly performance. International Journal of Hydrogen Energy. 2010;**35**(21):12029-12037

[84] Abdol Rahim AH, Tijani AS, Kamarudin SK, Hanapi S. An overview of polymer electrolyte membrane electrolyzer for hydrogen production: Modeling and mass transport. Journal of Power Sources. 2016;**309**:56-65

[85] Di Virgilio M, Basso Peressut A, Arosio V, Arrigoni A, Latorrata S, Dotelli G. Functional and environmental performances of novel electrolytic membranes for PEM fuel cells: A lab-scale case study. Clean Technology. 2023;**5**:74-93

[86] Ren K, Liu X, Rao Z. Effect of temperature and external electric field on proton transport in ordered and amorphous proton exchange membranes: A molecular dynamics study. International Journal of Hydrogen Energy. 2024;**54**:1422-1435

[87] Nicotera I, Simari C, Enotiadis A. Nafion-based cation-exchange membranes for direct methanol fuel cells. Cell Technology. 2020:13-36

[88] Kusoglu A, Weber AZ. New insights into perfluorinated sulfonic-acid ionomers. Chemical Reviews. 2017;**117**(3):987-1104

[89] Anensong CS, Vengatesan S, Tagne Kuate AC, Ngoune J, Ravichandran S, Vasudevan S. Distorted octahedral cobalt()-acylpyrazolone complex with a tunable lattice-strain structure - An efficient electrocatalyst for overall water splitting. Catalysis Science and Technology. 2023;**13**(5):2184-2200

[90] Shi L, Chen J, Zhao S, Du L, Ye S. Proton-exchange membrane water electrolysis: From fundamental study to industrial application. Chemical Catalysis. 2023;**3**:100734

[91] Pushkarev AS, Pushkareva IV, Bessarabov DG. Supported Ir-based

oxygen evolution catalysts for polymer electrolyte membrane water electrolysis: A mini review. Energy & Fuels. 2022;**36**(13):6613-6625

[92] Hrbek T, Kú P, Sutov TK, Veltrusk K, Dinhov TN, Dopita M, et al. Sputtered Ir–Ru based catalysts for oxygen evolution reaction: Study of iridium effect on stability. International Journal of Hydrogen Energy. 2022;**47**(89):21033-21043

[93] Lv H, Wang S, Li J, Shao C, Zhou W, Shen X, et al. Self-assembled RuO_2@IrO_x core-shell nanocomposite as high efficient anode catalyst for PEM water electrolyzer. Applied Surface Science. 2020;**514**:145943

[94] Bohm D, Beetz M, Gebauer C, Bernt M, Schröter J, Kornherr M, et al. Highly conductive titania-supported iridium oxide nanoparticles with low overall iridium density as OER catalyst for large-scale PEM electrolysis. Applied Materials Today. 2021;**24**:101134

[95] Yasutake M, Noda Z, Matsuda J, Lyth SM, Nishihara M, Ito K, et al. Ru-core Ir-shell electrocatalysts deposited on a surface-modified Ti-based porous transport layer for polymer electrolyte membrane water electrolysis. International Journal of Hydrogen Energy. 2024;**49**:169-183

[96] Kim IG, Lim A, Jang JH, Lee K-Y, Nah IW, Park S. Leveraging metal alloy-hybrid support interaction to enhance oxygen evolution kinetics and stability in proton exchange membrane water electrolyzers. Journal of Power Sources. 2021;**501**:230002

[97] Wang C, Lan F, He Z, Xie X, Zhao Y, Hou H, et al. Iridium-based catalysts for solid polymer electrolyte electrocatalytic water splitting. Chemistry Sustainability Chemistry. 2019;**12**:1576-1590

[98] Siracusano S, Van Dijk N, Payne-Johnson E, Baglio V, Arico AS. Nanosized IrOx and IrRuOx electrocatalysts for the O_2 evolution reaction in PEM water electrolysers. Applied Catalysis B: Environmental. 2015;**164**:488-495

[99] Oh H-S, Nong HN, Reier T, Gliech M, Strasser P. Oxide-supported Ir nanodendrites with high activity and durability for the oxygen evolution reaction in acid PEM water electrolyzers. Chemical Science. 2015;**6**:3321-3328

[100] Wu Z-Y, Chen F-Y, Li B, Yu S-W, Finfrock YZ, Meira DM, et al. Non-iridium-based electrocatalyst for durable acidic oxygen evolution reaction in proton exchange membrane water electrolysis. Nature Materials. 2023;**22**:100-108

[101] Hoang PT, Hong NN, Matej Z, Aram Y, Uta H, Martina R, et al. Reactivity and stability of reduced Ir-weight TiO_2-supported oxygen evolution catalysts for proton exchange membrane (PEM) water electrolyzer anodes. Journal of the American Chemical Society. 2024;**146**(46):31444-31455

[102] Muntean R, Pascal DT, Rost U, Holtkotte L, Näther J, Köster F, et al. Investigation of iridium nanoparticles supported on sub-stoichiometric titanium oxides as anodic electrocatalysts in PEM electrolysis. Part I: Synthesis and characterization. Topics in Catalysis. 2019;**62**:429-438

[103] Saha S, Rajbongshi BM, Ramani V, Verma A. Titanium carbide: An emerging electrocatalyst for fuel cell and electrolyser. International Journal of Hydrogen Energy. 2021;**46**(24):12801-12821

[104] Jiaqi K, Xingli W, Sebastian M, Shima F, Miklós MK, Johannes S, et al. Synthesis, molecular structure, and

water electrolysis performance of TiO2-supported Raney-IrO*x* nanoparticles for the acidic oxygen evolution reaction. ACS Catalysis. 2025;**15**:5435-5446

[105] Siracusano S, Fausta G, Antonino SA. Ag/Ti-suboxides as non-PGM anode electrocatalyst for PEM water electrolysis. Journal of Power Sources. 2023;**565**:232903

[106] Morozan A, Johnson H, Roiron C, Genay G, Aldakov D, Ghedjatti A, et al. Nonprecious bimetallic iron–molybdenum sulfide electrocatalysts for the hydrogen evolution reaction in proton exchange membrane electrolyzers. ACS Catalysis. 2020;**10**(4):14336-14348

[107] Liu J, Liu H, Yang Y, Tao Y, Zhao L, Li S, et al. Efficient and stable proton exchange membrane water electrolysis enabled by stress optimization. ACS Central Science. 2024;**10**:852-885

[108] Wan C, Deng R, Guo M, Zhang Q. Recent progress of advanced Co_3O_4-based materials for electrocatalytic oxygen evolution reaction in acid: From rational screening to efficient design. International Journal of Hydrogen Energy. 2023;**48**(82):31920-31942

[109] Peiyu H, Xu F, Changjian Z, Xuejun X, Guangjin W, Huawen H, et al. Na-doped cobalt spinel electrocatalysts prepared by programmed electrostatic spraying for improving OER activity and durability in acidic media. Chemical Engineering Journal. 2024;**500**:156846

[110] Qisheng JF, Wenjuan S, Wenzhe N, Zhuorong L, Kai S, Xiao Y, et al. Chromium-induced high covalent Co-O bonds for efficient anodic catalysts in PEM electrolyzer. Advanced Science. 2024;**11**(25):2402356

[111] Fanfan S, Huijie H, Peng L, Hairui C, Bei A, Xiaoqian L, et al. PO6 geometric configuration unit enhanced electrocatalytic performance of Co3O4 in acidic oxygen evolution. Journal of Colloid and Interface Science. 2023;**641**:329-337

[112] Ni J, Shi Z, Wang Y, Yang J, Wu H, Wang P, et al. Suppressing the lattice oxygen diffusion via high-entropy oxide construction towards stabilized acidic water oxidation. Nano. 2024;**17**(3):1107-1113

[113] Liu R-T, Xu Z-L, Li F-M, Chen F-Y, Yu J-Y, Ya Y, et al. Recent advances in proton exchange membrane water electrolysis. Chemical Society Reviews. 2023;**52**:5652-5683

[114] Ayers KE, Renner JN, Danilovic N, Wang JX, Zhang Y, Maric R, et al. Pathways to ultra-low platinum group metal catalyst loading in proton exchange membrane electrolyzers. Catalysis Today. 2016;**262**:121-132

[115] Martin S, Garcia-Ybarra P, Castillo J. Ten-fold reduction from the state-of-the-art platinum loading of electrodes prepared by electrospraying for high temperature proton exchange membrane fuel cells. Electrochemistry Communications. 2018;**93**:57-61

[116] Kannan A, Li Q, Cleemann LN, Jensen JO. Acid distribution and durability of HT-PEM fuel cells with different electrode supports. Fuel Cells. 2018;**18**(2):103-112

[117] Gabialti S, Baricci A, Casalegno A, Marchesi R. Degradation in phosphoric acid doped polymer fuel cells: A 6000 h parametric investigation. International Journal of Hydrogen Energy. 2013;**38**(15):6469-6480

[118] Pinar FJ, Cañizares P, Rodrigo MA, Ubeda D, Lobato J. Long-term testing of a high-temperature proton exchange

membrane fuel cell short stack operated with improved polybenzimidazole-based composite membranes. Journal of Power Sources. 2015;**274**:177-185

[119] Ramakrishna SUB, Srinivasulu RD, Kumar SS, Himabindu V. Nitrogen-doped CNTs supported palladium electrocatalyst for hydrogen evolution reaction in PEM water electrolyser. International Journal of Hydrogen. 2016;**41**(45):20447-20454

[120] Kim JH, Kim H, Kim J, Lee HJ, Jang JH, Ahn SH. Electrodeposited molybdenum sulfide as a cathode for proton exchange membrane water electrolyzer. Journal of Power Sources. 2018;**392**:69-78

[121] Sapountzi FM, Orlova ED, Sousa JPS, Salonen LM, Lebedev OI, Zafeiropoulos G, et al. FeP nanocatalyst with preferential [010] orientation boosts the hydrogen evolution reaction in polymer-electrolyte membrane electrolyzer. Energy & Fuels. 2020;**34**(5):6423-6429

[122] Kumar SS, Himabindu V. Boron-doped carbon nanoparticles supported palladium as an efficient hydrogen evolution electrode in PEM water electrolysis. Renewable Energy. 2020;**146**:2281-2290

[123] Wang T, Cao X, Jiao L. PEM water electrolysis for hydrogen production: Fundamentals, advances, and prospects. Carbon Neutrality. 2022;**1**(22):1-19

[124] Alkhaldi S, Aziz M, Amrite A, Prasad AK. Parametric study of PEM water electrolyzer performance. Journal of Applied Electrochemistry. 2025;**55**:327-343

[125] Ma L, Sui S, Zhai Y. Investigations on high performance proton exchange membrane water electrolyzer. International Journal of Hydrogen Energy. 2009;**34**(2):678-684

[126] Liso V, Savoia G, Araya SS, Cinti G, Kær SK. Modelling and experimental analysis of a polymer electrolyte membrane water electrolysis cell at different operating temperatures. Energies. 2018;**11**(12):3273

[127] Choi Y, Lee W, Na Y. Effect of gravity and various operating conditions on proton exchange membrane water electrolysis cell performance. Membranes. 2021;**11**(11):822

[128] Martin A, Trinke P, Stahler M, Stahler A, Scheepers F, Bensmann B, et al. The effect of cell compression and cathode pressure on hydrogen crossover in PEM water electrolysis. Journal of the Electrochemistry Society. 2021;**169**:014502

[129] Altus SJ, Inkson BJ, Hack J. Complementary X-ray and neutron imaging of water electrolysers for green hydrogen production. Journal of Material Chemistry A. 2024;**12**:23364-23391

[130] Ali M, Jin X, Liu F, Agar E, Wong H-W, Rashinkar S, et al. Investigating proton exchange membrane electrolyzer performance under variable renewable power: Insights into voltage and stack efficiency. Journal of Electroanalytical Chemistry. 2025;**988**:119130

[131] Zhou B, Gao R, Zou JJ, Yang H. Surface design strategy of catalysts for water electrolysis. Small. 2022;**18**(27):2202336

[132] Shaikh N, Sau S, Mitra S, Mukhopadhyay I, Ray A. A self-supported nickel pseudo-intermetallic alloy electrocatalyst for overall water splitting. Cell Reports Physical Science. 2025;**6**(1):102369

[133] Retuerto M, Pascual L, Torrero J, Mohamed AS, Tolosana Moranchel Á, Diego G, et al. Highly active and stable OER electrocatalysts derived from Sr2MIrO6 for proton exchange membrane water electrolyzer. Nature Communications. 2022;**13**(1):7935

[134] Yu G, Vetchinkin SI, Zimont SL, Ivleva IN, Yu MS. Nature of satellites in x-ray photoelectron spectra XPS of paramagnetic cobalt (II) compounds. Chemical Physics Letters. 1976;**42**(2):265-267

[135] Yin L, Adler I, Tsang T, Matienzo LJ, Grim SO. Paramagnetism and shake-up satellites in X-ray photoelectron spectra. Chemical Physics Letters. 1974;**24**(1):81-84

[136] Timoshenko J, Cuenya BR. In situ/operando electrocatalyst characterization by X-ray absorption spectroscopy. Chemical Reviews. 2021;**121**(2):882-961

[137] Kumari S, Thakur M, Kumar S, Devi M, Sharma S, Bhatt AK, et al. Synthesis, structural analysis, theoretical calculations, in silico prediction of antibacterial efficacy, toxicity assessment, in vitro antioxidant evaluation, and polymerization activity of cobalt(II) coordination complexes of para-biphenylol. Chemistry Africa. 2024;**7**:1-21

[138] Kumari S, Thakur M, Chauhan C, Kumari M. Synthesis, characterization, biological activity, and computation-based efficacy of cobalt(II) complexes of biphenyl-2-ol against SARS-CoV-2 virus. Journal of Biomolecular Structure Dynamics. 2023;**43**(1):483-497

[139] Kumari S, Thakur M, Kumar S. Design, spectroscopic analysis, DFT calculations, catalytic evaluation, and comprehensive in silico and in vitro bioactivity studies, including molecular docking, of novel Co(II) complexes of 2-hydroxy-5,3-(phenylallylidene) aminobenzoic acid. Journal of Inorganic and Organometallic Polymers and Materials. 2025;**35**:1535-1562

[140] Kumari S, Thakur M, Kumar S. Synthesis, spectroscopic analysis, and theoretical study of novel cobalt(II) complexes: Solar radiation-driven photocatalytic degradation of organic pollutants in wastewater. Journal of the Indian Chemical Society. 2025;**102**(2):101588

[141] Siracusano S, Baglio V, Stassi A, Merlo L, Moukheiber E, Arico AS. Enhanced performance and durability of low catalyst loading PEM water electrolyser based on a short-side chain perfluorosulfonic ionomer. Applied Energy. 2017;**192**:477-489

[142] Urbina LP, Liu J, Semagina N, Secanell M. Low loading inkjet printed bifunctional electrodes for proton exchange membrane unitized regenerative fuel cells. Journal of Power Sources. 2023;**580**:233448

[143] Lee C, Lee JK, George MG, Fahy KF, LaManna JM, Baltic E, et al. Reconciling temperature-dependent factors affecting mass transport losses in polymer electrolyte membrane electrolyzers. Energy Conversion and Management. 2020;**213**:112797

[144] Anantharaj S, Ede SR, Karthick K, Sam Sankar S, Sangeetha K, Karthik PE, et al. Precision and correctness in the evaluation of electrocatalytic water splitting: Revisiting activity parameters with a critical assessment. Energy & Environmental Science. 2018;**11**:744-771

[145] Li Y, Wang Z, Shi M, Fan R. Safety analysis of hydrogen leakage accident with a mobile hydrogen refueling station. Process Safety and Environmental Protection. 2023;**171**:619-629

[146] Kim Y. Hydrogen Tank Explosion in Gangneung, South Korea. Center Hydrogen Safety Conference; 2019. Available from: https://www.aiche.org/chs/conferences/international-centerhydrogen-safety-conference/2019/proceeding/paper/review-hydrogentank-explosion-gangneung-south-korea

[147] Tovini MF, Weiß AH, Gasteiger HA, El-Sayed HA. The discrepancy in oxygen evolution reaction catalyst lifetime explained: RDE vs MEA-dynamicity within the catalyst layer matters. Journal of the Electrochemical Society. 2021;**168**:014512

[148] Zerrougui I, Li Z, Hissel D. Investigating bubble impacts on PEM electrolysis performance through enhanced multiphysics modelling. International Journal of Hydrogen Energy;**98**:626-638

[149] Safetygram 33, the Hazards of Oxygen and Oxygen-Enriched Mixtures, Air Products and Chemicals, Inc., 1-10 (36202) 900-13-101-US, 2014

[150] Panchawadkar D. Review of Opportunities to Valorise the Oxygen Generated from Water Electrolysis. Degree Project in Chemical Engineering for Energy and Environment TRITA-CBH-GRU-2024. Stockholm, Sweden: KTH Royal Institute of Technology; 2024. pp. 1-58

[151] Principi G, Agresti F, Maddalena A, Lo RS. The problem of solid state hydrogen storage. Energy. 2009;**34**(12):2087-2091

[152] Xu Y, Zhou Y, Li Y, Ding Z. Research progress and application prospects of solid-state hydrogen storage technology. Molecules. 2024;**29**(8):1767

[153] Dilshan RADP, Perera MSA, Matthai SK. Effect of mechanical weakening and crack formation on caprock integrity during underground hydrogen storage in depleted gas reservoirs – A comprehensive review. Fuel. 2024;**371**:131893

[154] Schlapbach L, Züttel A. Hydrogen-storage materials for mobile applications. Nature. 2001;**414**:353-358

[155] Kalibek MR, Ospanova AD, Suleimenova B, Soltan R, Orazbek T, Makhmet AM, et al. Solid-state hydrogen storage materials. Discover Nano. 2024;**19**:195

Chapter 5

Perspective Chapter: Research into Obtaining Green Hydrogen from Today's Wind Turbines

Saban Pusat and Yasin Karagoz

Abstract

The present study aims to examine the relationship between price changes in the day-ahead electricity market and changes in the amount of electricity generated from wind energy, and to assess the implications of this relationship for green hydrogen production and storage strategies. The present study is chiefly concerned with the analysis of the relationship between day-ahead electricity market price data and wind energy production data. Analyses conducted for different countries have determined that both electricity prices and wind electricity fluctuate significantly throughout the day. Following a thorough evaluation of the situation, it was determined that the implementation of a green hydrogen production and storage strategy could be a viable solution from both economic and environmental standpoints.

Keywords: green hydrogen, wind power, electricity market, electorlysis, green energy, energy storage

1. Introduction

The present moment is characterized by an unprecedented global imperative to combat climate change and transition toward a sustainable, carbon-neutral energy system. However, the intermittent nature of dominant renewable sources such as wind and solar power poses a significant challenge to grid stability and energy security. This challenge has thus been instrumental in prompting the search for effective large-scale energy storage and sector-coupling solutions. In this context, green hydrogen, produced *via* water electrolysis powered exclusively by renewable electricity, has emerged as a cornerstone of the future energy landscape. It is capable of decarbonizing hard-to-abate sectors such as heavy industry, shipping, and long-haul transport [1].

It is evident that, within the domain of renewable energy sources, wind energy has attained a notable degree of technological maturity and cost-effectiveness. Modern wind turbines, both onshore and offshore, represent a significant source of clean electricity generation. However, the variability of their output is intrinsic and frequently incongruent with energy demand patterns. This has the potential to result in grid congestion, wasted resources that could otherwise be utilized [2]. This paradox underscores the necessity for innovative approaches to facilitate more effective integration of wind energy into the energy value chain.

The direct integration of contemporary wind turbines with electrolysis systems to produce green hydrogen exhibits a compelling synergy. This integration of wind turbines with hydrogen production represents a paradigm shift in the conventional function of wind turbines, transforming them from mere electricity generators into integrated renewable hydrogen production facilities. The solution is multifaceted in nature. It provides a valuable outlet for excess wind generation, mitigates grid balancing issues, and creates a storable energy vector that can be used flexibly across various sectors [3]. This concept, frequently termed "power-to-gas," is undergoing rapid evolution from a theoretical model to a practical engineering and economic pursuit.

Electricity prices in the day-ahead market, where wind power plants predominantly supply electricity, exhibit high levels of volatility. When considered alongside the significant fluctuations in electricity generated by wind power plants throughout the day, this situation renders planning extremely difficult. The strategic management of hydrogen reserves emerges as a pivotal solution to this planning challenge.

This research paper explores the practicalities and potential of obtaining green hydrogen from wind turbines. The present study investigates the change of day-ahead electricity prices and wind power production in the context of storage strategy. The present study aims to provide a comprehensive assessment of how today's wind energy assets can be immediately leveraged to become a foundational pillar of the emerging green hydrogen economy, thereby enhancing their value and accelerating the clean energy transition. To this end, the study will examine real-world operational data and existing market structures.

2. General perspective

Presently, the energy sector is undergoing a fundamental transformation process with the objective of breaking free from its dependence on fossil fuels and combating climate change. The cornerstones of this transformation are wind energy, energy storage systems, and green hydrogen. The synergy between these three components is pivotal in establishing a sustainable energy ecosystem.

The advent of technological advancements over the past two decades has resulted in wind energy becoming one of the most cost-effective renewable energy sources. The role of onshore and offshore wind farms in energy production is growing, as their capacities are increasing. However, the intermittent nature of wind energy requires energy storage systems to increase efficiency.

Energy storage technologies have been identified as a solution to the intermittent production issue, which is considered to be the most significant drawback of wind energy. Lithium-ion batteries are well-suited to meeting short-term storage requirements, while green hydrogen is poised to become a leading option for large-scale and long-term energy storage. Excess electricity generated by wind turbines can be converted into hydrogen through water electrolysis, thereby minimizing energy loss.

Green hydrogen has been proposed as a means of facilitating the integration of wind energy with storage systems. It is imperative to acknowledge the pivotal role this will play in achieving decarbonization in sectors such as steel production, heavy transport, shipping, and aviation. Electrolysis plants established in regions abundant in wind energy have the potential to meet the local industry's hydrogen needs and contribute to international hydrogen trade.

In conclusion, wind energy, energy storage, and green hydrogen are complementary and reinforcing technologies. The coordination of investments in these three areas has the potential to accelerate the clean energy transition and play a critical role in achieving zero-emission targets. The future of energy systems is predicated on the intelligent integration of these three components.

2.1 Wind energy

Wind energy is one of the fastest-growing and most technologically mature systems among renewable energy sources. The growing recognition of the environmental challenges posed by fossil fuels and the escalating threat of climate change has led to a concomitant increase in the importance of wind energy within the global energy system. The historical development of wind energy can be traced back millennia. It is established that the earliest windmills were utilized by the Persians circa 2000 BCE for the purpose of grinding grain [4]. During the medieval period, windmills became a common sight throughout Europe. These devices were utilized for a variety of purposes, including water pumping and various industrial activities. The first wind turbine to generate electricity in the modern sense was constructed in 1887 by the Scottish Professor James Blyth [5]. The oil crisis of the 1970s represented a significant turning point, accelerating the development of modern wind turbine technology.

Contemporary wind energy technologies can be categorized into two primary classifications: onshore and offshore systems. Onshore wind turbines generally possess a capacity of 3–5 MW and a tower height of 80–120 meters [6]. The most significant advantages of these systems are their relatively low installation costs and ease of maintenance. Nevertheless, the implementation of these measures may encounter resistance at the local level, primarily due to concerns regarding noise pollution and land use. In contrast, offshore wind farms are characterized by turbines with capacities in excess of 10 MW, accompanied by floating platforms installed in deep waters [7]. These systems facilitate greater access to more robust and reliable wind resources, while concomitantly exerting a lesser visual and environmental impact in comparison with onshore systems. Nevertheless, the elevated expenses associated with installation and maintenance constitute a notable disadvantage of offshore systems.

Global wind energy capacity is increasing at a rapid rate. According to data from 2023, the global installed capacity of wind power has surpassed 900 GW [6]. China is the global leader in this field, with 365 GW of installed capacity, followed by the United States with 140 GW and Germany with 66 GW. The approximately 70% decrease in wind energy costs over the past decade has significantly increased the economic competitiveness of this technology [8]. This decline in costs can be attributed primarily to large-scale projects and technological innovations.

The outlook for the future of wind energy is highly promising. The utilization of floating wind turbines in deep seas has the potential to facilitate energy generation. The integration of wind, solar, and storage systems in hybrid energy systems has the potential to offer significant advantages in terms of uninterrupted energy provision. The utilization of wind-powered electrolyzers on a large scale is poised to play a pivotal role in the development of a hydrogen economy that is devoid of carbon emissions [9]. However, alongside these positive developments, there are also some obstacles to the widespread adoption of wind energy. In consideration of the variability inherent in its capacity as an energy source, the necessity arises for the development of energy storage solutions, thereby ensuring grid stability. Furthermore, the promotion of public-private sector collaborations is of paramount importance in order to enhance investment levels.

Wind energy plays a pivotal role in the clean energy transition, and this role is expected to expand further in the future. In order to achieve the full potential of wind energy, there is a necessity for increased investment in offshore wind projects, the development of energy storage and smart grid technologies, and the reduction of costs through international cooperation. Should these measures be implemented, it is projected that wind energy's contribution to the global energy landscape will undergo significant expansion, thereby solidifying its role as a pivotal instrument in the global effort to combat climate change.

Despite its status as one of the fastest-growing technologies among renewable energy sources, wind energy poses significant challenges for integration into energy systems due to its inherent intermittency and unpredictability [10]. This phenomenon can be attributed to the temporal and spatial variability of wind-based electricity production. The irregular nature of wind energy can be examined in three ways:

- Short-term variability: Sudden fluctuations in wind speed have been shown to result in changes of up to 50% in production within minutes or hours [11]. This poses a significant threat to grid frequency stability.

- Daily and seasonal variations: It is evident that wind patterns demonstrate marked variations throughout the diurnal cycle and across seasonal transitions. For instance, in numerous regions, wind production exhibits an increase during nocturnal hours, while concurrently decreasing during diurnal periods of maximum demand [12].

- Geographical differences: The variability of wind resources from region to region has been identified as a significant complicating factor in energy planning [13].

A range of solutions has been developed to address the variable nature of wind energy. The utilization of advanced forecasting systems, employing artificial intelligence and machine learning algorithms, has yielded a commendable level of accuracy in wind speed predictions, ranging from 85 to 90% [14]. The implementation of these systems has been demonstrated to engender substantial enhancements in the domain of energy production planning.

Energy storage solutions are of critical importance in ensuring the balancing of fluctuating levels of wind energy production. Lithium-ion batteries are employed for short-term fluctuations, while hydrogen storage systems are utilized for long-term and seasonal storage requirements [15]. These technologies ensure a more stable integration of wind energy into the grid.

The geographical diversification strategy is predicated on the principle of the dispersal of wind farms across a wide geographical area. The diversity of wind patterns in different regions has been demonstrated to enable more balanced energy production across the system [16]. This approach serves to mitigate the impact of local weather conditions on production fluctuations.

Wind energy is a good addition to other renewable energy sources when used together in a hybrid energy system. In particular, wind-solar-storage combinations have been shown to significantly mitigate the intermittent nature of wind energy alone [17]. The distinguishing characteristic of these integrated systems is their capacity to deliver uninterrupted energy supplies.

These technological advancements and strategic approaches have enabled the more effective management of the intermittent and unpredictable nature of wind energy.

Research and development is making it possible for wind energy to play a central role in future energy systems [6]. These advancements are accelerating the clean energy transition and contributing to the construction of a sustainable energy future.

2.2 Energy storage systems

Energy storage systems have been identified as playing a critical role in modern power grids, particularly in the context of the increasing share of variable renewable energy sources, such as wind and solar, in the generation mix. The intermittent nature of these renewable sources gives rise to challenges with regard to grid stability and reliability, thus rendering energy storage essential for balancing supply and demand, providing grid services, and ensuring continuous power availability [15]. Energy storage can be considered to comprise a diverse range of technologies, each of which is suited to different applications based on capacity, discharge duration, and scalability.

Battery Energy Storage (BESS): The preponderance of lithium-ion batteries in the market is attributable to three key factors: their high energy density, efficiency, and declining costs [18]. These devices are predominantly employed for brief-duration storage (ranging from minutes to hours) in grid stabilization, electric vehicles, and residential applications. Emerging alternatives, such as solid-state and sodium-ion batteries, promise improved safety and lower material costs [19].

Pumped Hydro Storage (PHS): PHS is the most advanced and largest-capacity storage technology, making up over 90% of the world's storage capacity [20]. The operational mechanism entails the pumping of water to an elevated reservoir during periods of low demand, and the subsequent release of the water through turbines to generate electricity when demand is high. However, geographical constraints and high capital costs limit its expansion.

Thermal energy storage (TES): TES systems are capable of storing energy in the form of heat or cold, frequently in conjunction with concentrated solar power (CSP) plants. A notable illustration of this is the potential of molten salt storage to facilitate the dispatch of electricity by solar plants even after sunset hours [21].

Hydrogen and power-to-X (P2X): When there is more renewable electricity than is needed, it can be used to make hydrogen. This is done by using electrolysis, which separates hydrogen and oxygen. The hydrogen can then be stored and used to power things, or it can be used in industry [22]. The prospect of long-duration hydrogen storage is particularly promising for the purpose of seasonal energy balancing.

Notwithstanding the rapid advancements witnessed in the field, energy storage is confronted with numerous challenges, including but not limited to:

- The financial implications and the potential for expansion must be given due consideration. Despite a significant decrease in lithium-ion battery prices (~89% since 2010), long-duration storage remains costly [23].

- The reliance on critical minerals (e.g., lithium, cobalt) has given rise to concerns regarding sustainability and geopolitical risks [20].

- Outdated market structures frequently fail to provide adequate compensation for storage facilities in their role in stabilizing the grid [24].

The combination of batteries with hydrogen or thermal storage is a strategy employed to address the requirement for multi-hour to multi-day storage. As posited by Ahmadi et al. [25], the repurposing of EV batteries for stationary storage has been identified as a means to reduce costs.

The ability to store energy is imperative for the successful integration of high levels of renewable energy into the energy mix. The ongoing innovation, favorable policies, and investments in diverse storage technologies are identified as pivotal in overcoming the current limitations and facilitating the transition to a fully decarbonized energy system.

2.3 Hydrogen

Hydrogen has emerged as a pivotal energy carrier in the global transition to decarbonized energy systems. The usual way to make hydrogen uses fossil fuels. But green hydrogen, made using water electrolysis with renewable electricity, is a better option as it causes no carbon emissions [26]. This section examines hydrogen's role in the energy transition, focusing on production methods, current applications, and future potential. There are different ways to make hydrogen, and these are put into groups based on how much carbon they use:

Gray hydrogen: Gray hydrogen is produced *via* steam methane reforming (SMR) of natural gas. It accounts for approximately 95% of current global production but emits 9–12 kg of CO_2 per kg of H_2 [27].

Blue hydrogen: Similar to gray hydrogen but incorporates carbon capture and storage (CCS), reducing emissions by 50–90% [28]. However, CCS technology faces scalability and cost challenges.

Green hydrogen: Green hydrogen is a sustainable energy source that is produced through water electrolysis using renewable energy sources such as wind, solar, and hydro power. It does not emit any emissions, making it an environmentally friendly option.

Current electrolyzer technologies are Proton Exchange Membrane (PEM) Electrolyzers, Alkaline Electrolyzers (AEL), and Solid Oxide Electrolyzers (SOEC). AEL is a mature technology with lower costs but limited flexibility [29]. PEM Electrolyzers are more efficient and responsive but require expensive catalysts [30]. SOECs are high-temperature systems with superior efficiency but durability challenges [31].

The use of green hydrogen is set to play a crucial role in the decarbonization of sectors that currently lack viable electrification pathways. In the industrial sector, it has the potential to substitute for hydrogen derived from fossil fuels in the production of ammonia, steel, and chemicals [32]. In the transport sector, there is potential for its use in fuel cell electric vehicles for heavy-duty vehicles, trucks, and ships [33]. In the domain of energy storage, hydrogen-energy systems present a significant solution for the purpose of long-term storage and grid balancing [34]. The utilization of power-to-fuel (P2X) technologies facilitates the production of synthetic fuels (e-fuels), which find applications in the domains of aviation and maritime transport [35].

However, there are still several things that make it difficult for most people to start using green hydrogen. The high production costs (currently between \$3 and \$6 per kilogram) make it hard to compete with gray hydrogen, which typically costs less than \$2 per kilogram [36]. The limitations in infrastructure, including the scarcity of pipelines, storage facilities, and refueling stations, represent significant impediments to the effective functioning of the system [37]. Moreover, in order to compete with hydrogen derived from fossil fuels, policy support in the form of subsidies and carbon pricing is required [38].

Future developments in this field may focus on innovations in electrolyzer technology, such as the reduction of iridium usage in PEM systems and the scaling up of AEL/SOEC technologies [39]. When it comes to using renewable energy sources, putting electrolyzers together with offshore wind farms could mean lower costs [40]. Moreover, the establishment of hydrogen supply chains between regions with abundant renewable energy sources and those with high energy-intensity demands offers a substantial opportunity for global trade [41].

3. Market for wind power and hydrogen

The escalating global demand for energy, coupled with concerns regarding climate change, underscores the imperative for sustainable and clean energy solutions that will diminish reliance on fossil fuels. In this context, renewable energy sources, particularly wind energy, have begun to account for a significant share of electricity generation. Nevertheless, the variable and intermittent nature of wind energy gives rise to challenges in terms of grid stability and meeting energy demand. One of the most promising solutions to overcome these challenges and maximize the potential of wind energy is the production of green hydrogen [1].

Green hydrogen is made by splitting water using electricity, and the electricity used for this process comes entirely from clean energy sources. This carbon-emission-free process is regarded as a pivotal technology for energy storage, transportation, and decarbonization [42]. The purpose of this section is to examine the technological infrastructure, integration models, advantages, challenges, and future prospects of green hydrogen production from wind energy.

The electrolyzer is pivotal in the process of converting wind energy into hydrogen. The integration of wind farms with electrolyzers can be achieved in a number of ways.

Grid-connected systems: The electrolyzer is connected to the grid at the same point. The wind farm's primary function is to sell energy to the grid. The electrolyzer is a device that uses energy to make hydrogen when there is excess energy or when electricity prices are low. It can draw power from the grid or directly from the farm. The model has the capacity to provide grid balancing services and to prevent curtailment of the farm [3].

Off-grid systems: The electrolyzer is directly connected to the wind farm and has no connection to the grid. This model is particularly well-suited for remote areas lacking grid infrastructure or for offshore wind farms. The hydrogen produced is compressed for storage and transported to shore/industrial consumers by tanker or pipeline [43]. However, the intermittent nature of wind power generation may result in low utilization of the electrolyzer capacity, thereby posing a risk to the system as a whole.

Hybrid systems: The system may comprise solar panels or a small grid connection in addition to wind energy. Using different energy sources like this means the electrolyzer can work better and cost less to make hydrogen [2].

3.1 Electricity market for wind power plants

Wind power plants sell their electricity on the free market through three main channels: (1) spot markets (Day-Ahead and Intraday), (2) bilateral agreements/PPAs (physical or virtual), (3) support/certificate mechanisms (e.g., YEKDEM, YEK-G). The Day-Ahead Market (DAM) is the primary reference market in which supply and demand are matched for each hour of the following day through a closed auction process, taking into account system constraints; the result is the "market clearing price." It is evident that imbalances are managed within the intraday and balancing/balance markets.

The day-ahead market is the market where wind power plants conduct the most transactions. The process is characterized by a centralized optimization procedure, whereby bids are amassed for each hour of the subsequent day. This is accomplished through the utilization of a closed auction method, a technique that is contingent upon transmission constraints. The market clearing price (MCP) is ultimately determined [44, 45].

The day-ahead market in Europe functions in an integrated manner under the Single Day-Ahead Coupling (SDAC) framework. A total of 27 countries are subject to a uniform clearing mechanism, with transmission capacity constraints being a key consideration [44, 46]. In contrast, the Intraday Market functions on a continuous basis across the continent, utilizing the SIDC/XBID infrastructure [47]. State-supported CfD auctions and corporate PPA agreements are also very common in wind investments [48]. Green certificates are important for corporate buyers [49].

Within the United States, there are binding Day-Ahead and Real-Time markets in ISO/RTO regions (PJM, ERCOT, ISO-NE, etc.). While regions such as PJM also have a capacity market, ERCOT only operates an energy market [50].

In Turkey, wind power sales are facilitated by EPİAŞ. For the day-ahead market, participants submit hourly price-quantity bids for the following day, and EPİAŞ determines the MCP through centralized optimization [51]. YEKDEM provides a TL-based, inflation-indexed purchase guarantee for renewable energy projects, thus remaining a significant financing mechanism for wind investments [52].

The day-ahead market is the primary sales channel for wind power. However, due to the uncertainty surrounding production forecasts, Intraday Market and balancing costs are of critical importance. Market liquidity has increased in Europe thanks to integration (SDAC/SIDC). Within the Turkish context, the EPİAŞ structure, in conjunction with the YEKDEM and YEK-G mechanisms, offers wind producers a dual income stream, encompassing both regulated and market-based revenue opportunities.

3.2 Hydrogen market

The objectives of combating climate change and ensuring energy supply security are compelling global energy systems to undergo a fundamental transformation. In this transformation, there is a need for an energy carrier that will balance the intermittent nature of clean energy sources and decarbonize various sectors.

It is evident that more than 30 countries worldwide have announced their national hydrogen strategies. These strategies encompass investments in research and development, infrastructure development, and mechanisms for market creation.

The European Union (EU) is pursuing an assertive policy agenda with the objective of establishing itself as a global leader in the field of green hydrogen. In the context of the European Green Deal and the REPowerEU Plan, the EU has signaled its commitment to increasing green hydrogen, with a view to reducing its reliance on Russian fossil fuels for energy supply and achieving its objective of achieving carbon neutrality by 2050. The EU has announced that it aims to produce a total of 20 million tons of renewable hydrogen by 2030, with 10 million tons to be produced domestically and 10 million tons through imports [53]. In order to support this objective, financial resources such as the "Hydrogen Bank" and regulatory frameworks such as the Carbon Border Adjustment Mechanism are being established to mandate that imports conform to environmental standards.

The USA offers attractive incentives to green hydrogen producers through the Inflation Reduction Act (IRA), with a maximum value of $3 per kilogram. This has led to a rise in global competition, with the US becoming a major recipient of investment [54].

China makes and uses the most hydrogen of all countries. Most of the hydrogen it makes comes from coal. Nevertheless, there is an increasing trend of investment and targets for renewable hydrogen projects. Japan and South Korea regard hydrogen as a pivotal component of their national energy security strategies, with both countries demonstrating notable leadership in the field of hydrogen transport technologies, particularly in the domain of fuel cell technologies.

The Middle East and North Africa (MENA) regions are of particular interest in this regard. Saudi Arabia, the UAE, Morocco, and Oman have announced their intention to establish large-scale green hydrogen production centers and export to Europe by utilizing their abundant solar and wind potential [55].

Turkey's geographical location, robust industrial infrastructure, and expanding renewable energy capacity position it as a significant contender in the green hydrogen economy. Turkey is currently a country that produces and consumes gray hydrogen, particularly in refineries and the fertilizer industry. The Ministry of Energy and Natural Resources has outlined its strategic vision for the development of hydrogen technologies in Turkey. By the close of the 2030s, the strategy aims to achieve electrolyzer capacity of 2 GW. Moreover, by the year 2050, the ministry envisions the incorporation of green hydrogen into the nation's final energy consumption [56]. It is evident that there is considerable potential for the generation of renewable energy. Turkey's wind and solar energy capacity is increasing steadily on an annual basis. This will provide cheap and abundant renewable electricity, which is the basic input for green hydrogen. The proximity of Turkey to Europe's largest green hydrogen market has the potential to establish the country as a logistical hub for hydrogen exports *via* pipelines or sea routes. Making changes to the natural gas pipelines so that they can be used to mix hydrogen at certain ratios, as well as replacing the hydrogen used in industry with green hydrogen, will make it much easier to set up a domestic market.

Despite the fact that the global green hydrogen market is still in its infancy, climate targets and energy geopolitics are rendering it an inevitable future. While Europe exerts its influence as the predominant demand center, significant competition has emerged from the US and the MENA region in terms of supply.

Turkey has the opportunity to function not only as a consumer but also as a producer and a transit hub, exporting to Europe in this new and massive global market. In order to capitalize on this opportunity, it is critical to increase R&D incentives for technology development and localization, support pilot projects, develop international partnerships, and implement a clear and long-term policy framework that will attract private sector investment. The potential of green hydrogen in Turkey is twofold: firstly, as a source of energy, and secondly, as a high-value-added export product. Furthermore, it is imperative to recognize the pivotal role that green hydrogen could play in catalyzing a green industrial transformation.

4. Market dynamics and shifts

The inherent difficulties and costs associated with the storage of electrical energy necessitate its real-time balancing, rendering it a commodity of significant importance. The day-ahead market constitutes a fundamental wholesale market mechanism, wherein electricity prices and quantities are determined for the subsequent 24-hour period. This enables the transmission system operator to operate the system safely.

Electricity prices in the electricity market exhibit significant variations depending on the time of day. When this situation is considered alongside the difficulty in predicting the electricity that a wind farm will generate the following day, the economic profit margin is considerably limited. **Figure 1** provides a detailed analysis of the fluctuations in electricity prices within the day-ahead market across various countries.

Electricity prices demonstrate significant fluctuations both diurnally and geographically. A thorough examination of the available data reveals that the range of electricity prices is from a minimum of €24/MWh to a maximum of €136/MWh. It is noteworthy that both of these values are observed in Romania. For instance, a decline in electricity prices is observed in France between 11 am and 7 pm, while in Romania, this decrease occurs between 10 am and 5 pm. It has been observed that electricity tariffs tend to be elevated during the morning and evening periods.

The analysis conducted for the sample day revealed that the highest standard deviation was €30/MWh for Romania, while the lowest standard deviation was €8/MWh for Italy. It is evident that Italy's hourly electricity price exhibits significantly less volatility compared to other countries.

Analogous to the shift in hourly electricity prices, the variation in electricity derived from wind has also been examined. For instance, **Figure 2** illustrates the 15-minute wind electricity price for Germany and France on 20/08/2025, which is analogous to the electricity price.

The capacity of wind power generation in France ranges from 2873 MW to 4743 MW. In Germany, wind power generation varies between 5296 MW and 12,823 MW. The higher standard deviation is elevated for Germany. It is evident that there is an increase in wind power generation after 17:00. For France, the standard deviation is minimal.

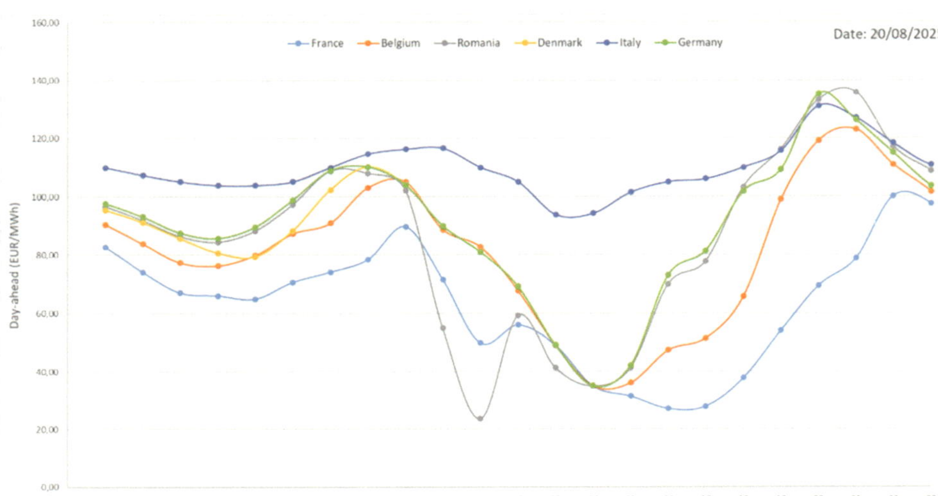

Figure 1.
Change in day-ahead electricity prices for different countries. Data taken from the ENTSO-E transparency platform [57].

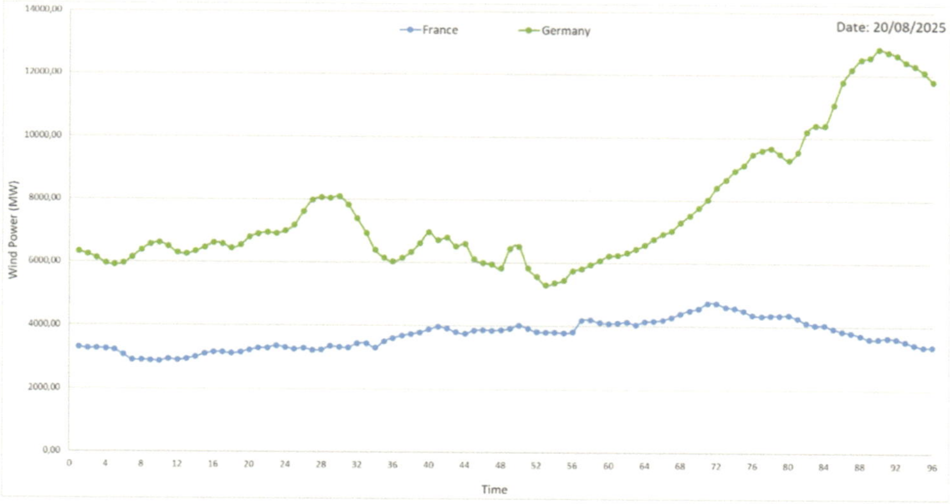

Figure 2.
Change of wind power for Germany and France. Data taken from the ENTSO-E transparency platform [57].

A consideration of wind power generation and hourly electricity prices reveals that they do not exhibit a similar pattern. It is not a realistic approach to expect a highly fluctuating energy source such as wind to align with electricity consumption. Consequently, the efficacy of storage options in optimizing revenue generation from wind electricity is of paramount importance.

When evaluating the potential for the production of green hydrogen from wind power in terms of future projections, it emerges as a significant option. A cost analysis of hydrogen, categorized as "green" or "gray," reveals that the former has a cost of approximately $2–3/kg, whereas the latter is around $3–6/kg [58, 59]. However, when the imposition of a carbon tax is also taken into account, the price of gray hydrogen can rise to the range of 4–7 $/kg [60, 61]. Consequently, when considering carbon emissions, there is no substantial discrepancy in the costs of gray and green hydrogen. Conversely, the heightened level of activity and growth in the green hydrogen market serves to further emphasize the significance of this option.

The present study examined the relationship between the day-ahead market electricity price and wind-generated electricity in Turkey (**Figure 3**). An assessment was conducted utilizing hourly data from 07/07/2025 to 13/07/2025.

As demonstrated in **Figure 3**, there is no discernible correlation between periods of wind power generation and periods of high electricity prices. This finding indicates that the revenue generated from wind power is minimal during periods when electricity prices are low. Consequently, the implementation of a hydrogen storage strategy during periods of low electricity prices may prove to be a more economically viable option. Conversely, the mitigation of emissions from green electricity sources could also have a beneficial effect on the environment.

Normalization has been performed to ensure easier comprehension using sample data taken for Turkey. The normalization data is presented in **Figure 4**. As can be seen, it is clearly evident that electricity prices remain low during periods of high wind power generation. It has been assessed that higher revenues could be achieved by producing hydrogen using the wind power generated during these periods.

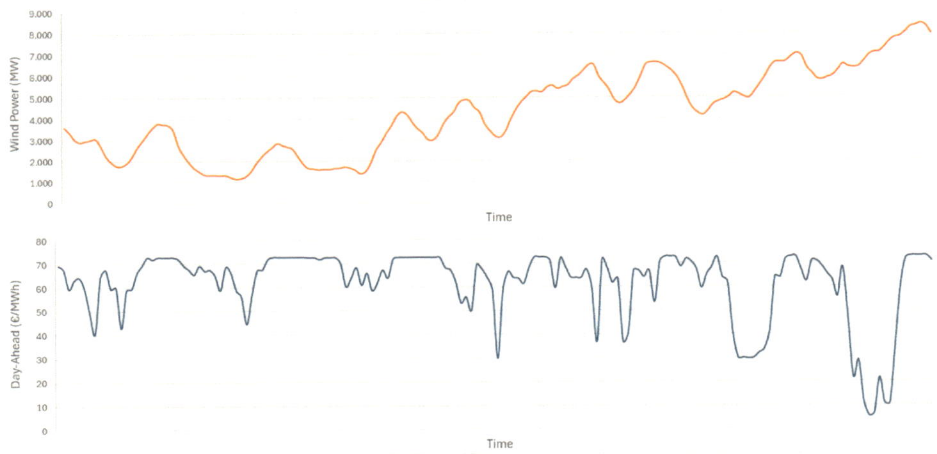

Figure 3.
Hourly change of wind power and electricity price for Turkey (07/07/2025–2113/07/2025). Data taken from the EPİAŞ transparency platform [62].

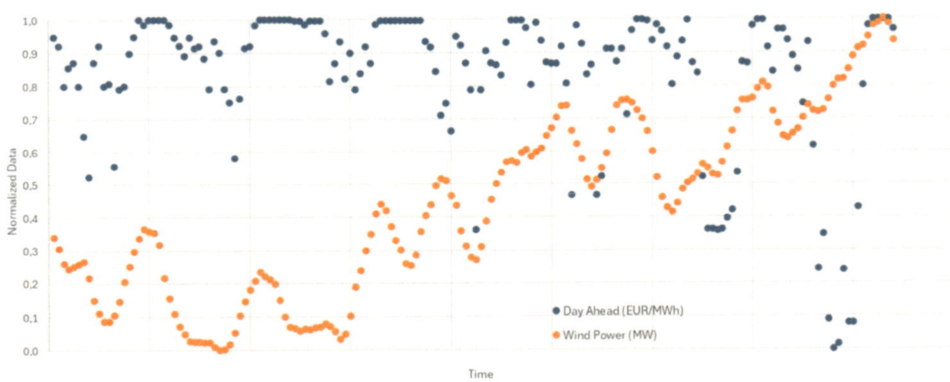

Figure 4.
Hourly normalized data for Turkey (07/07/2025–2113/07/2025). Data taken from the EPİAŞ transparency platform [62].

5. Conclusion

The present study investigates the relationship between the price of electricity in the day-ahead market and the fluctuation of electricity generated from wind. This relationship has been the basis of an analysis of green hydrogen production. The challenges associated with predicting wind energy production and the volatility of electricity prices have led to a renewed focus on storage strategies.

The market price of electricity is subject to considerable variation depending on the time of day. A recent study of electricity price fluctuations in Romania revealed a range of €24/MWh to €136/MWh, with a standard deviation of €30/MWh, as observed in specific city samples. In a similar manner, wind-generated electricity was found to exhibit a daily variable profile. A sample analysis for Germany found that wind power fluctuated between 5296 MWh and 12,823 MWh.

Analyses of a similar nature conducted for Turkey determined that there is no correlation between electricity prices and the amount of electricity generated from wind power.

The findings of the present study demonstrate that there is an absence of a complete correlation between electricity prices in the day-ahead market and electricity generated from wind power. In this particular instance, it has been determined that the implementation of hydrogen storage applications has the potential to enhance the profitability and environmental sustainability of wind power. It has been determined that the operation of a power plant that is more economically and environmentally efficient is possible through the production and storage of green hydrogen when electricity prices are low.

Conflict of interest

The authors declare no conflict of interest.

Author details

Saban Pusat* and Yasin Karagoz
Yildiz Technical University, Istanbul, Turkey

*Address all correspondence to: spusat@yildiz.edu.tr

References

[1] IRENA. Green Hydrogen: A Guide to Policy Making. Masdar City, Abu Dhabi: International Renewable Energy Agency; 2020

[2] Gökçek M, Kale C. Optimal design of a hydrogen-based hybrid system for a remote community in Turkey. International Journal of Hydrogen Energy. 2018;**43**(23):10721-10732

[3] González A, McKeogh E, Gallachóir BÓ. The role of hydrogen in high wind energy penetration electricity systems: The Irish case. Renewable Energy. 2003;**29**(4):471-489

[4] Spera DA. Wind Turbine Technology: Fundamental Concepts of Wind Turbine Engineering. New York, NY: American Society of Mechanical Engineers; 2009

[5] Price TJ. James Blyth: Britain's first modern wind power Pioneer. Wind Engineering. 2005;**29**(3):191-200

[6] Global Wind Energy Council (GWEC). Global Wind Report 2023. Belgium: GWEC; 2023

[7] International Renewable Energy Agency (IRENA). Offshore Renewable Energy: Technologies, Policies, and Economics. Masdar City, Abu Dhabi: IRENA; 2022

[8] Lazard. Levelized Cost of Energy Analysis. New York, NY: Lazard; 2023

[9] International Energy Agency (IEA). The Future of Wind Energy in a Net-Zero World. Paris, France: IEA; 2023

[10] IEA. World Energy Outlook 2022. Paris, France: IEA; 2022

[11] Potter CW, Negnevitsky M. Very short-term wind forecasting for Tasmanian power generation. IEEE Transactions on Power Systems. 2006;**21**(3):965-972

[12] Archer CL, Jacobson MZ. Historical trends in the jet streams. Journal of Geophysical Research. 2007;**112**(D12):1252-1259

[13] IRENA. Future of Wind: Deployment, Investment, Technology, Grid Integration and Socio-Economic Aspects. Masdar City, Abu Dhabi: IRENA; 2021

[14] Wang H et al. Taxi trajectory data based fast-charging facility planning for urban electric taxi systems. Applied Energy. 2021;**286**:116541

[15] Dunn B et al. Electrical energy storage for the grid: A battery of choices. Science. 2011;**334**(6058):928-935

[16] Kempton W et al. Energy efficient vehicles: Technologies and challenges. Proceedings of the IEEE. 2010;**98**(6):1180-1197

[17] Denholm P, Hand M. Grid flexibility and storage required to achieve very high penetration of variable renewable electricity. Energy Policy. 2011;**39**(3):1815-1821

[18] Larcher D, Tarascon JM. Towards greener and more sustainable batteries for electrical energy storage. Nature Chemistry. 2015;**7**(1):19-29

[19] Tapia-Ruiz N et al. High voltage structural evolution and enhanced Na-ion diffusion in P2-Na2/3Ni1/3-xMgxMn2/3O2 (0 ≤x ≤ 0.2) cathodes from diffraction, electrochemical and ab initio studies. Energy & Environmental Science. 2021;**14**(4):1924-1933

[20] IEA. The Role of Critical Minerals in Clean Energy Transitions. Paris, France: IEA; 2021

[21] Gil A et al. State of the art on high temperature thermal energy storage for power generation. Part 1—Concepts, materials and modellization. Renewable and Sustainable Energy Reviews. 2010;**14**(1):31-55

[22] Glenk G, Reichelstein S. Economics of converting renewable power to hydrogen. Nature Energy. 2019;**4**(3):216-222

[23] BNEF. Battery Price Survey. London, UK: BNEF; 2023

[24] Sioshansi R et al. Impact of policy instruments in the implementation of renewable sources of energy in selected European countries. Energy Policy. 2022;**164**:112881

[25] Ahmadi L et al. A multi-objective optimization evaluation framework for integration of distributed energy resources. Journal of Energy Storage. 2020;**27**:101037

[26] IEA. The Future of Hydrogen. Paris, France: IEA; 2021

[27] Bloomberg NEF. Hydrogen Economy Outlook. London, UK: Bloomberg NEF; 2023

[28] Global CCS Institute. Blue Hydrogen Report. Melbourne, Australia: Global CCS Institute; 2022

[29] Carmo M et al. A comprehensive review on PEM water electrolysis. International Journal of Hydrogen Energy. 2013;**38**(12):4901-4934

[30] Ayers K et al. Research advances towards low cost, high efficiency PEM electrolysis. ECS Transactions. 2019;**58**(1):3-15

[31] Laguna-Bercero MA. Recent advances in high temperature electrolysis using solid oxide fuel cells: A review. Journal of Power Sources. 2012;**203**:4-16

[32] IRENA. Green Hydrogen for Industry: A Guide to Policy Making. Masdar City, Abu Dhabi: IRENA; 2022

[33] Staffell I et al. The role of hydrogen and fuel cells in the global energy system. Energy & Environmental Science. 2019;**12**(2):463-491

[34] Buttler A, Spliethoff H. Current status of water electrolysis for energy storage, grid balancing and sector coupling via power-to-gas and power-to-liquids: A review. Renewable Energy. 2018;**123**:802-820

[35] Fasihi M et al. The value of ammonia towards integrated power and heat system decarbonisation. Applied Energy. 2019;**233-234**:854-893

[36] Hydrogen Council. Hydrogen Insights Report. Rotter, Netherlands: Hydrogen Council; 2023

[37] JRC. Hydrogen Infrastructure Development. Belgium: JRC; 2020

[38] ETC. Making the Hydrogen Economy Possible. London, UK: ETC; 2021

[39] Shiva Kumar S, Himabindu V. Hydrogen production by PEM water electrolysis – A review. Materials Science for Energy Technologies. 2019;**2**(3):442-454

[40] Reuß M et al. Community compensation in the context of carbon capture and storage: Current debates and practices. Energy Policy. 2019;**132**:1092-1100

[41] IRENA. Global Hydrogen Trade to Meet the 1.5°C Target. Masdar City, Abu Dhabi: IRENA; 2022

[42] Dawood F, Anda M, Shafiullah GM. Hydrogen production for energy: An overview. International Journal of Hydrogen Energy. 2020;**45**(7):3847-3869

[43] Schiebahn S, Grube T, Robinius M, Tietze V, Kumar B, Stolten D. Power to gas: Technological overview, systems analysis and economic assessment for a case study in Germany. International Journal of Hydrogen Energy. 2015;**40**(12):4285-4294

[44] EPEX SPOT – Day-Ahead Market. Epex Markets. Available from: www.epexspot.com/en/market-results

[45] Nord Pool – Day-Ahead Market Clearing

[46] ACER & ENTSO-E, SDAC Report

[47] ENTSO-E, SIDC/XBID Intraday Market Report

[48] "EU renewable energy financing mechanism" Report

[49] EPİAŞ, "YEK-G Sertifika Piyasası" Information Note

[50] FERC & PJM/ISO-NE/ERCOT Market Design Documents

[51] EPİAŞ, "Gün Öncesi Piyasası (GÖP)" Guide

[52] Republic of Türkiye Ministry of Energy and Natural Resources "Annual report 2021". Available from: www.enerji.gov.tr

[53] European Commission. REPowerEU Plan. COM(2022) 230 Final. Belgium: EU; 2022

[54] U.S. Department of the Treasury. Inflation Reduction Act of 2022: Hydrogen Production Tax Credit. Washington, D.C: U.S. Department of the Treasury; 2023

[55] IRENA. *Global Hydrogen Trade to Meet the 1.5°C Climate Goal: Part II - Technology Review of Hydrogen Carriers*. Masdar City, Abu Dhabi: International Renewable Energy Agency; 2022

[56] Republic of Türkiye Ministry of Energy and Natural Resources. Türkiye Ulusal Hidrojen Teknolojileri Stratejisi ve Yol Haritası. Anlara, Turkiye: Republic of Türkiye Ministry of Energy and Natural Resources; 2023

[57] European Network of Transmission System Operators for Electricity ENTSO-E Transparency Platform. 2025. Available from: https://transparency.entsoe.eu

[58] BloombergNEF (BNEF). 1H 2023 Hydrogen Levelized Cost Update. New York: Bloomberg Finance L.P; 2023

[59] IRENA. Green Hydrogen Cost Reduction: Scaling up Electrolysers to Meet the 1.5°C Climate Goal. Masdar City, Abu Dhabi: International Renewable Energy Agency; 2020

[60] International Energy Agency (IEA). Global Hydrogen Review 2023. Paris: IEA Publications; 2023

[61] Hydrogen Council. Hydrogen Insights 2023. Brussel: Hydrogen Council; 2023

[62] EPİAŞ Transparency Platform. 2025. Available from: https://seffaflik.epias.com.tr/

Section 2

Advances in Hydrogen Energy: Storage and Utilization

Chapter 6

Computational Insights into Advanced Materials for Solid-State Hydrogen Storage

Bilal Ahmed, Muhammad Bilal Tahir, Muhammad Rafique, Tahir Iqbal and Ghulam Nabi

Abstract

The world is moving toward sustainable and low-carbon energy systems, which means we need hydrogen storage solutions that are safe, scalable, and cost-effective. Among several methods, solid-state hydrogen storage has become a promising alternative to traditional gaseous and liquid storage methods because it has a higher gravimetric capacity, is more stable, and is safer. This chapter delineates a computational analysis of perovskite hydrides XYH_3 (X = Li, Na, K) employing first-principles density functional theory (DFT). The study thoroughly investigates their structural, electrical, optical, and mechanical properties to assess their viability for hydrogen storage applications. $LiYH_3$, $NaYH_3$, and KYH_3 have lattice constants of 4.42 Å, 4.30 Å, and 4.41 Å, and their gravimetric hydrogen storage capacities are 3.06 wt%, 2.64 wt%, and 2.31 wt%, respectively. Thermodynamic study showed that the formation enthalpies were negative (−16.19 eV, −13.54 eV, and − 9.13 eV), which confirmed their stability. The Gibbs free energy values at 1000 K were also still good (−0.51 eV, −0.69 eV, and − 1.48 eV). The mechanical investigation showed that Young's moduli were 27.59 GPa, 52.19 GPa, and 49.29 GPa, and all of the compounds were brittle ($B/G < 1.75$). Optical tests revealed significant absorption peaks between 26 and 29 eV, underscoring their suitability for optoelectronic applications. These findings underscore that perovskite hydrides are promising candidates for next-generation hydrogen storage devices, providing computationally informed design principles that enhance experimental initiatives. This combination of simulation and materials engineering helps create solid-state hydrogen storage solutions that are efficient, long-lasting, and cheap, which is important for the growth of the hydrogen economy.

Keywords: solid-state hydrogen storage, perovskite hydrides, density functional theory, structural stability, computational materials science, energy storage materials

1. Introduction

The world economy is booming and the population is growing continuously, which is putting a lot of stress on current energy resources [1]. According to the

International Energy Agency (IEA), using fossil fuels mostly meets this need [2]. Burning a wide range of fossil fuels releases carbon dioxide, which is a well-known dangerous substance that pollutes the environment and speeds up global warming [3]. However, the increasing global demands for energy and climate change have led academics to conduct new experiments to find energy storage technologies that are both sustainable and good for the environment [4]. In this view, pumped hydro, hydrogen storage, and battery storage are all ways to store energy [5]. Batteries have a finite lifespan and need certain materials, while pumped hydro needs to be in a certain place [6]. Hydrogen storage is the greatest choice because it can be expanded and can hold a lot of hydrogen for a long time [7].

Preparation, storage, and delivery make up the three primary links in the design chain of hydrogen energy applications [8]. Traditional methods of storing data, such as gaseous or liquid-state storage, have many limitations in terms of both cost and storage density, despite their widespread use in business [9]. Consequently, it is critical that all links in the chain settle on a more secure and effective method of hydrogen storage [7]. As it stands, one viable option for storing hydrogen energy is the solid hydrogen storage technique [10]. It has a lot of benefits, such as being able to absorb and release hydrogen very well, being more space-efficient, and being safer [11]. As a result, studies on solid hydrogen storage materials have also received a lot of interest in the last several years. Perovskite hydrides have become a focus of hydrogen storage research among the many solid-state materials that have been studied [12, 13]. These materials have great structural stability, a lot of room for hydrogen storage, and good physicochemical properties. ABH_3-type perovskite hydrides, which are made from main group and transition metal elements, are very fascinating because they have high gravimetric hydrogen densities and can be desorbed and reabsorbed [14]. This means they could be used in hydrogen storage systems of the next generation. A lot of these compounds also meet the Department of Energy's objective of having 5.5 wt% hydrogen storage capacity by 2025 [15, 16]. The results are between 1.2 wt% and 6 wt% [17]. Because of this, simple hydride perovskite materials are still one of the most investigated types of compounds for storing hydrogen [18]. Computational materials science has lately emerged as a transformative approach for discovering and enhancing novel materials for solid-state hydrogen storage [19, 20]. Density Functional Theory (DFT) is one of the best ways to use computers to predict the structural, electronic, thermodynamic, and kinetic properties of hydride systems with a lot of accuracy and without much experimental data [21]. DFT enables us see how materials behave at the atomic level, such as how much energy it takes to bond hydrogen, how much energy it takes to create a molecule, how charge density is distributed, and how electronic band structures work. These are vital for figuring out if a substance can hold hydrogen.

This chapter seeks to deliver a thorough examination of the current state of solid-state hydrogen storage materials from a computational standpoint. Although numerous prior investigations have examined various perovskite hydrides, including $MgCrH_3$ [22], $LiScH_3$ [23], and $XSrH_3$ (X = K, Rb) [13], for hydrogen storage purposes, XYH_3 hydrides have received scant attention, notwithstanding their capacity to fulfill the gravimetric capacity objectives established by the U.S. DOE. Prior studies predominantly concentrated on structural stability or restricted property subsets, resulting in a deficiency in the systematic correlation of structural, thermodynamic, mechanical, electrical, optical, and hydrogen storage attributes within this category of materials. In this paper, we fill this gap by using first-principles DFT to look at XYH_3 (X = Li, Na, K) and give numerical information about their formation energies, gravimetric hydrogen capacities, phonon stability, mechanical behavior, and optical properties. Our study not only adds to what is known about perovskite-type hydrides, but it also sets up computationally guided design concepts that work well with experimental work. This closes the gap between theoretical predictions and real-world hydrogen storage applications.

2. Overview of solid-state hydrogen storage materials

There are a numerous of different materials that can be employed in hydrogen storage, such as carbon allotropes, zeolites, metal-organic frameworks, and metal hydrides [24]. A short summary of these contents will follow. The section will go into great detail about metal hydrides, which are the main focus of this paper.

The process of adsorption happens when one species, the adsorbate, sticks to the surface of another species, the adsorbent [25]. If the connection between the adsorbate and the adsorbent is weak and involves intermolecular forces (like Van der Waals forces), the adsorption is physical [26]. If the forces are ionic or covalent, the adsorption is chemical. In absorption, the same thing happens, except this time the whole object is involved, not just the surface [27]. Absorption, like adsorption, can be either physical or chemical. Because they have a lot of surface area, carbon allotropes are some of the most investigated materials for solid-state hydrogen storage. In this respect, the most well-known examples include graphene, fullerenes, and carbon nanotubes (both single-walled and multi-walled) [27]. Zeolites are porous aluminosilicates that can store hydrogen because they have a large surface area [28]. Hydrogen can stick to zeolites at very low temperatures and pressures (cryogenic temperatures and low pressures) or at normal temperatures and high pressures (encapsulation). Metal-organic frameworks (MOFs) are materials that have a very high specific surface area. They are formed of metal ions or clusters that are linked together by organic ligands. These materials only work well at cryogenic temperatures when it comes to gravimetric capabilities [29].

Hydrides are a type of material that is made up of hydrogen and one or more other elements [30]. Hydrides can be put into distinct groups based on the elements that are bonded to H and the type of bonding that is used [31]. The broad classification of hydrides is presented here, followed by a specific classification of metal hydrides. Hydrogen may make binary compounds with almost all elements, except for He, Ne, Ar, Kr, Pm, Os, Ir, Rn, Fr, and Ra [27]. Based on the type of bond that H has with the other element, hydrides can be divided into three groups: ionic hydrides, covalent hydrides, and interstitial hydrides (also called metallic hydrides) [32, 33]. In ionic hydrides, hydrogen is attached to an electropositive element and behaves as a hydride ion, H^-. An alkali or alkaline-earth metal is normally the electropositive element; however some lanthanides can also form ionic hydrides. For the later members of the alkali and alkaline-earth metals groups, the ionic nature is more obvious, while for the smaller elements, the covalent nature is more obvious: Be, in particular, exhibits polymeric covalent behavior. H and the elements in groups 12 to 17 (including Be) make covalent hydrides [34]. Hydrides are polymeric for Be and groups 12 and 13 (elemental hydrides from group 12 are frequently seen as a middle ground between covalent polymeric hydrides and metallic hydrides), while they are volatile for groups 14 to 17 [35].

Transition metals, several lanthanides, and actinides create interstitial hydrides; hydrogen is integrated inside the tetrahedral and/or octahedral interstices of the

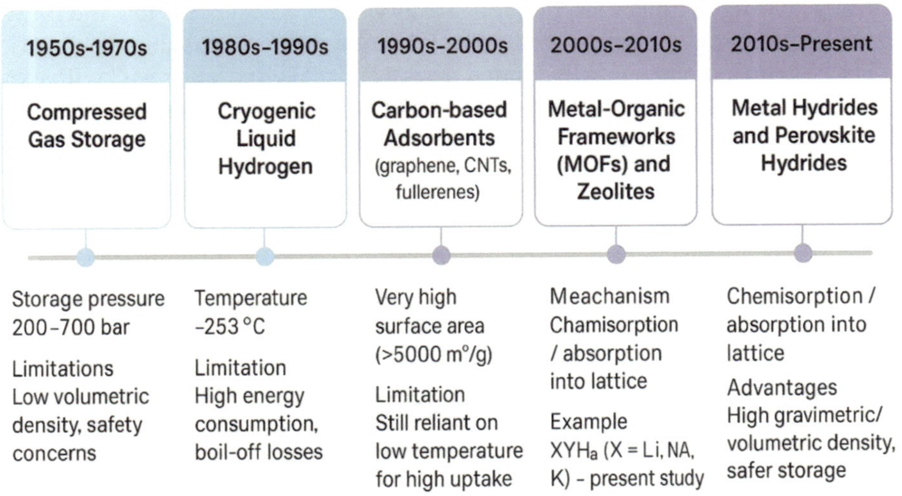

Figure 1.
Evolution of hydrogen storage technologies from physical methods to advanced solid-state materials

metal lattice via metal bonds, resulting in little alterations to the metallic structure, although the occurrence of some distortions. In general, ionic and covalent hydrides (also called non-interstitial hydrides) can hold more hydrogen by weight than interstitial hydrides [36]. They are also more stable, therefore the hydrogen release happens at higher temperatures. The lower atomic weight of the elements that usually make non-interstitial hydrides and the stronger ionic and covalent bonds compared to metallic bonds explain these traits (**Figure 1**) [37].

3. Computational methods in hydrogen storage research

To make good solid-state hydrogen storage materials, we need to know a lot about the basic atomic and electrical processes that control how hydrogen is absorbed, diffused, and released. Experimental methods yield essential insights; yet, they are frequently hindered by constraints in spatial resolution, temporal scales, and the challenges associated with investigating dynamic processes at the atomic level. In this context, computational materials science has become an essential complement to experimental methods, providing predictive and interpretive functions that greatly speed up the search for and improvement of materials for storing hydrogen.

Density functional theory (DFT) is the most important computational method for modeling contemporary quantum materials [38]. It enables highly accurate predictions of the structural, electrical, and thermodynamic characteristics of the ground state derived from quantum mechanics. In hydrogen storage research, DFT is commonly applied to examine structural stability through the computation of total energies and relaxed geometries and to investigate electronic properties including band structures, charge density distributions, and density of states [21, 39]. Also, DFT makes it possible to figure out formation enthalpies and the energies needed to adsorb or desorb hydrogen, which are important for figuring out if hydrogen storage processes will work. Using DFT, researchers may logically figure out how doping, replacement, or functionalization changes the performance of storage. This lets them make materials at the atomic level that can hold more, go back to their original state, and stay stable at high temperatures.

Ab initio molecular dynamics (AIMD) adds time evolution and temperature effects to DFT. This enables scientists see how systems that contain hydrogen act at different temperatures [40, 41]. AIMD simulations use Newtonian dynamics that are changed by quantum phenomena. This lets you see how atoms move at temperatures that are not zero. This method is especially good for assessing the structural integrity of materials while they are in operation, studying how hydrogen diffuses, and studying dynamic phase changes as hydrogen is absorbed and released. AIMD gives us vital information about kinetic limitations and thermally activated phenomena that static DFT computations do not show by simulating real-time interactions at high temperatures [42]. Along with electrical and thermal impacts, vibrational characteristics are highly significant in hydrogen storage systems. You can find vibrational modes, zero-point energy, and vibrational entropy by doing phonon calculations. These are commonly done with density functional perturbation theory or finite displacement methods [43]. These traits change the total Gibbs free energy and help us learn about dynamic stability by looking at phonon dispersion curves. The absence of imaginary modes in the phonon spectrum confirms the dynamical stability of the proposed hydride phase. Phonon analysis also helps us estimate thermodynamic properties like heat capacity and thermal conductivity, and it helps us understand how vibrations affect hydrogen diffusion and desorption.

The combination of various computational methods, which include first-principles quantum computations, large-scale statistical modeling, and data-driven prediction, makes it possible to gain a full and multi-scale picture of hydrogen storage materials [44]. They give us a lot of information on processes that are hard or impossible to measure in a lab, like how hydrogen atoms migrate via defect sites or how electronic structure affects desorption energetics. Also, more and more, computational tools are being employed alongside experimental work to help with material synthesis, make sense of experimental data, and improve theoretical models based on feedback from real-world experiments. As the hydrogen economy grows, computational modeling will play an even bigger role in finding, designing, and improving solid-state hydrogen storage materials. This will help make hydrogen storage systems that are safer, more efficient, and more cost-effective.

4. Design principles for high-performance hydrogen storage materials

A set of basic design rules that make sure the materials are useful, efficient, and reliable throughout time lead to the creation of next-generation hydrogen storage materials. One of the most important things is to get both high volumetric and gravimetric hydrogen densities [45, 46]. For a material to meet energy density standards, it needs to be able to carry a lot of hydrogen by weight and still be small enough to be useful in real life, like when you store fuel in a car. To do this, you need to use light elements like Li, Mg, or B and crystal structures that can accommodate more than one hydrogen atom without breaking down.

Another key thing to think about is how hydrogen absorbs and releases heat, which is commonly quantified by the enthalpy of formation [47]. Hydrogen should be able to be absorbed and released in a fashion that can be reversed at normal temperatures and pressures. It could be harmful if materials with very low enthalpy let hydrogen out too quickly. Materials with a lot of enthalpy, on the other hand, need a lot of energy to let hydrogen go. So, a moderate enthalpy of formation of 20 to 50 kJ/mol H_2 is usually the ideal way to establish a balance between stability and convenience of usage [48]. Hydrogen storage devices need to be able to swiftly adsorb and desorb hydrogen, in addition to being thermodynamically stable. Kinetic barriers can slow down the process of taking in or letting out hydrogen molecules. This makes response times longer and affects efficiency. For materials to perform well, hydrogen needs to be able to travel fast through the lattice. This is based on the kind of chemical bond, the presence of diffusion channels, and the presence of catalytic or structural promoters. Reversibility and cyclability are also very crucial for finding out how effectively hydrogen storage materials will operate over time. A good material should be able to hold hydrogen for many cycles without breaking down or turning into other phases [49, 50]. This means that it needs to be stable both mechanically and thermally while it goes through repeated cycles of heating, cooling, and hydrogenation. For materials to last, they must not sinter, crack, or break down with time. You should also consider about the economy and the environment. The optimum material for storing hydrogen should be cheap, easy to find on the Earth, and safe for people. When putting energy systems in place on a large scale, this is especially vital to think about because sustainability and the availability of materials are very significant.

Computational approaches are useful for making these design criteria better and more precise. DFT calculations can help us understand how substituting host atoms with lighter or more electropositive elements affect the stability and energy of hydrogen bonds. Doping procedures, which inject small amounts of atoms from outside the material, can modify the electrical structure or create active patches that make it simpler for hydrogen to diffuse and desorb. Nano-structuring, which includes making particles smaller or modifying the form of surfaces, can considerably increase kinetics by making the surface area bigger and the diffusion routes shorter. You can even introduce vacancies, interstitials, or dislocations on purpose to affect how hydrogen bonds and moves about via defect engineering. These computational insights make it easier to systematically develop materials with certain properties. This reduces the need for trial-and-error experimentation and speeds up the search for good hydrogen storage solutions.

5. Case studies of promising material classes

A targeted comprehension of solid-state hydrogen storage materials is optimally illustrated through concentrated case studies that emphasize the application of computational methods to assess their potential. This section discusses some types of materials that have gotten a lot of attention because they can be changed, hold a lot of hydrogen, and can be used in a variety of ways. These materials include perovskite-type hydrides. This case study examines the structural, electronic, optical mechanical and hydrogen storage properties derived from computer analyses, illustrating how these insights facilitate material optimization and selection.

5.1 Theoretical approach

In the present research we are looking into the space group Pm3m as a possible cubic structure for XYH_3 (X = Li, Na, K). The atomic coordinates for the X are (0.0, 0.0, 0.0). The atomic coordinates for the elements Y are at (0.5, 0.5, 0.5), whereas the atomic coordinate for H is at (0.5, 0.5, 0.0). This simulation used the ultra-soft pseudo potential USP plane wave method to guess all of its characteristics. Perdew, Burke, and Ernzerhof (PBE) came up with the exchange correlation functional for the generalized gradient approximation (GGA). The CASTEP software [51], grounded on Density Functional Theory (DFT), utilizes the Kohn-Sham Equations [52]. In this study, we utilized the USP (ultra-soft pseudo potential) plane wave method [53] to accurately determine all the physical parameters. This strategy lets us fast run simulations without having to know anything about the orbitals' geometry beforehand. The nuclei's interaction with the inner shell electrons creates a cluster of ions in the center. So, the ionic core and the valence electrons are quite close to each other. Because of this interaction, the electron-ion potential quickly reaches a point of convergence. In this case, each atom has energy of 2.0×10^{-5} electron volts (eV). After optimizing the shape, the leftover atomic forces are 0.2 eV/atom. We set the energy cutoff for the complete Brillouin zone at 489.8 eV after doing k-integration with a $6 \times 6 \times 6$ k-points mesh on the Monkhorst pack grid.

5.2 Results and discussions

5.2.1 Structural properties

The geometry of each compound was initially optimized. By using the lattice parameters and internal coordinates to optimize the geometry, the structures are fully relaxed. After conducting geometry optimization, the computed lattice parameters of $LiYH_3$, $NaYH_3$, and KYH_3 compounds are listed in **Table 1**. **Figure 2** depicts the crystal structures of the $LiYH_3$, $NaYH_3$, and KYH_3 compounds. The Murnaghan's State Equation was used to determine these parameters. All the compounds possess a cubic morphology and are categorized within the Pm-3 m space group (#221). The electron configurations for the atoms in XYH_3 (X = Li, Na, K) are as follows: lithium (Li): $1s^2 2s^1$, sodium (Na): $1s^2 2s^2 2p^6 3s^1$, potassium (K): $1s^2 2s^2 2p^6 3s^2 3p^6 4s^1$, yttrium (Y): $1s^2 2s^2 2p^6 3s^2 3p^6 4s^2 3d^{10} 4p^6 5s^2 4d^1$, and hydrogen (H): $1s^1$.

Hydride	a (Å)	V (Å)^3	Cwt%
$LiYH_3$ (present study)	4.42	86.35	3.06
$NaYH_3$ (present study)	4.30	79.51	2.64
KYH_3 (present study)	4.41	85.76	2.31
$MgCrH_3$ [54]	3.45	41.06	3.77
$LiScH_3$ [23]	3.89	58.86	5.7
$LiTiH_3$ [23]	3.76	53.15	5.5
$LiVH_3$ [23]	3.24	34.02	4.4
$MgFeH_3$ [54]	3.02	27.54	3.60
$MgMnH_3$ [54]	3.35	37.59	3.64
$KSrH_3$ [55]	3.92	60.23	2.71
$LiSrH_3$ [55]	4.30	79.50	2.38
$KCaH_3$ [55]	4.55	94.19	3.27
$LiCaH_3$ [55]	4.18	73.03	2.73

Table 1.
Lattice constant, volume, and Cwt% of XYH_3 (X = Li, Na, K).

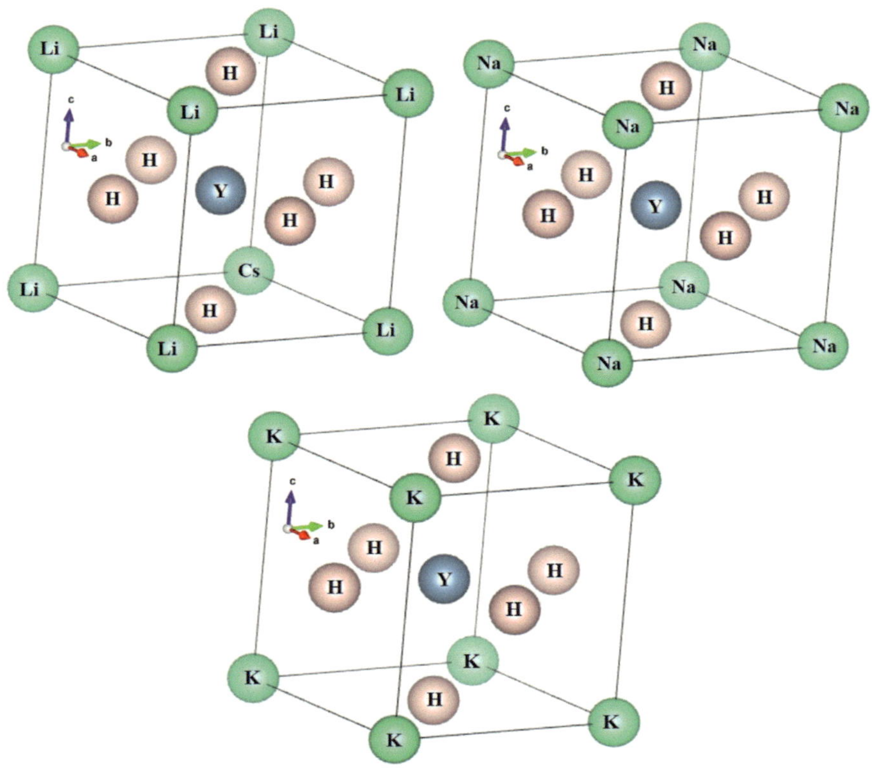

Figure 2.
Optimized crystal structure of XYH_3 (X = Li, Na, K).

We utilized the following formula to calculate the enthalpy of formation (ΔH) while accounting for the energy difference in order to offer more proof of the thermodynamic stability [56]:

$$\Delta H = E_{total}\left(XYH_3\right) - aE_X - bE_Y - cE_H \tag{1}$$

Individual atomic energies inside their intrinsic lattice per unit cell volume are represented by E_X, E_Y, and E_H, respectively, whereas the overall energies of XYH_3 are denoted by Etotal (XYH_3). The amounts of Li, Na, K, Y, and H atoms per unit cell volume are represented by the symbols a, b, and c, respectively. The computed formation energies for $LiYH_3$, $NaYH_3$, and KYH_3 are −16.19 eV/atom, −13.54 eV/atom, and − 9.13 eV/atom, respectively. Thermodynamically stable materials have negative enthalpy of production values. Materials that are thermodynamically stable have negative values for enthalpy of production. We have also added phonon dispersion curves for all of the compounds to show that they are dynamically stable. **Figure 3** shows that the compounds are dynamically stable as there are no imaginary frequencies.

5.2.2 Thermodynamic properties

A thermodynamic analysis was conducted to comprehend the thermal stability of XYH_3 (X = Li, Na, K) hydride compounds, focusing on enthalpy, entropy, and Gibbs free energy, as illustrated in **Figure 4**. This study employed temperatures from 0 to

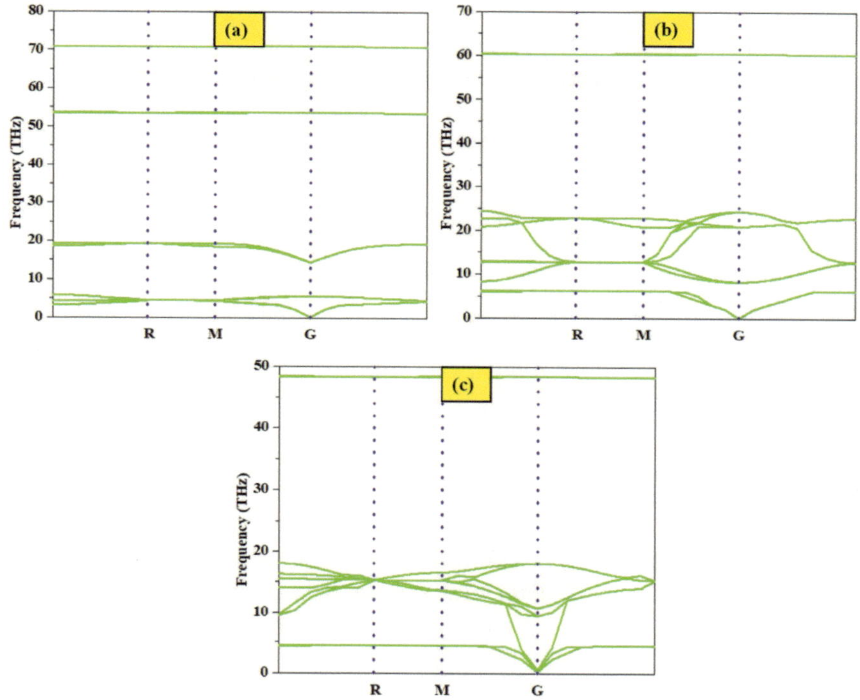

Figure 3.
Phonon dispersion of (a) $LiYH_3$, (b) $NaYH_3$, and (c) KYH_3.

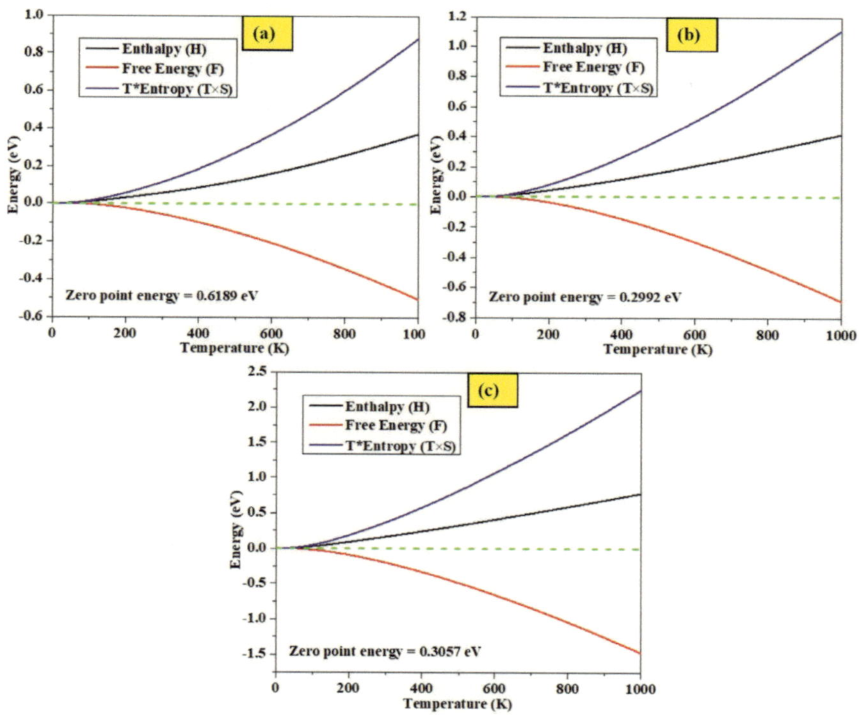

Figure 4.
Thermodynamic parameter of (a) $LiYH_3$, (b) $NaYH_3$, and (c) KYH_3.

1000 K. Also, (ΔH) stands for the amount of heat that is transported when hydrogen is absorbed and released from the perovskite lattice structure [57]. **Figure 4** illustrates that the enthalpy of the XYH_3 compounds goes up in a straight line as the temperature rises. This means that the system's internal energy is slowly rising. At 1000 K, the compounds $LiYH_3$, $NaYH_3$, and KYH_3 showed maximum enthalpy values of 0.37 eV, 0.41 eV, and 0.77 eV, respectively.

Figure 3 shows that the Gibbs free energy (F) calculations demonstrate that the value becomes closer to zero as the temperature rises. These hydrides are stable from a thermodynamic point of view, as shown. At 1000 K, the Gibbs free energy values for $LiYH_3$, $NaYH_3$, and KYH_3 were – 0.51 eV, −0.69 eV, and – 1.48 eV, respectively. **Figure 4** demonstrates that as the temperature goes up, the entropy (ΔS) goes up quicker than the enthalpy and Gibbs free energy. The highest values for $LiYH_3$, $NaYH_3$, and KYH_3 at 1000 K were 0.87 eV, 1.10 eV, and 2.25 eV, respectively.

5.2.3 Electronic properties

In the Brillouin zone, the high symmetry directions are typically the focus of the electronic property analysis. To determine the electronic band structures and density of states of the XYH_3 (X = Li, Na, K) compounds, the PBE + GGA approximation

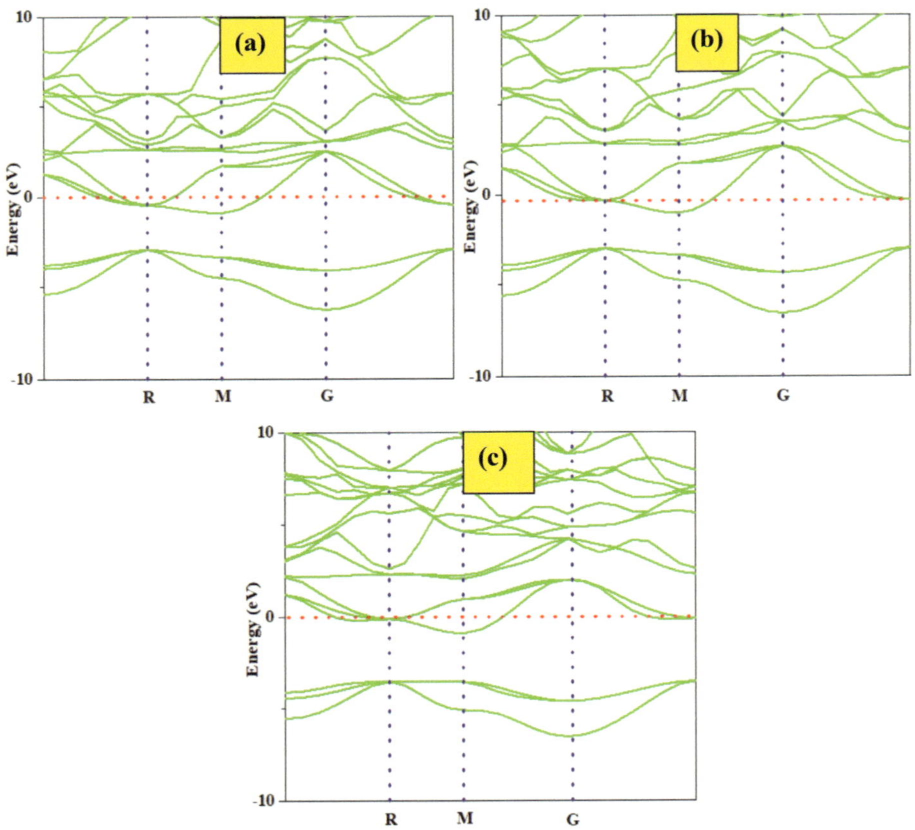

Figure 5.
Band structure of (a) $LiYH_3$, (b) $NaYH_3$, and (c) KYH_3.

[58] were utilized. Based on their Total Density of States (TDOS), the electrical characteristics of XYH_3 (X = Li, Na, K) were examined. **Figure 5** shows the band structures of these materials in (a), (b), and (c). The conduction band (CB) is a feature of the electronic band structure that shows the possible energy levels for electrons and the range of energies in which they are not present. Without taking temperature into account and presuming a Fermi level of 0 eV, the computations were carried out at a temperature of absolute zero. The Fermi energy divides the energy levels into two distinct bands: the conduction band and the valence band. These compounds' metallic properties are made abundantly obvious by the band structure data, which does not distinguish between the lowest and greatest energy levels of the conduction band and the valence band, respectively. The Fermi level is where all the compounds are found. It is possible to show the band structure findings for these compounds by applying density of state data. Both the total and partial densities of states for the composites under study are displayed in **Figure 6**. Total density of states spectra for $LiYH_3$ has prominent peaks at −23.44 eV and − 3.71 eV, for $NaYH_3$ at −24.91 eV, 23.17 eV, and − 3.85 eV, and for KYH_3 at −23.61 eV, −14.99 eV, and 7.60 eV. The PDOS screen mainly shows the valence and conduction bands. The s-state is the most prominent in the PDOS spectra's valance band, with only a small amount of contribution from the p- and d-states. The s-, p-, and d-states predominate in the conduction band.

5.2.4 Optical properties

One can obtain a complete grasp of the interactions between light and matter by studying optical characteristics. Materials' optical characteristics are determined by electron transitions and excitations. Excitations can be produced by plasmons, phonons, or electrons. Every optical quality is related in a methodical way. The dielectric function can be used to compute optical characteristics. The following describes how these characteristics are related to each other [59].

$$\varepsilon(\omega)=\varepsilon_1(\omega)+i\varepsilon_2(\omega) \tag{2}$$

$$n(\omega)=\left[\varepsilon_1(\omega)/2+\left\{\varepsilon_1^{\,2}(\omega)+\varepsilon_1^{\,2}(\omega)\right\}^{1/2}/2\right]^{1/2} \tag{3}$$

$$L(\omega)=-I'm\left(\varepsilon(\omega)^{-1}\right)=\varepsilon_2(\omega)/\varepsilon_1(\omega)^2+\varepsilon_2(\omega)^2 \tag{4}$$

$$I(\omega)=2^{1/2}\omega\left[\left\{\varepsilon_1^{\,2}(\omega)+\varepsilon_1^{\,2}(\omega)\right\}^{1/2}-\varepsilon_1(\omega)\right]^{1/2} \tag{5}$$

$$R(\omega)=(n+ik-1)/(n+ik+1) \tag{6}$$

To ensure consistency, all computations of optical properties were done under the same steady-state conditions and starting values as the theoretical model said they should be. This makes sure that $LiYH_3$, $NaYH_3$, and KYH_3 compounds can all be compared in the same way, which lets us fairly judge how they respond to light under the same simulated settings.

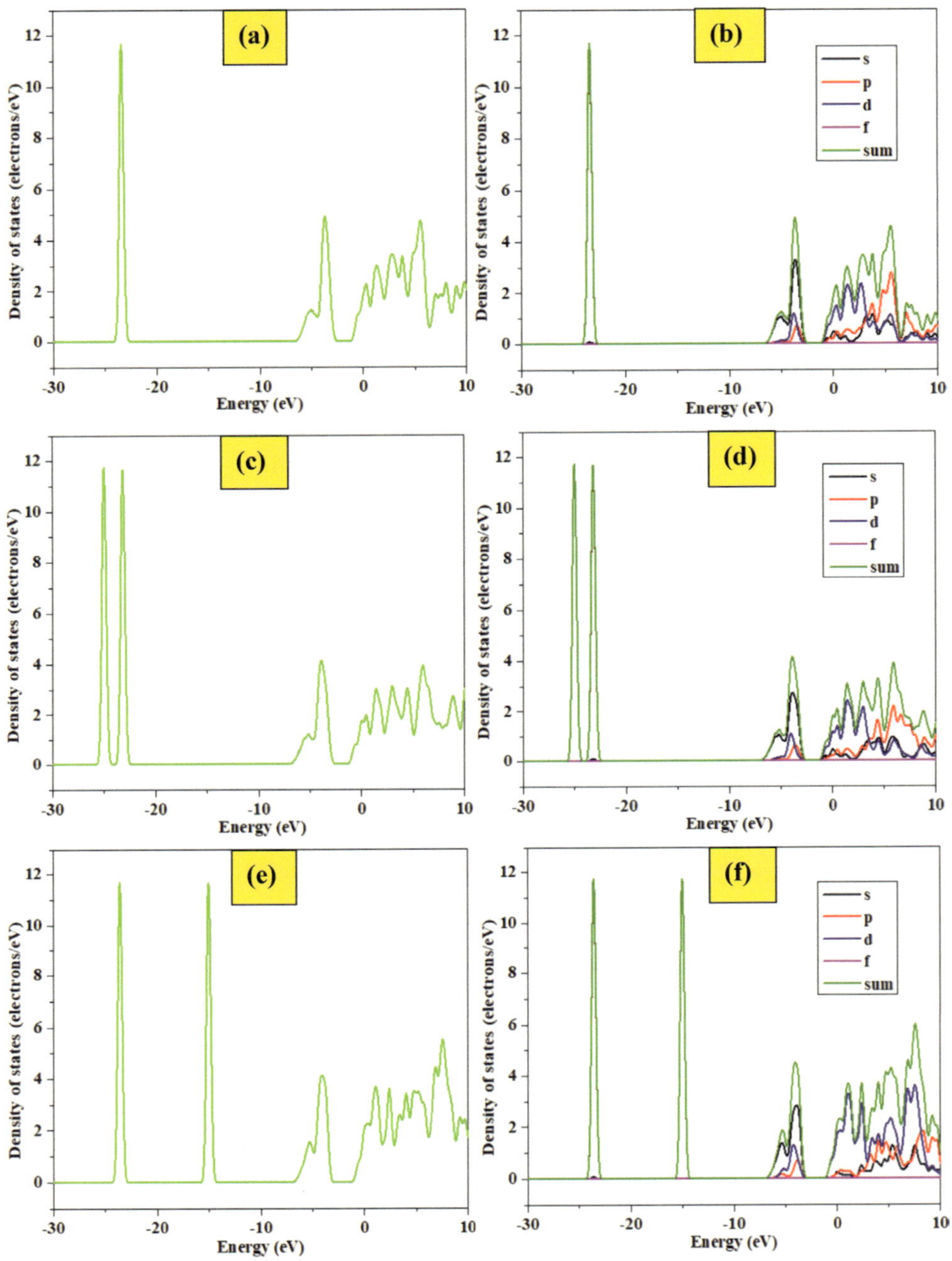

Figure 6.
(a) Total density of states. (b) Partial density of states of $LiYH_3$. (c) Total density of states. (d) Partial density of states of $NaYH_3$. (e) Total density of states. (f) Partial density of states of KYH_3.

The dielectric function, denoted as ε(ω) in Eq. (3), comprises two components: $\varepsilon_1(\omega)$ and $\varepsilon_2(\omega)$, which correspond to its real and imaginary sections, respectively. The dielectric function real component $\varepsilon_1(\omega)$ indicates polarization, and an imaginary component $\varepsilon_2(\omega)$ represents energy dissipation. The components can be seen in **Figure 7(e-f)**. The dielectric function governs the relationship between the material and the external electromagnetic field, determining the behavior of all other

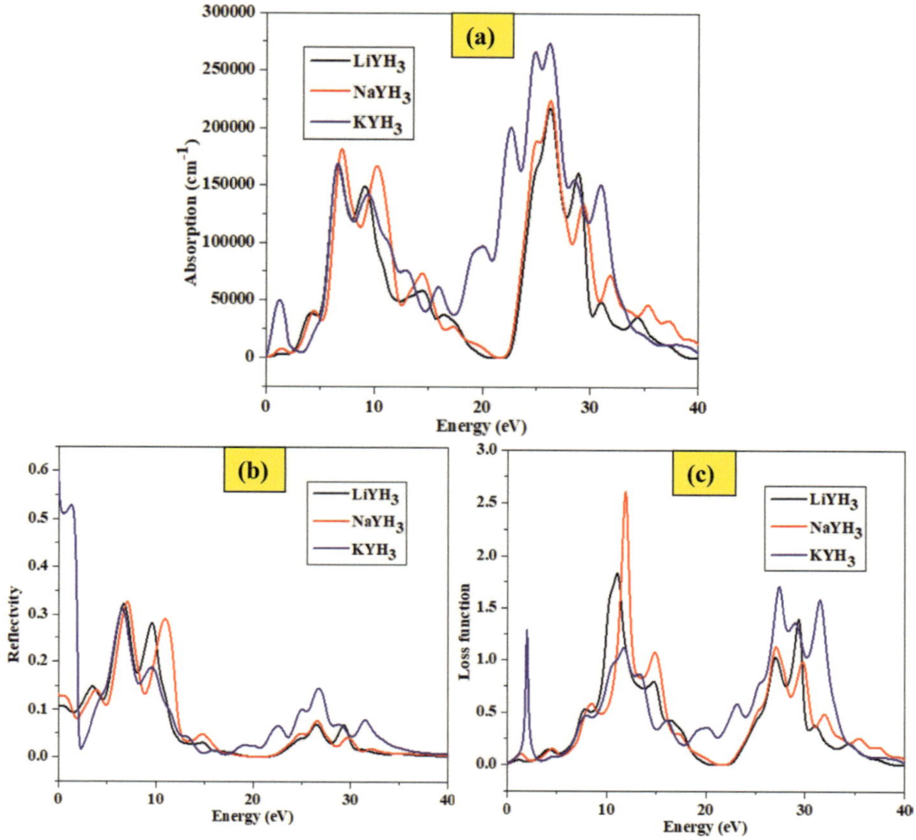

Figure 7.
(a) Loss function (b) Reflectivity (c) of XYH3 (X = Li, Na, K).

parameters in the system. The main peaks in the real part of dielectric function appeared at 6.63 eV and imaginary component peak at 6.21 eV for all the investigated compounds. All three chemicals exhibit a distinct primary peak at 6.90 eV in the reflectivity spectra. The consistent peak position observed indicates a similarity in how these substances interact with electromagnetic waves. This similarity may be due to a common electrical structure or bonding property that affects their reflectivity. To further understand these materials' optical characteristics; we can look at their refractive index spectra, which include both the real and imaginary parts. The real section of the refractive index spectra shows prominent peaks at 5.95 eV for all three substances. The uniform peak position suggests that the phase velocity of light is identical when it passes through these materials, showing that their electronic polarizability is equal. The imaginary section of the refractive index spectra has prominent peaks at 6.73 eV across all substances. This statement describes the absorption losses that occur in certain materials. The peaks indicate specific energy ranges where the substance absorbs light very effectively, resulting in greater extinction coefficients. Each compound exhibits separate primary peaks in the real section of the optical conductivity spectra. Specifically, $LiYH_3$ has a peak at 29.45 eV, $NaYH_3$ has a peak at 10.73 eV, and KYH_3 has a peak at 26.74 eV. These peaks indicate the frequency at which these materials have the highest level of electrical conductivity when exposed to optical stimulation. The variances in peak positions imply variations in the electronic band structures of the materials, which could impact their appropriateness for different optoelectronic applications. The absorption spectra of $LiYH_3$, $NaYH_3$, and KYH_3 exhibit prominent peaks at 26.28 eV. Out of these options, the peak for KYH_3 exhibits the greatest intensity, indicating that KYH_3 has a greater inclination to absorb light at this specific energy level in comparison to $LiYH_3$ and $NaYH_3$. This attribute could be beneficial in applications that necessitate a high level of absorption efficiency. The energy loss function tells us how much energy is wasted when light interacts with a substance because electrons move so quickly. Eq. (5) shows how this event can be described mathematically. Plasma oscillations occur due to the displacement of rapidly moving electrons from their lattice sites. The primary peaks in the loss function spectra for the compound $LiYH_3$ are observed at 11.05 eV and 29.36 eV; for $NaYH_3$, they are at 11.95 eV; and for KYH_3, they are at 27.44 eV and 31.45 eV (**Figure 8**).

5.2.5 Mechanical properties

The mechanical characteristics of the XYH_3 (X = Li, Na, K) can be studied by determining their elastic constants. To comprehend mechanical stability and bonding forces, elastic constants are crucial. We know the elastic constants (C_{11}, C_{12}, and C_{44}) from studying the strain-volume relationship. At birth, cubic compounds are determined to be stable according to the following rules [60, 61];

$$\begin{cases} C_{11} - C_{12} > 0 \\ C_{11} > 0, C_{44} > 0 \\ C_{11} + 2C_{12} > 0 \end{cases} \tag{7}$$

The elastic constants satisfy the conditions for Born stability, as given in **Table 2**. The x-direction compression resistance is given by the coefficient C_{11}. C_{44} provides similar information regarding the capacity to resist shear deformation caused by shear force. The C_{44}'s primary function is to measure the material's hardness. Equations for determining the elastic anisotropy (A) and the moduli of elasticity (Young's modulus, bulk modulus, and shear modulus) using the elastic constants are as follows [62].

$$\mathrm{A} = \frac{2C_{44}}{C_{11} - C_{12}} \tag{8}$$

$$\mathrm{B} = \frac{(C_{11} + 2C_{12})}{3} \tag{9}$$

$$G_v = \frac{(C_{11} - C_{12} + 3C_{44})}{5} \tag{10}$$

$$G_R = \frac{5C_{44}(C_{11} - C_{12})}{4C_{44} + 3(C_{11} - C_{12})} \tag{11}$$

$$\mathrm{G} = \frac{(G_V + G_R)}{2} \tag{12}$$

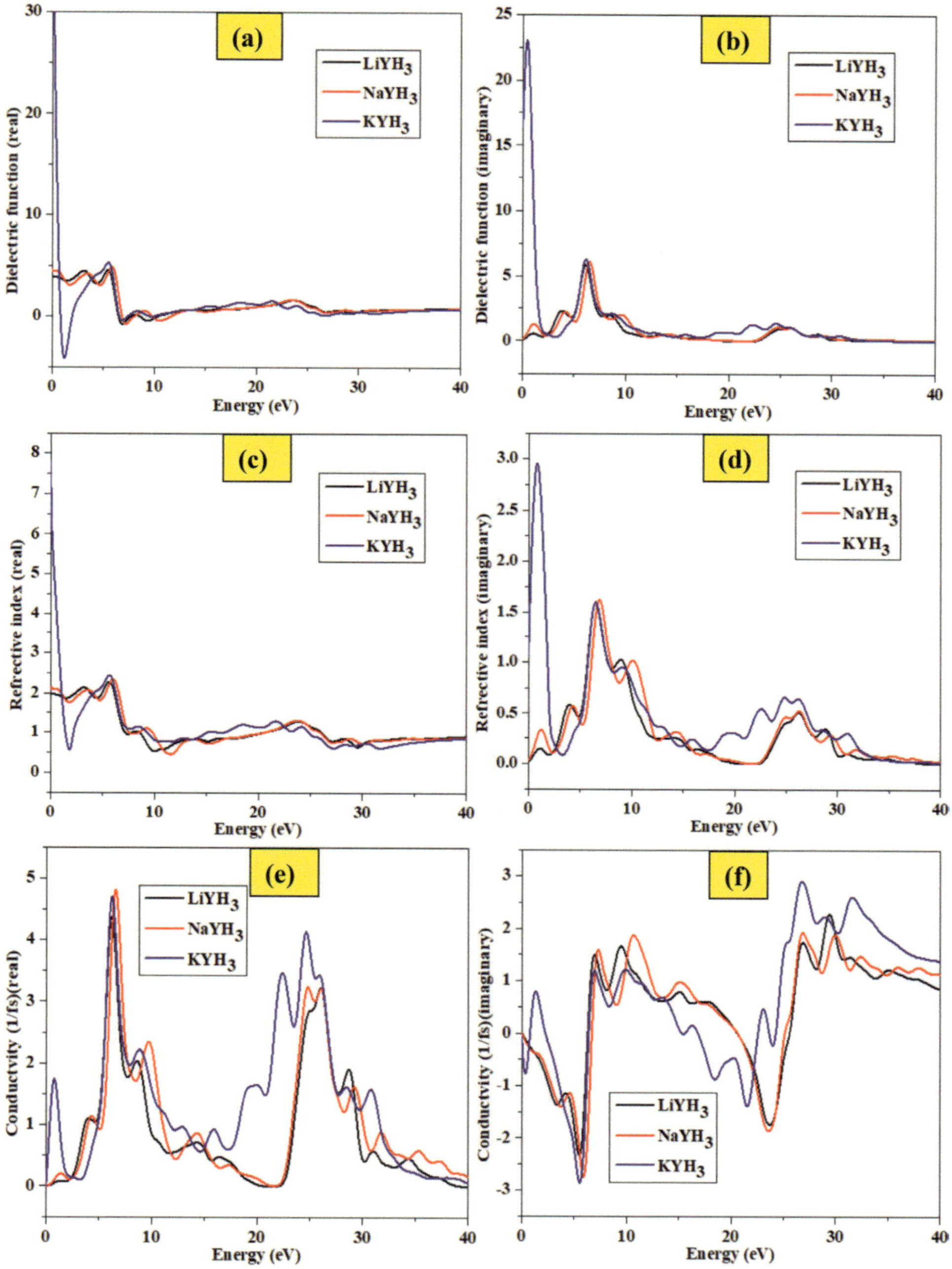

Figure 8.
Optical properties of XYH_3 (X = Li, Na, K).

$$E = \frac{(9BG)}{(3B + G)} \tag{13}$$

Eq. (5)'s elastic constants help with the computation of the resistance to changes in volume under pressure, which is defined by the component B. While E predicts a material's stiffness, G evaluates its resistance to permanent deformation under shear stress and

Parameters	$LiYH_3$	$NaYH_3$	KYH_3
C_{11}	30.05	90.25	62.66
C_{12}	5.65	5.36	15.68
C_{44}	11.59	12.58	17.83
A	0.95	0.29	0.76
B	13.78	33.65	31.34
G_v	11.82	24.52	20.09
G_R	11.83	17.50	19.73
G	11.83	21.02	19.91
E	27.59	52.19	49.29
B/G	1.16	1.60	1.57
C_{11}-C_{12}	24.40	84.89	46.98
T_m	586.397 K ± 300 K	584.683 K ±300 K	645.684 K ± 300 K

Table 2.
The estimated mechanical parameters.

provides a more all-encompassing picture of its hardness than B. A comprehensive set of the calculated elastic moduli is shown in **Table 2**. The results demonstrated that $NaYH_3$ exhibited higher Young's modulus and shear modulus compared to $LiYH_3$ and KYH_3. The $NaYH_3$ is more resistant to damage than $LiYH_3$ and KYH_3. Composites are evaluated for their overall ductility or brittleness using Pugh's ratio (B/G). A ductile composite has a B/G ratio greater than 1.75, whereas a brittle compound has a ratio less than 1.75 [63, 64]. In this case, all the compounds exhibit the property of brittle. Importantly, the anisotropic ratio (A) indicates how much elastic anisotropy a material possesses. For isotropic materials, where the characteristics are same in all directions, A = 1. But A is not equal to 1 for anisotropic materials, which have varying properties in different directions. According to the findings, A is not equal to 1, which means that the hydrides being studied exhibit anisotropy. The investigated composites' dynamic stability has been assessed by computing the value of G. According to **Table 2**, all compounds have a positive shear constant, which means they are dynamically stable. We also figured out the Cauchy pressure (C_{12}-C_{44}) so we could look at how the hydrides were bonded together. Cauchy pressure values that are positive denote ionic bonding while values that are negative denote covalent bonding. Positive Cauchy pressure values for $LiYH_3$, $NaYH_3$, and KYH_3 indicate that these compounds include ionic bonds. The perovskites' melting point has been determined using an equation that takes into account the elastic constant data. In **Table 2**, you can see the calculated results.

$$T_m(K) = \left[553(K) + 5.911C_{12}\right]\text{Gpa} \pm 300\,K \tag{14}$$

5.2.6 Hydrogen storage capabilities

An important barrier to the general adoption of hydrogen as a fuel is the absence of technology capable of storing hydrogen fuel with gravimetric ratios (Cw%) that are sufficient for practical applications. A promising approach to address this issue is the synthesis of molecules capable of efficiently storing hydrogen at high densities. The maximum hydrogen storage capacity of hydride perovskites XYH_3 (X = Li, Na, K) has been examined and analyzed. The gas release of these compounds will also demonstrate their hydrogen storing capabilities. An investigation will be conducted to explore the potential use of perovskite hydride XYH_3 (X = Li, Na, K) for hydrogen storage, focusing on the materials' ability to release gas. The gravimetric ratios (Cw%) of the substances under investigation are ascertained by quantifying their hydrogen content per unit mass. The following correlation demonstrates the methodology employed to examine the General Relativity theory for the item under investigation [65]:

$$\mathrm{Cwt\%} = \frac{\mathrm{pMH}}{\mathrm{Mcom} + \mathrm{pMH}} \times 100\% \tag{15}$$

The letter p stands for the ratio of hydrogen to compound atoms, the letter Mcom stands for the molar mass of the material, and the letter MH stands for the molar mass of hydrogen. We have found that $LiYH_3$, $NaYH_3$, and KYH_3 can hold 3.06 Cw%, 2.64 Cw%, and 2.31 Cw% of hydrogen, respectively. As a result of the preceding line of reasoning, it is possible to draw the conclusion that perovskite hydrides XYH_3 (X = Li, Na, K) are suitable combinations for hydrogen storage systems due to the fact that they possess high gravimetric hydrogen storage ratios. Because of the qualities that they possess, all of these molecules are stable and can be put through the synthesis process.

6. Conclusion

First-principles approaches have been used to look at the structural, electrical, optical, and mechanical properties of XYH_3 (X = Li, Na, K) for hydrogen solid-state storage. The lattice parameters of $LiYH_3$, $NaYH_3$, and KYH_3 are 3.42 Å, 4.30 Å, and 4.41 Å, respectively. Their structural characteristics show that these compounds have a cubic shape. The band structure of the hydrides we looked at showed that they have metal-like electronic properties. The PDOS and TDOS analyses confirmed this finding even more. We used elastic constants to look at the mechanical properties of these hydrides and found that they met the Born stability requirements. As a result, these hydrides are classified as compounds with mechanical stability. All ternary hydrides are fragile because their modulus ratio is less than 1.75. They also have ionic bonding, which is shown by the positive values of Cauchy pressures. The findings of this latest study provide novel and relevant information to the existing body of literature on solid-state hydrogen storage. But it's apparent that perovskite hydrides could change the way we store hydrogen for sustainable energy in a big way. As science and technology continue to improve, it is becoming easier to find ways to store hydrogen safely, efficiently, and affordably using perovskite hydrides. The ongoing efforts in this field are essential for addressing global energy challenges and transitioning to a sustainable hydrogen-based energy economy in the future.

7. Integration with experimental work

The combination of computer modeling and experimental research has become a key part in making better hydrogen storage materials [19]. Computational methods are very important for screening and predicting how materials will behave before they are synthesized in the lab [66]. This saves a lot of money, time, and work compared to the old trial-and-error methods. By simulating a large selection of materials in the lab, researchers can focus on the best ones. These candidates should have the best hydrogen storage capacity, the right thermodynamic properties, and the best structural stability. These predictions act as a strategic filter that helps scientists focus their experiments on the chemicals that are most likely to work. One of the most important things that computational studies can do is anticipate structural and spectroscopic signatures, like X-ray diffraction (XRD) patterns, infrared (IR) and Raman spectra, and UV-Vis optical absorption. These generated fingerprints are very useful for experimentalists who want to be sure they have successfully made a target phase or find unknown or intermediate structures. For example, DFT-based phonon simulations can predict the vibrational modes of complex hydrides. These modes can then be compared directly with experimental IR or Raman spectra to check the bonding arrangements and see if there are any vibrations associated to hydrogen. Simulated optical properties, like the dielectric function and absorption coefficient, can also help us understand UV-Vis data and figure out electronic transitions.

In addition to characterization, computational techniques can also offer ways to dope, replace chemicals, or make nanostructures that can improve the kinetics or thermodynamic profile of hydrogen storage materials [67]. For instance, theoretical predictions about how transition metal doping would affect MgH_2 and $LiBH_4$ have been confirmed by experiments that showed higher desorption temperatures and better cycling stability. Also, first-principles calculations can help choose the best synthesis routes by finding reaction pathways and precursor phases that are thermodynamically stable [68]. This prediction information lets you choose the right synthesis settings, including temperature, pressure, and environment, to get the material phase you want with as few by-products as possible. Computational insights are especially useful for in situ and operando experimental studies, where structural changes and hydrogen release mechanisms are watched in real time. Theoretical models elucidate minor changes observed during dynamic processes, such as phase transitions or the generation of intermediate species, by providing atomistic interpretations that are challenging to obtain experimentally. A significant instance is the theoretical forecasting of intermediate dehydrogenation stages in complex hydrides, including alanates and borohydrides, subsequently validated through in situ neutron diffraction and infrared experiments [69].

8. Outlook and future directions

The combination of computational materials science and experimental hydrogen storage research is likely to have a bigger and bigger impact on the creation of new energy systems. As the need for decarburization becomes more urgent and the demand for energy around the world grows, it is evident that solid-state hydrogen storage technologies need to be discovered, improved, and used more. Future study will probably focus on the multistage modeling of hydrogen storage systems. This means linking atomic-scale simulations with models at the musicale and macro scale to better predict how the system would work in real life. This entails integrating first-principles computations with finite element models to evaluate thermal, mechanical, and chemical stability in operational conditions. One of the most promising options is to employ machine learning (ML) and artificial intelligence (AI) to speed up the process of finding new materials. When used with high-throughput DFT databases, ML models may quickly find correlations between structure and property and guess how well thousands of new compounds can store hydrogen. At the same time, automated workflows and materials informatics will make the screening process even easier. This will let researchers not only find materials but also plan synthetic routes and processing conditions that are likely to work. The investigation of innovative material categories, including high-entropy hydrides, hybrid organic–inorganic frameworks, and low-dimensional systems (e.g., functionalized 2D materials), will broaden the spectrum of prospective possibilities. These materials have special qualities, like stability that comes from configurationally entropy, a large surface area, and adjustable hydrogen binding energies. For more precision, their theoretical assessment will rely more and more on beyond-DFT methods, such as hybrid functionals, random phase approximation (RPA), and quantum Monte Carlo (QMC) approaches.

In the future, it will be more usual to use computational models in real-time experimental feedback loops. This will make adaptive experimentation possible, where simulations help with continuous synthesis and characterization processes. Because these two systems are so closely linked, they will be able to iterate more quickly and build more useful storage systems more quickly. In the end, scalable, safe, and efficient hydrogen storage solutions will depend on more than just improvements in individual approaches. They will also depend on a comprehensive, cross-disciplinary strategy that combines computation, experimentation, and systems engineering into one development pipeline.

9. Conclusion

We used first-principles methods to look into the structural, thermodynamic, electrical, optical, and mechanical features of XYH_3 (X = Li, Na, K) hydrides for solid-state hydrogen storage. The optimized lattice constants for $LiYH_3$, $NaYH_3$, and KYH_3 were 4.42 Å, 4.30 Å, and 4.41 Å, respectively. This shows that they have cubic symmetry. The gravimetric hydrogen storage capacities that were computed were 3.06 wt%, 2.64 wt%, and 2.31 wt%. Formation enthalpies (−16.19 eV, −13.54 eV, and − 9.13 eV) and Gibbs free energy values at 1000 K (−0.51 eV, −0.69 eV, and − 1.48 eV) revealed that the system was stable thermodynamically. Phonon dispersion curves provided additional confirmation of dynamical stability. Electronic band structure research showed that all the compounds had metallic properties. The mechanical analysis showed that the material met the Born stability criterion, with Young's modulus values of 27.59 GPa, 52.19 GPa, and 49.29 GPa. The B/G ratio (< 1.75) also revealed that the material was brittle. Optical experiments revealed pronounced absorption peaks within the 26–29 eV range, indicating potential applicability in optoelectronic applications.

The current results indicate that XYH_3 (X = Li, Na, K) perovskite hydrides are viable candidates for next-generation hydrogen storage devices due to their stability, sufficient gravimetric capacities, and synergistic optical and mechanical features. Future studies ought to concentrate on customizing these materials by doping, substitution, and nanostructuring to enhance kinetics and reversibility. It will also be important to combine computer predictions with experimental synthesis and characterization in order to confirm theoretical ideas and speed up the search for useful storage materials. New methods including high-throughput screening, machine learning, and multistate modeling should also be looked into to make the search area bigger and improve design techniques. Perovskite hydrides can help make hydrogen-based energy sources more sustainable in this way.

Author details

Bilal Ahmed[1*], Muhammad Bilal Tahir[1*], Muhammad Rafique[2], Tahir Iqbal[3] and Ghulam Nabi[3]

1 Institute of Physics, Khwaja Fareed University of Engineering and Information Technology, Rahim Yar Khan, Pakistan

2 Department of Physics, University of Sahiwal, Sahiwal, Pakistan

3 Department of Physics, University of Gujrat, Gujrat, Pakistan

*Address all correspondence to: raisbilalahmed@gmail.com and m.bilaltahir@kfueit.edu.pk

References

[1] Ali A et al. A DFT study of novel perovskite hydrides $RbXH_3$ (X= Cu, Zn) for hydrogen storage applications. Physica B: Condensed Matter. 2025;**715**:417608

[2] Guo Y, Guo R. DFT study to explore the physical properties of Fe-based perovskite hydrides for hydrogen storage application. International Journal of Hydrogen Energy. 2025;**148**:150066

[3] Patil T et al. Sustainable methanol production from carbon dioxide: Advances, challenges, and future prospects. Environmental Science and Pollution Research. 2024;**31**(32):44608-44648

[4] Elkhatat A, Al-Muhtaseb S. Climate change and energy security: A comparative analysis of the role of energy policies in advancing environmental sustainability. Energies. 2024;**17**(13):3179

[5] Awan AB et al. Performance analysis of various hybrid renewable energy systems using battery, hydrogen, and pumped hydro-based storage units. International Journal of Energy Research. 2019;**43**(12):6296-6321

[6] Islam S et al. Diversifying the materials and Technologies for the Future of energy storage. Energy & Fuels. 2025;**39**(18):8369-8390

[7] Sun H et al. Advancements in hydrogen storage technologies: Enhancing efficiency, safety, and economic viability for sustainable energy transition. International Journal of Hydrogen Energy. 2025;**105**:10-22

[8] Waqdim A et al. Lithium-based hydride perovskites $LiXH_3$ (X= Mo, Tc, Rh) for hydrogen storage applications: A DFT study. International Journal of Hydrogen Energy. 2025;**153**:150065

[9] Fang L et al. Economic analysis of compressed gaseous hydrogen, liquid hydrogen, and cryo-compressed hydrogen storage methods for large-scale storage and transportation. International Journal of Hydrogen Energy. 2025;**162**:150725

[10] Zacharia R, Rather SU. Review of solid state hydrogen storage methods adopting different kinds of novel materials. Journal of Nanomaterials. 2015;**2015**(1):914845

[11] Xu X et al. Electrochemical hydrogen storage materials: State-of-the-art and future perspectives. Energy & Fuels. 2024;**38**(9):7579-7613

[12] Siddique A et al. Structures and hydrogen storage properties of $AeVH_3$ (Ae= Be, Mg, Ca, Sr) perovskite hydrides by DFT calculations. International Journal of Hydrogen Energy. 2023;**48**(63):24401-24411

[13] Raza HH et al. First-principle investigation of $XSrH_3$ (X= K and Rb) perovskite-type hydrides for hydrogen storage. International Journal of Quantum Chemistry. 2020;**120**(24):e26419

[14] Bakar A et al. Ab-initio insights of $BeXH_3$ (X= In, Sn, Sb) hydrides for structural, phonon, electronic, thermodynamic and hydrogen storage properties. Journal of Energy Storage. 2025;**132**:117817

[15] Nguyen NN. Prospect and challenges of hydrate-based hydrogen storage in the low-carbon future. Energy & Fuels. 2023;**37**(14):9771-9789

[16] Simanullang M, Prost L. Nanomaterials for on-board solid-state hydrogen storage applications. International Journal of Hydrogen Energy. 2022;**47**:29808-29846

[17] El-Adawy M et al. Review of sustainable hydrogen energy processes: Production, storage, transportation, and color-coded classifications. Energy & Fuels. 2024;**38**(23):22686-22718

[18] Rafique A et al. Investigation of structural, electronic, mechanical, optical and hydrogen storage properties of cobalt-based hydride-perovskites $XCoH_3$ (X= In, Mn, Sr, Sn, Cd) for hydrogen storage application. Journal of Physics and Chemistry of Solids. 2023;**181**:111559

[19] Yang X et al. Crystal structure prediction and performance assessment of hydrogen storage materials: Insights from computational materials science. Energies. 2024;**17**(14):3591

[20] Zhou P et al. Machine learning in solid-state hydrogen storage materials: Challenges and perspectives. Advanced Materials. 2025;**37**(6):2413430

[21] Shevlin S, Guo Z. Density functional theory simulations of complex hydride and carbon-based hydrogen storage materials. Chemical Society Reviews. 2009;**38**(1):211-225

[22] Fatima R et al. First principles study to investigate physical properties of $MgXH_3$ (X= Cr, Cu) hydrides for hydrogen storage applications. Materials Chemistry and Physics. 2025;**44**:131072

[23] Siddique A et al. Structural, electronic, mechanical and dynamical stability properties of $LiAH_3$ (A= Sc, Ti & V) perovskite-type hydrides: A first principle study. Chemical Physics. 2023;**568**:111851

[24] Le TH et al. Recent developments in materials for physical hydrogen storage: A review. Materials. 2024;**17**(3):666

[25] Alardhi SM. "We are here!" ethanethiol in crude oil, natural gas, and fuels: Removal by adsorption process, a mini review. Chemical Papers. 2025;**79**:1-17

[26] Aljamali NM, Khdur R, Alfatlawi IO. Physical and chemical adsorption and its applications. International Journal of Thermodynamics and Chemical Kinetics. 2021;**7**(2):1-8

[27] Scarpati G et al. A comprehensive review on metal hydrides-based hydrogen storage systems for mobile applications. Journal of Energy Storage. 2024;**102**:113934

[28] Paizullakhanov M et al. Hydrogen storage on porous absorbers with a zeolite composition. Applied Solar Energy. 2024;**60**(4):604-608

[29] Jimenez-Lopez L, Ospino RM, de Araujo LG, Celzard A, Fierro V. Latest developments in the synthesis of metal–organic frameworks and their hybrids for hydrogen storage. Nanoscale. 2025;**17**:6390-6413

[30] Azhar U et al. First principle investigation on the physical properties of rhodium-based $XRhH_3$ (X= Na, Cs or Sr) perovskites hydride for hydrogen storage applications. International Journal of Hydrogen Energy. 2025;**101**:1448-1459

[31] Grabowski SJ. Hydrogen bond types which do not fit accepted definitions. Chemical Communications. 2024;**60**(49):6239-6255

[32] Norton JR, Sowa J. Introduction: Metal hydrides. Chemical Reviews. 2016;**116**(15):8315-8317

[33] Nemukula E, Mtshali C, Nemangwele F. Metal hydrides for sustainable hydrogen storage: A review. International Journal of Energy Research. 2025;**2025**(1):6300225

[34] Mandal SK, Roesky HW. Group 14 hydrides with low valent elements for activation of small molecules. Accounts of Chemical Research. 2012;**45**(2):298-307

[35] Aldridge S, Downs AJ. Hydrides of the main-group metals: New variations on an old theme. Chemical Reviews. 2001;**101**(11):3305-3366

[36] Moury R, Gigante A, Hagemann H. An alternative approach to the synthesis of NaB3H8 and $Na_2B_{12}H_{12}$ for solid electrolyte applications. International Journal of Hydrogen Energy. 2017;**42**:22417-22421

[37] Kozhakhmetov Y et al. Boron-based compounds for solid-state hydrogen storage: A review. Crystals. 2025;**15**(6):536

[38] Gusarov S. Advances in computational methods for modeling photocatalytic reactions: A review of recent developments. Materials. 2024;**17**(9):2119

[39] Ahangari MG, Mashhadzadeh AH. Density functional theory based molecular dynamics study on hydrogen storage capacity of C_{24}, $B_{12}N_{12}$, Al_{12} N_{12}, $Be_{12}O_{12}$, $Mg_{12}O_{12}$, and $Zn_{12}O_{12}$ nanocages. International Journal of Hydrogen Energy. 2020;**45**(11):6745-6756

[40] Roldan A. Frontiers in first principles modelling of electrochemical simulations. Current Opinion in Electrochemistry. 2018;**10**:1-6

[41] Nunzi F, De Angelis F. Modeling titanium dioxide nanostructures for photocatalysis and photovoltaics, Chemical Science. 2022;**13**:9485-9497

[42] Ashbrook SE et al. Understanding dynamics and mechanisms: General discussion. Faraday Discussions. 2025;**255**:553-582

[43] Baroni S, Giannozzi P, Isaev E. Density-functional perturbation theory for quasi-harmonic calculations. Reviews in Mineralogy and Geochemistry. 2010;**71**(1):39-57

[44] Banerjee T et al. Machine learning driven search of hydrogen storage materials. arXiv preprint arXiv:2503.04027; 2025

[45] Çolak AB. Optimizing hydrogen storage in Li-decorated borophene slit pores: A novel machine learning study on volumetric and gravimetric capacities. International Journal of Hydrogen Energy. 2025;**144**:912-923

[46] Wu J et al. Techno-economic analysis on low-temperature and high-pressure cryo-adsorption hydrogen storage. Fuel. 2025;**381**:133532

[47] Laghdas A et al. First principles analysis of hydrogen storage and release performance of the perovskite type $XCrH_3$: Structural, mechanical, optical and thermodynamic behavior. Journal of Energy Storage. 2025;**132**:117862

[48] Adametz P, Müller K, Arlt W. Energetic evaluation of hydrogen storage in metal hydrides. International Journal of Energy Research. 2016;**40**:1820-1831

[49] Graetz J. New approaches to hydrogen storage. Chemical Society Reviews. 2009;**38**(1):73-82

[50] Dewangan SK et al. A comprehensive review of the prospects for future hydrogen storage in materials-application

and outstanding issues. International Journal of Energy Research. 2022;**46**(12):16150-16177

[51] Ratul Hasan M et al. Impact of applied pressure on tin-based cubic halide perovskite $ASnX_3$ (A= Li, Na and X= Cl, Br, and I) in reference to their optoelectronic applications. International Journal of Energy Research. 2024;**2024**(1):8213804

[52] Ahmed B et al. An Ab-initio simulation of boron-based hydride perovskites XBH_3 (X= Cs and Rb) for advance hydrogen storage system. Computational and Theoretical Chemistry. 2023;**1225**:114173

[53] Shafiq M et al. DFT screening of Ga-dopped $ScInO_3$ perovskite for optoelectronic and solar cell applications. Inorganic Chemistry Communications. 2024;**161**:112054

[54] Ur Rehman Z et al. A DFT study of structural, electronic, mechanical, phonon, thermodynamic, and H_2 storage properties of lead-free perovskite hydride $MgXH_3$ (X= Cr, Fe, Mn). Journal of Physics and Chemistry of Solids. 2024;**186**:111801

[55] Song Y et al. Theoretical prediction of perovskite ARH_3 (A= K, Li, Rb; RCa, Sr) hydride materials for hydrogen storage applications: A DFT investigation. International Journal of Hydrogen Energy. 2024;**79**:1472-1482

[56] Ahmed B et al. DFT insights on structural, electronic, optical and mechanical properties of double perovskites X_2FeH_6 (X= Ca and Sr) for hydrogen-storage applications. International Journal of Hydrogen Energy. 2024;**50**:316-323

[57] Azdad A, Boutahar A, Lemziouka H. A DFT study on the structural, electronic, mechanical, optical, and dynamic properties of double perovskite-type hydrides K_2BeXH_6 (X= Mg and Ca) for hydrogen storage. Journal of Energy Storage. 2025;**132**:117980

[58] Ahmed B et al. First-principles study to investigate effect of pressure on electronic and optical properties of $KCdCl_3$ for improved solar cells and optoelectronic applications. Emergent Materials. 2023;**6**(5):1697-1705

[59] Hoat D, Silva JR, Blas AM. First principles study of structural, electronic and optical properties of perovskites $CaZrO_3$ and $CaHfO_3$ in cubic phase. Solid State Communications. 2018;**275**:29-34

[60] Al S, Yortanlı M, Mete E. Lithium metal hydrides (Li_2CaH_4 and Li_2SrH_4) for hydrogen storage; mechanical, electronic and optical properties. International Journal of Hydrogen Energy. 2020;**45**:18782-18788

[61] Al S. Theoretical investigations of elastic and thermodynamic properties of $LiXH_4$ compounds for hydrogen storage. International Journal of Hydrogen Energy. 2019;**44**(3):1727-1734

[62] Pugh S. Relations between the elastic moduli and the plastic properties of polycrystalline pure metals. The London, Edinburgh, and Dublin Philosophical Magazine and Journal of Science. 1954;**45**(367):823-843

[63] Ahmed B et al. An ab-initio insight computation of structural, electronic and optical properties of Na-doped $CsSrF_3$ fluoro-perovskite for optoelectronic applications. Inorganic Chemistry Communications. 2023;**156**:111151

[64] Tariq A et al. Lithium doping effect on structural, electronic and optical properties of $KCaF_3$ fluoroperovskite: DFT insights using GGA-PBE.

Inorganic Chemistry Communications. 2023;**158**:111475

[65] Baysal M et al. The effect of hydrogen on the electronic, mechanical and phonon properties of $LaMgNi_4$ and its hydrides for hydrogen storage applications. International Journal of Hydrogen Energy. 2018;**43**(52):23397-23408

[66] Ludwig A. Discovery of new materials using combinatorial synthesis and high-throughput characterization of thin-film materials libraries combined with computational methods. npj Computational Materials. 2019;**5**(1):70

[67] Attia NF, Elashery SE, Nour MA, Policicchio A, Agostino RG, Abd-Ellah M, et al. Recent advances in sustainable and efficient hydrogen storage nanomaterials. Journal of Energy Storage. 2024;**100**:113519

[68] Szymanski NJ et al. Quantifying the regime of thermodynamic control for solid-state reactions during ternary metal oxide synthesis. Science. Advances. 2024;**10**(27):eadp3309

[69] Majzoub EH, Rönnebro EC. Methodology of materials discovery in complex metal hydrides using experimental and computational tools. Materials Science and Engineering: R: Reports. 2012;**73**(2):15-26

Chapter 7

Electrical and Optomechanical Sensors for Hydrogen Monitoring: Sensing Strategies and Applications

Nancy Anna Anasthasiya Anthuvan, Karuppiah Veerathangam, João Piteira and Hugo M. Oliveira

Abstract

This chapter discusses the role of electrical and optomechanical sensors in hydrogen monitoring. To ensure safety in a simple, quick, and cost-effective manner, the emerging electrical and optomechanical sensors, along with their various sensing strategies, are described and discussed in detail, along with the current state of the art. The nanostructured sensing element plays a crucial role in detecting H_2 under various ambient conditions. Hence, the synthesis and deposition of nanostructured materials, including semiconducting metal oxides, metals, conducting polymers, and unique layered/dimensional structures, are crucial to ensure enhanced gas sensing properties. Finally, some examples of sensing systems applied in real-world scenarios are reviewed, providing the current ground where future developments of electrical and optomechanical systems will play a key role.

Keywords: hydrogen, electrical, optomechanical, nanomaterials, sensor array

1. Introduction

Hydrogen sensors have garnered significant attention in the research community because they are essential for the processes of H_2 production, transport, storage, and utilization. The majority of H_2 was used as a reducing agent and fuel due to its distinctive nature compared to other gases, namely, high energy density and clean combustion. H_2 has low density (0.0899 kg/m^3), small molecular size, low boiling point (20.39 K), minimum ignition energy (0.017 mJ), high heat of combustion (142 kJ/g H_2), wide flammable range (4–75% in air) [1], high diffusion coefficient (0.61×10^{-4} m^2/s) [2], high propagation velocity (3.46 m/s) [3], high propensity for flame acceleration, and high sensitivity to detonation [1]. Diatomic hydrogen does not have a toxic nature to humans; at very high concentrations in the air, it acts as an asphyxiant gas, which causes hypoxia. H_2 has a wide explosive range in terms of concentration, from 4 to 75% in the air; also, leakage cannot be identified by the human senses since it is colorless, odorless, and tasteless. Hence, it needs to be monitored for safety considerations.

Hence, this chapter addresses the current quest for H_2 sensing, focusing on detecting H_2 in the gaseous state using electrical and optomechanical sensors (**Figure 1**).

IntechOpen

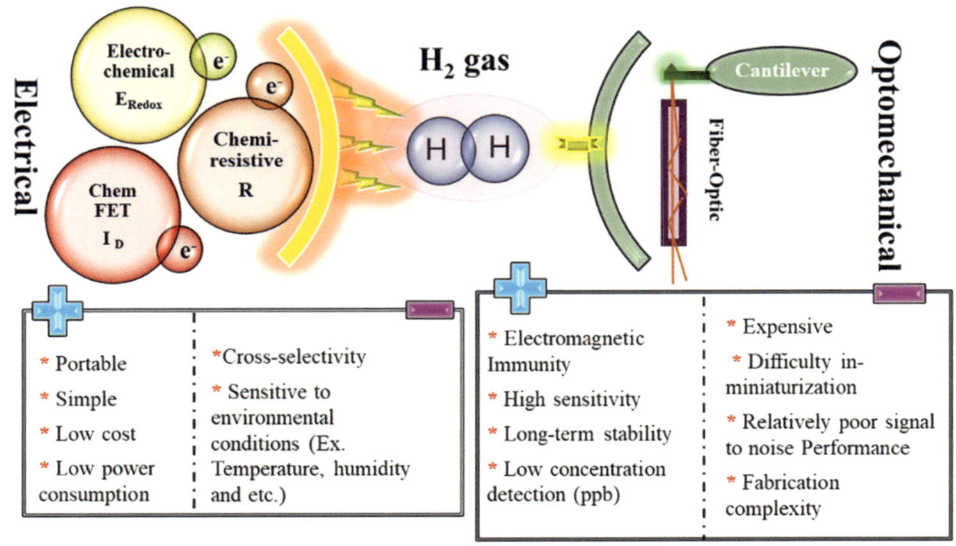

Figure 1.
Electrical- and optomechanical-based H_2 gas sensors and their major features.

These types of sensors play an important role where the rapid and in situ monitoring of H_2 is required (e.g., in leakage situations). More specifically, this discussion will cover electrical sensors, including chemi-resistive, chemFET, and electrochemical sensors, as well as optomechanical sensors. In the latter group, we will cover fiber optic sensors and cantilever sensors. On one hand, the electrical-based sensors are portable, cost-effective, and low-power-consuming devices; quick detection is possible, and they do not require an operator [4–6]. On the other hand, optomechanical sensors are resistant to electromagnetic interference, suitable for corrosive environments, and do not induce any electrical sparks [3]. In general, the optomechanical sensors mainly use the Pd-based sensing element due to its greater affinity toward H_2, and it has remarkable reversibility in the PdH_x (with $x < 0.015$) [2]. To build the literature review presented in this chapter (**Table 1**), we considered the publications describing the most relevant sensors in terms of analytical performance published over the last 5 years. Due to the high number of published articles in the field of H_2 gas sensors, we selected three relevant examples for each kind of sensor.

In this context, the objective of the chapter is to highlight the role of electrical and optomechanical sensors for H_2 monitoring. A brief introduction to their working principles, as well as the sensing mechanism for semiconductors and Pd metal, will be provided. The major focus will be on the synthesis of various sensing elements, namely metal oxide, palladium metal, conductive polymer, carbon-based materials, and MXene, due to their major features for H_2 sensing. Selected examples of reported sensing systems for real-time H_2 identification and conclusion are presented at the end of this chapter.

1.1 Electrical sensor

Three of the most relevant electrical-based sensing strategies for hydrogen monitoring comprise the chemi-resistive, chemFET, and electrochemical sensors.

In chemi-resistive gas sensors [5], the resistance change in the absence and presence of H_2 is measured. The sensor consists of a substrate, an electrical contact, a

Sensing element	Method	Detection limit	OT (°C)	Response/ sensitivity	Response time & recovery time (s)	Ref.
$Pd_{0.62}$ $Au_{0.38}$	Chemi-resistive	0.0002–5%	RT	~2.7% @ 1%	9.5 & 139.3	[7]
$AgInO_2$		1–10,000 ppm	RT	66 @ 10,000 ppm	~6 & ~24	[8]
Ag–Pd/Cr		800 ppm–4%	200	2.13% @ 4%	80 & 170	[2]
CNT	ChemFET	0–500 ppm	70	~80 mV ppm^{-1}	22 & n.a.	[9]
Pd-Ag		1–5000 ppm	245	n.a.	16 & 14	[10]
Pd-gated AlGaN/GaN		50 ppb–20,000 ppm	120	19,137% 500 ppm	2.6 & 100	[11]
Pt-Pd/ rGO-CNT/SPE	Electrochemical	0.8–4%	RT	1.336 μA $\%^{-1}$	6 & 6	[12]
Nafion and Pt/C-Pt		20–100%	−40	0.1 mV $\%^{-1}$	3.5 & n.a	[13]
IL/Pt–Ni		0.05–1.25%	RT	0.230 μA $cm^{-2}\%^{-1}$	n.a	[14]
$PDMS/Pt/WO_3$	FPI	0.5–2.9%	RT	2.664 nm $\%^{-1}$	51 & 73	[15]
Pd nanofilm		0.5–3.5%	RT	0.397 nm $\%^{-1}$	8 & 16	[16]
$Pt/WO_3/PDMS$		0.25–1.75%	RT	5.255 nm $\%^{-1}$	n.a	[17]
Pd	Cantilever	0.2–1.0%	RT	0.48 mm^{-1} @ 1%	n.a	[18]
Cu/Pd		0.4–2%	RT	2825@ 0.4%	1 & 1	[19]
Pd/GaN		180 ppb–1000 ppm	RT	n.a.	10 & 90	[20]
Graphene–Au–Pd	FPI and cantilever	0–1000 ppm	RT	30.3 pm ppm^{-1}	31.5 & n.a	[21]
Polymer/Pd		0–4.5%	23.5	−2 nm $\%^{-1}$	13.5 & n.a	[22]
Au/Pd		1–3.5%	25	0.153 nm $\%^{-1}$	12 & 1	[23]

Table 1.
Survey on electrical (chemi-resistive, chemFET, and electrochemical) and optomechanical (Fabry-Perot interferometer, FPI, and cantilever)-based hydrogen sensing devices. (n.a.: not available and OT: operating temperature).

sensing element, and a heater (when high operation temperatures are needed). The interdigitated electrode (IDE) is a popular design for chemi-resistive gas sensors because it improves the movement of electric charge.

Regarding the chemFET [24], it is a chemically sensitive field-effect transistor. The changes in the drain current in the presence of the H_2 gas are used to detect it. Based on the structure, it is classified as a back-gate FET, a top-gate FET, a dual-gate FET, and a floating-gate FET. It consists of source, gate, and drain electrodes along with an insulating gate oxide, a semiconductor material channel, and a substrate.

Concerning the electrochemical sensor-based H_2 detection [25], it can rely on conductometric-, amperometric-, or potentiometric-based measurements. It consists of an electrode, electrolyte, membrane, and substrate. During the H_2 gas detection, H_2 interacts with the working electrode (anode), and it forms H^+ and electrons through an anode oxidation reaction. The H^+ moves to the counter electrode via the proton-conducting membrane, and the cathode reduction reaction will occur. This process will change the potential of the overall electrochemical cell, which can be correlated with the concentration of H_2. It is reported that electrochemical H_2 sensors can offer higher selectivity than the chemi-resistive semiconducting sensor [12].

The substrate provides a platform for the sensing material, and the adhesion between the sensor and substrate plays a crucial role in ensuring long-term stability. This primarily depends on the thermal expansion coefficient and surface roughness. Based on the elastic limit, substrates are classified as rigid and flexible. In general, the rigid substrate is used for relatively high operating temperatures (above 100°C), whereas the flexible substrate is used for low operating temperatures (e.g., wearable sensor applications). Both can be used for integrated wireless technology in end-user applications. Glass, conducting glass (fluorine-doped tin oxide and indium tin oxide), Si wafers, sapphire, and Al_2O_3 are widely used as a rigid platform for various sensing elements. Si, sapphire, and Al_2O_3 are used for sensors operating above 200°C due to their high chemical and mechanical stability. However, these substrates are expensive and less flexible compared to polymer-based substrates [6]. Nowadays, flexible substrates such as polyethylene terephthalate (PET), polyimide (PI), polydimethylsiloxane (PDMS), polyethylene naphthalate (PEN), polyurethane (PU), Kapton, cloth, and paper are utilized for wearable applications due to their stretchability, amenability, biocompatibility, low weight, and high degree of bending and tilting nature [6]. Yet these substrates are unsuitable for high temperatures and relatively humid environments. Gold (Au) is widely used as an electrode and heater for electrical-based H_2 gas sensors.

1.2 Optomechanical sensor

The different sensing schemes based on fiber optic sensors to detect H_2 are reported in **Figure 2**, and the performance comparison between each measurement method is given in **Table 2**. All the mentioned fiber optic and cantilever-based H_2 sensors consist of a thin sensitive layer mainly made up of Pd or Pd-based material. This thin layer changes its physicochemical property in the presence of H_2 gas, and it induces the variation in the intensity, wavelength, and phase of the light. The interaction between the Pd and the H_2 is discussed in Section 2. In the cantilever-based sensor, the resonance frequency of the cantilever changes due to the mass change in the Pd-sensitive layer [18–20]. The reported detection methods in Pd-based cantilevers are Fabry-Perot interferometry, piezoresistivity, plasmonic amplification, quartz crystal microbalance, frequency detection, and interferometry.

Instead of using the optical and cantilever sensors separately to detect H_2, Xiong et al. [21] integrated the cantilever with FBG fiber to construct H_2 sensors.

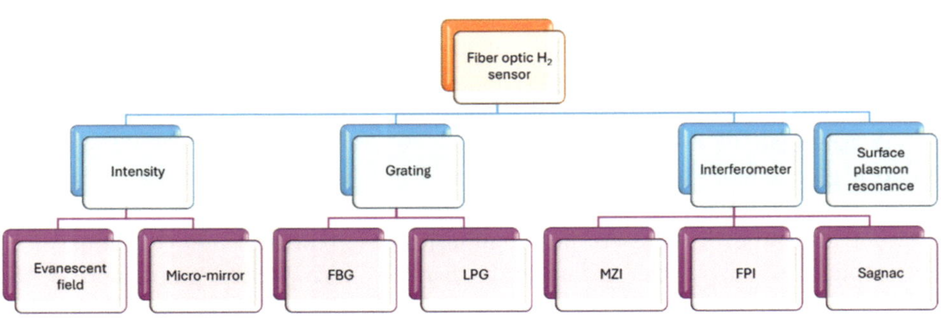

Figure 2.
Schematic representation of various types of fiber optic H_2 sensors (FBG—fiber Bragg grating, LPG—long-period grating, MZI—Mach-Zehnder interferometer, and FBI—Fabry-Perot interferometer).

Performances	Intensity sensors	FBG sensors	LPG sensors	Interferometer sensors	SPR sensors
Structural complexity	Simple	Relatively simple	Relatively simple	Moderate	Complex
Sensor cost	Low	Relatively low	Relatively high	Relatively high, but can be improved	High
Sensor size	Small	Small	Moderate	Relatively large, but can be improved	Moderate
Fabrication difficulty	Low	Low	Relatively high	Moderate	High
Sensitivity	Low	Low	Moderate	Relatively high	High
Precision	Low	Moderate	Moderate	Relatively high	High
Measurement range	Relatively narrow	Wide	Restricted	Restricted	Moderate
Response rate	Slow	Slow	Relatively slow	Moderate	Quick
Multiplexed measurement	Unable	Able	Unable	Unable	Able
Mechanical strength	Fragile	Fragile	High	Depend on the specific structure	Moderate

FBG—fiber Bragg grating, LPG—long-period grating, and SPR—surface plasma resonance.
Reused with permission from the reference [26], copyright © 2017 Elsevier.

Table 2.
Performance comparison of different optical fiber H_2 sensors.

These sensors were downscaled to a portable format and have a reduced noise floor (caused by the environment of the measurements), which is a major drawback of both optical-based and cantilever-based sensors. For that Pd-coated polymer, a μ-cantilever was printed on the end face of a single-mode fiber by a femtosecond laser-induced two-photon polymerization process. The sensing setup consists of a broadband light source (BBS), an optical spectrum analyzer (OSA), a microchannel, a gas flow meter, a carrier gas, and a hydrogen generator. The induced interference from the fiber end face, as well as from the upper and lower surfaces of the cantilever signal, was taken into account. These optomechanical sensors can detect concentrations from 0 to 4.5% (v/v); the reported sensitivity and response time were – 2 nm $\%^{-1}$ and 13.5 s at 4%, respectively. In 2023, Zhong et al. [21] described a cantilever made from a thin film of graphene-Au-Pd combined with an FBG interferometer; it could measure concentrations from 0 to 1000 ppm and had a sensitivity of 30.3 pm per ppm.

In the optomechanical detection of H_2, the cantilever is covered with two distinct materials, with the upper layer often being palladium (Pd). The absorption of hydrogen causes a volumetric expansion of the palladium (top layer), generating differential tensions between the upper and lower surfaces of the cantilever, which leads to its bending, as expressed in the eq. (1) [21].

$$Z = 3\left(\propto_{Pd} - \propto_f\right)\left(\frac{t_{Pd} + t_f}{t_f^2 k}\right) CL^2 \qquad (1)$$

In this context, L represents the cantilever length, while t_{Pd}, t_f, α_{Pd}, and α_f denote the respective layer thicknesses and thermal expansion coefficients. The subscript f denotes the cantilever foundation, C represents the concentration of hydrogen atoms in the Pd film, Z indicates the bending deflection of the cantilever, and C signifies the hydrogen concentration. The cantilever and the fiber end facet function as two partially reflective mirrors, creating a Fabry-Pérot cavity. The correlation between the reduction in cavity length (ΔL) and the shift in dip wavelength (Δλ) is expressed as $\Delta\lambda/\lambda = \Delta L/L$, where λ represents the wavelength of the observed dip and L denotes the cavity length. [22]. Consequently, the H_2 concentration represented by cantilever deflection can be determined by directly measuring the dip wavelength shift of the FP cavity. When the external hydrogen concentration diminishes, the process is inverted. Aside from the materials, the dimensions, layer thickness, and production significantly influence sensitivity. Furthermore, reduced stiffness is a crucial factor for preserving temperature cross-sensitivity.

Based on the background compiled in **Table 1**, it is inferred that the optomechanical sensor shows adequate sensitivity even at room temperature, and the chemFET sensor is a more suitable candidate for measuring H_2 at low concentrations. The chemi-resistive and electrochemical sensors have better stability, and they are simple to fabricate. The reliability of the response and recovery time comparison based on the detection method is questionable; it highly depends on factors such as the catalytic nature of the material, the ad-atom effect, morphology, dimensions, and others. A major limitation of electrical-based sensors is their sensitivity to environmental conditions, such as humidity, the influence of other interfering gases, and temperature. The mechanical properties of the sensor material can be influenced by changes in temperature and humidity, which can affect its dimensions and resonant frequencies. These properties are frequently employed to estimate physical parameters. Additionally, humidity can cause interference with light-based measurements by influencing the optical properties of materials, including the refractive index [18, 21].

2. Sensing mechanisms

Hydrogen acts as the reducing gas when it interacts with semiconducting metal oxides (MOs), metal chalcogenides (MCs), or conducting polymers (CPs). The electrical charge transfer can be explained by two models, namely the oxygen vacancy mechanism (**Figure 3(a)**) and the direct adsorption mechanism (**Figure 3(b)**). The expected reaction in each model is given in eqs. 2 and 3 [8, 28].

$$\textit{Oxygen vacancy model}: O^-_{ads} + H_2 \rightarrow H_2O + e^- \left(\textit{sensing elements}\right) \tag{2}$$

$$\text{Direct adsorption model}: H_2 \rightarrow 2H^+ + 2e^- \tag{3}$$

The sequence of steps involved in every response is as follows: (i) exposure of the sensing element in air, (ii) chemisorption of oxygen leading to the formation of a depletion layer (DL), (iii) interaction between adsorbed oxygen and H_2, and (iv) desorption of reacted product during the exposure to air/removal of H_2 gas. Unsaturated surface states in the sensing element led to the chemisorption of oxygen in the form of O_2^-, O^-, and O^{2-} ions when the operating temperatures are below 100°C, 100–300°C and above 300°C, respectively [29]. This leads to the formation of electron depletion in n-type and hole accumulation in p-type sensing elements.

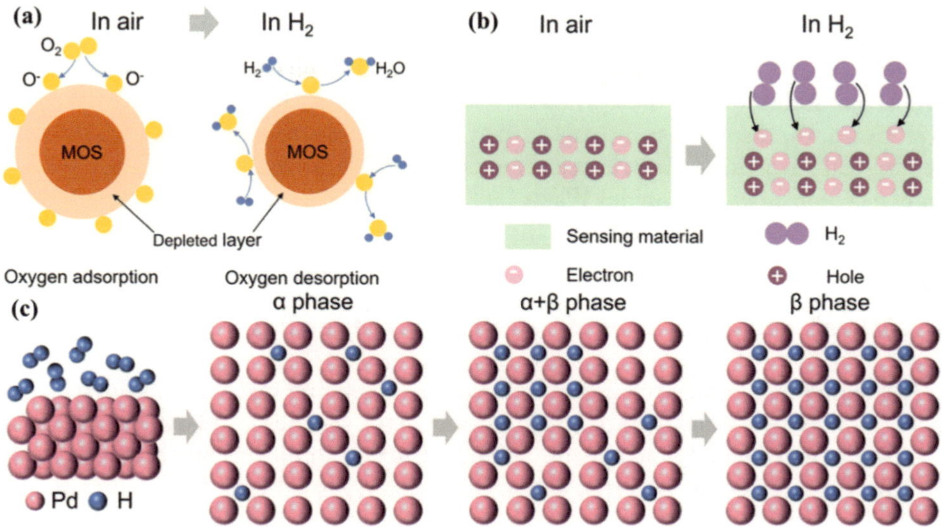

Figure 3.
(a) Schematic of hydrogen diffusion into the Pd lattice. (b) Schematic of chemisorption based on MOSs. (c) Schematic of physisorption (Reused with permission from the reference [27] ©The Royal Society of Chemistry 2024).

When the sensing element is exposed to H_2 gas, the H_2 acts as a donor, given in eq. (2). Hence, for the n-type material, the electrical conductivity will increase, and for the p-type material, it will decrease [29].

According to the direct adsorption/desorption model, many materials undergo redox reactions when exposed to H_2 gas without the influence of adsorbed oxygen ions on the surface. Based on this mechanism, the gas directly interacts with the sensing element, given in eq. (3), resulting in a change in electrical conductivity.

Another important mechanism is the hydrogen diffusion mechanism [30]. It is generally suitable for Pd-based sensing elements (**Figure 3(c)**). In this case, the first step is the chemisorption of hydrogen on the Pd surface, followed by dissociation of H_2. The second step is the diffusion of dissociated hydrogen atoms into the Pd lattice, which subsequently occupies the interstitials of the lattice (this step is also called the dissolution of hydrogen in the Pd lattice), given in eq. 4 [19].

$$H_2 \text{ diffusion mechanism}: Pd + \tfrac{x}{2}H_2 \leftrightarrow PdH_x \tag{4}$$

The hydrogen diffusion coefficient in the Pd is on the order of 10^{-7} cm^2 s^{-1} at ambient temperature [30]. The H diffuses into the FCC lattice of Pd, where it acts as a scattering center for charge carrier transport, thereby increasing the bulk resistance of the Pd. Hence, it is used to detect hydrogen using an electrical sensor. Along with the resistivity changes, the formation of α and β palladium hydride (PdH_x) phase induces mechanical stress due to lattice expansion, which leads to the optomechanical detection of H_2 gas [26]. Depending on the availability of hydrogen and temperature, α and β phases of PdH_x can form, and both have higher resistivity and larger unit cell volume than that of pure Pd. The α-PdH_x and plateau region in the phase diagram of the Pd-H system are suitable for sensor application due to the reversibility in the structural changes when the H_2 flow is removed. The β-PdH_x region is unsuitable for the sensing application due to the incomplete recovery to its baseline after the removal of H_2 flow, and the potential delamination of the film from the substrate [2].

In the case of carbon-based materials, such as graphene oxide (GO), MXene, and carbon nanotube (CNT), they primarily employ hydrogen bonding or electrostatic attraction at their surface terminals in the charge transfer process, which facilitates the physical adsorption of gases onto their surfaces. These materials engage with target gas molecules via Van der Waals forces or acceptor-donor systems, resulting in alterations in electrical resistance [31].

3. Synthesis and deposition of nanostructured sensing elements

Regarding the intrinsic characteristics of the H_2 sensing mechanisms described above, a major research effort has been placed on the synthesis and deposition of the sensing element/layer. Using appropriate preparation conditions and methods, one can alter the size, shape, porosity, and grain boundary to achieve the enhancement in "3S" (sensitivity, selectivity, and stability) of the sensor. **Table 3** provides a list of top-down and bottom-up approaches for the synthesis/deposition of sensing elements. The top-down approach refers to a subtractive process in which a bulk starting material is divided into smaller ones of the nanoscale. Bottom-up approach refers to an additive process that starts with precursor atoms (or molecules), which combine to form a nanosized structure. The term "top-down approach" describes a size reduction method that involves reducing a bulk starting material to nanoparticles. In a bottom-up technique, atoms or molecules used as precursors are added together to create a nanoscale structure.

3.1 Semiconducting metal oxide, metal, and conducting polymers

So far, semiconducting metal oxides (SMOs) have been extensively reported as sensing elements for H_2. Among them, ZnO [32], SnO_2 [33], In_2O_3 [34], TiO_2 [35], MoO_3 [36], V_2O_5 [37], and WO_3 [38] are well-known n-type SMOs, and NiO [39], CuO [40], and Fe_2O_3 [41] are p-type SMOs utilized to detect H_2 at different operating temperatures. While comparing the sensing performance of the SMOs with that of metal chalcogenides (MCs), the SMOs show better stability and repeatability in the air ambient, which is the main reason to be approached in the present chapter. Along with SMOs, palladium metals/palladium-based oxides are also reported for the detection of H_2 [30].

For room temperature detection, conducting polymers (CPs) are the potential candidates, due to the presence of a π-conjugation structure that leads to the p-type behavior toward H_2 gases. Polyaniline (PANI) [42], polypyrrole (PPy) [43], polythiophene (PT) [44], poly(para-phenylene) (PPP), poly(3,4-ethylenedioxythiophene) (PEDOT) [45], and their derivatives are widely studied materials for room temperature H_2 gas sensor applications. To enhance the sensing performance, these materials are prepared in nanostructured form, as detailed in **Table 3**. This table summarizes various top-down and bottom-up approaches, their methods, modes, advantages, and drawbacks for the synthesis of metal and semiconducting metal oxides in nanostructure forms [46, 47]. When the sensing element is prepared in a nanostructured form rather than as a bulk material, it takes advantage of the surface-to-volume ratio of the nanostructures that leads to a higher number of active sites and a larger exposure area for interaction.

Method	Mode	Merit	Demerit
Etching	Wet	Low costSimple process	Poor repeatabilityChemical contamination
Grinding	Mechanical energy	Low-costLarge-scale production	The shape of nanomaterials is irregular.Contamination occurs from the grinding/milling medium.
Milling			
Etching	Dry	High feature resolutionHigh reproducibilityEasy to control	High costPower-consuming process
Ablation	Pulsed laser	High-quality thin film formationEnvironmentally friendly processHigh-quality multilayer formationThickness can be controlled	High capital costThe rate of deposition is quite slowRequirement of controlled vacuumLarge-area deposition is the major limitationNot easy to scale up
Evaporation	Thermal and E-beam		
Sputtering	Ions		
Ion plating	Ions		
Co-precipitation	Chemical reaction under controlled temperature, pH, concentration, and volume	Simple, easy, and low-cost synthesis methodProduction of high-quality nanostructures by controlling pHBetter yield	Usage of additives (e.g., NaOH/KOH/NH_3/thiourea/citric acid/glycine, etc.) in high concentrations to achieve the desired morphology/reaction
Sol-gel			
Combustion			
Hydrothermal			
Solvothermal			
Spray pyrolysis	Deposition of materials over the surface through chemical reaction/thermal decomposition	Simple and suitable for large-area depositionCost-effective processEasy to scale upSuitable for all types of substrates	The deposition rate and thickness control are possible.Avoiding oxide formation will be difficult.
Chemical bath deposition			
SILAR			
Electrochemical	Chemical reaction under the control of applied potential	Production of high-quality 1D nanostructureAn ordered and aligned structure can be obtained using an anodized Al_2O_3 substrate.	Use of conductive substrateMass production is limited
Electrospinning	Nanofiber formation under the control of applied potential	Controllable fiber formation	Use of toxic solventsLimited control of the structure porosityLow cell-density controllability

Method	Mode	Merit	Demerit
Chemical vapor deposition	Thermal/LASER/Plasma	Uniform film formationHigh purity and stabilityGood yield and reproducibility	Highly expensive instrumentsEmit toxic gaseous byproducts
Atomic layer deposition	Thermal/Plasma	High-quality filmUltimate stoichiometric and thickness accuracyExcellent adhesion	Economically not a viable methodVery high materials and energy waste

Table 3.
Preparation methods for the formation of metal and metal oxide nanostructures.

The top-down approach involves breaking, grinding, and milling bulk materials into nanostructured powders. It also includes thin film deposition techniques, such as ablation, evaporation, sputtering, etching, and lithography. Conversely, the buildup of nanomaterials from their atomic and molecular precursors corresponds to the bottom-up approach. Conducting polymer can be synthesized by oxidation of corresponding monomers using chemical and electrochemical methods. The deposition of conducting polymer film can be done using electrochemical deposition, spin coating, thermal evaporation, vapor polymerization, layer-by-layer self-assembly, and electrochemical polymerization [46–48]. The aforementioned synthesis methods yield the sensing elements in the form of powders, which need to be coated onto electrodes using techniques, such as spin coating, screen printing, drop casting, and dip coating to form thick film sensors. Conversely, the material should also be made into pellets. When comparing the performance and sample requirements, thick/thin film sensors generally require a small amount of sensing material, and they can show good sensitivity and quick recovery due to their high surface-to-volume ratio, porosity, and low-dimensional nature [48].

3.2 Unique layered/dimensional structures

In addition to the SMOs and conducting polymers, the widely reported H_2 sensing elements are carbon-based materials, such as graphene, graphene oxide, carbon nanotubes, and carbon nanofibers due to their textural properties, such as high specific surface area, porosity, and pore size distribution, as well as their chemical characteristics [49].

Exfoliation of the graphene or graphene oxide from graphite is the top-down approach that involves mechanical, chemical, or electrochemical exfoliation. The scotch tape method is a simple way to produce graphene from a carbon sheet. However, the major limitation lies in controlling the size, shape, and thickness of the graphene. The widely used chemical exfoliation method is the modified Hummer's method [50]. First, the graphite is oxidized with a strong acid and oxidant, and then the layers are isolated, resulting in the formation of graphene oxide (GO). Finally, the graphene is produced through the reduction process using thermal, chemical, or laser treatment [51]. The yield obtained through the modified Hummer's method is high, but major attention is needed to convert the graphene oxide to graphene if high-quality graphene is needed. Reduced graphene oxide (rGO) or graphene can also be produced using the electrochemical exfoliation method [52], which provides good quality and yield but requires an external electric field. Electrolytes used for the exfoliation process are ionic liquids, aqueous acids, and inorganic salts.

Chemical vapor deposition (CVD) can yield high-quality graphene [53], but it has major drawbacks, including relatively low yield, high power consumption, and a time-consuming process. In CVD, gases and carbon sources, such as acetylene and methyl methacrylate, are decomposed in the presence of a catalyst to form graphene. Desorption of Si atoms from the SiC surface at higher temperature and vacuum leads to the epitaxial growth of graphene layers [54]. LASER irradiation is another method that provides a 3D porous graphene structure using the carbonaceous precursors [55]. Carbon nanofibers (CNFs) [56] can be made using electrospinning, chemical vapor deposition, and template methods. The electrospinning technique deposits fibers onto the collecting plate, followed by the carbonization process in a nitrogen atmosphere to form CNFs. While depositions can be performed at room temperature, they require high voltage, post-treatment, and a carbonization process. CVD involves the decomposition of a hydrocarbon composite in the presence of a metal catalyst at high or low temperature with plasma to form carbon nanofibers. While it offers high-quality CNFs, it has significant disadvantages, including high cost and energy consumption. Porous anodic alumina membranes are used as a template for the deposition of CNFs [56], but the removal of these templates needs considerable attention.

An alternative to carbon-based materials is MXene. It is a 2D transition metal carbide/nitride/carbonite discovered by Naguib et al. [57] in 2011 through selective chemical etching of the Al layer from the MAX phase using hydrofluoric acid (HF), which results in fluorine-containing functional groups. MXenes are represented by the general formula Mn + 1XnTx, where M denotes transition metals (Ti, Mo, Ta, W, V, Cr, Y, Zr, Sc, Hf, Ta, and Nb), X stands for carbide/nitride/carbonites (C/N/CN), and T represents the surface functionality (O, OH, F, Cl, I, Br, NH_2, S, Se, and Te). There are different stoichiometries of MXenes discovered to date, including M_2AX_2, M_3AX_2, M_4AX_3, and M_5AX_4. Even though this method provides high-quality MXene sheets, there is a need for an alternative etchant to replace HF, due to its toxic and corrosive nature. The hydrochloric acid (HCl) and lithium fluoride (LiF) mixture (LiF/HCl) and other fluoride-containing salts (NaF, KF, NH_4F, FeF_3, CsF, and NH_4HF_2) with HCl are some of the proposed alternatives [57]. Nevertheless, they still generate HF gas during the process, which raises safety issues. Other methods adopted are non-fluoride-based etching (NaOH and H_2SO_4), molten fluoride salt etching, electrochemical etching, alkali etching, Lewis acidic molten salt etching, and ionic liquid-based etching. Non-fluoride-based etching using potassium hydroxide as an alkaline etchant results in oxygen-containing functional groups [58]. Sun et al. [59] reported the electrochemical etching of the MAX phase to form Ti_2CT_x in a diluted HCl electrolyte. The major drawback of this method is over-etching, and it can be controlled by optimizing the etching voltage and etching time. In 2019, Li et al. [60] proposed the Lewis acid etching method for $Ti_3C_2Cl_2$ and Ti_2CCl_2 preparation. Husmann et al. [61] reported an acid-free synthesis using 1-butyl-3-methylimidazolium hexafluorophosphate (BMIMPF6) and 1-ethyl-3-methylimidazolium tetrafluoroborate ($EMIMBF_4$) for $Ti_3C_2T_x$ and Ti_2CT_x preparation from Ti_3AlC_2 and Ti_2AlC, respectively. The above-mentioned methods are top-down approaches. Xu et al. [62] deposited the ultrathin Mo_2X, WC, and TaC MXenes on Cu/Mo foil. However, to date, wet etching is a widely used and effective synthesis method for various MXenes [57]. These materials attracted attention as gas sensors due to their tailored surface functionality, higher surface area, and low dimensionality [58]. The performance of SMOs, metal, conducting polymer, and unique layered/dimensional structures shows remarkable electrical and optomechanic property changes during H_2 detection [13, 18, 42, 44, 63–66].

To improve the sensing performance, namely the sensitivity, a sensing element prepared in a nanoscale morphology can contribute to the quantum size effect, higher surface area, better charge transfer and transport mechanisms (such as hopping, tunneling, ballistic transport, and hybrid transport), and the thickness of the depletion layer formation [67]. Another method to enhance sensitivity involves doping, modifying surface defect states, and forming heterojunctions. Hence, these actions improve or tune the electrical and optomechanical properties. The expected electrical changes include a shift in the Fermi energy level due to bond bending, and an adjustment of the carrier concentration, both of which influence the baseline resistivity of the sensing element. In the case of optomechanical sensors, the cantilever material and its dimensions need to be designed in such a way that it has adequate withstanding ability even at large bending forces, and a robust mechanical structure, without compromising the sensitivity. In this scenario, the design of the fiber should include low insertion loss, high contrast, and high finesse [22].

In more detail, **Figure 4(a)** shows the influence of size on the depletion layer formation [68]. The CeO_2 quantum dots measuring approximately 3 nm exhibited an enhanced sensor response, which diminished for CeO_2 nanoballs and nanoparticles sized at 15 ± 3 and 30 ± 12 nm, respectively. This indicates that optimizing the size of nanomaterials is crucial for achieving improved sensor performance in CeO_2 nanostructure-based sensors. The reduced size of CeO_2 quantum dots, smaller than the Debye length, was identified as the primary reason for the roughly 4-fold increase in sensor responsiveness and selectivity toward reducing hydrogen gases compared to oxidizing gases at 1–10 ppm levels [68]. Additionally, **Figure 4(b)** schematically illustrates the impact of the Pd layer functioning as a gate on the Si channel. Upon the deposition of Pd on the Si surface, an electron depletion layer was established in the Si channel due to the elevated work function of the deposited Pd nanoparticles, increasing resistance. The numerical simulation of a semiconductor device simulator indicates that the electron concentration altered following the deposition of a Pd layer on the Si surface, which possesses a work function of 5.12 eV [69]. A bifunctional sensing mechanism is presented to elucidate the enhanced sensing performance of $SnO_2 - ZnO$ composite nanofibers toward H_2 gas. The bifunctional sensing system comprised some elements: (1) diminishing the gas interaction among adjacent SnO_2 nanograins and (2) a metallization phenomenon between the ZnO and SnO_2 nanograins. The initial mechanism (m1) pertains to a grain-boundary-dependent event observed in SnO_2 homojunctions (**Figure 4(c)**, m1). The second mechanism (m2) is depicted in **Figure 4(c)**, m2. In this case, the $ZnO - SnO_2$ heterojunctions

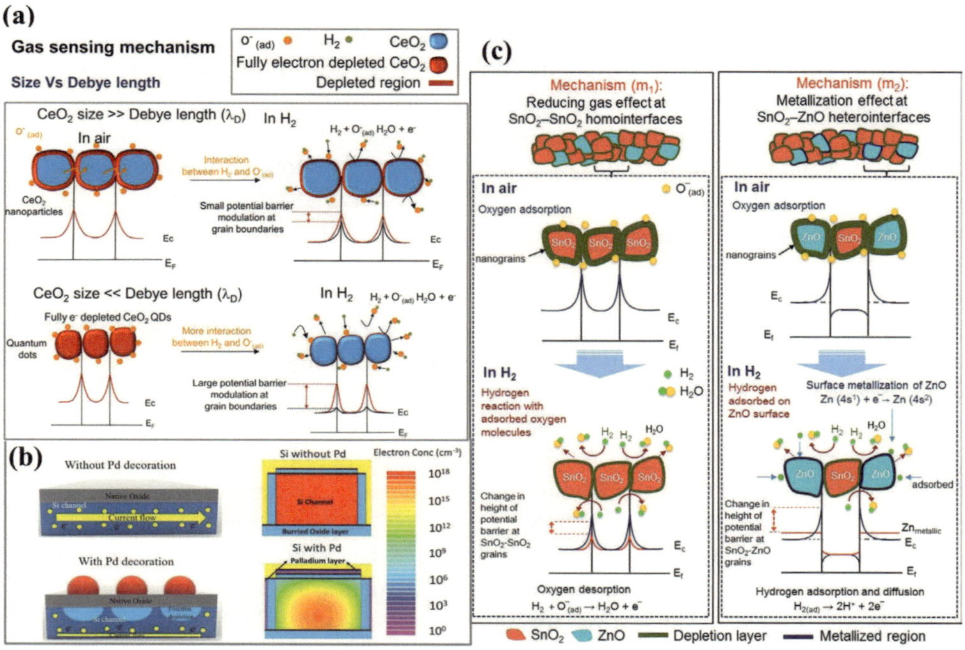

Figure 4.
(a) Size effect (reused with permission from the reference [68] ©2024 Elsevier Ltd.), (b) metal doping effect (reused with permission from the reference [69] © 2018 WILEY-VCH Verlag GmbH & Co. KGaA, Weinheim), and (c) nanocomposite effect (reused with permission from the reference [70] © 2015, American Chemical Society).

boost the sensor response. $SnO_2 - ZnO$ heterojunctions are characterized by a significant barrier at the junctions and a semiconductor-to-metallic transition on the surfaces of the ZnO nanograins following exposure to H_2 gas [70].

4. Sensing systems for real-time H_2 identification

The current state of the art reports several strategies to sense H_2 in different contexts. For example, Song et al. [71] reported a self-powered integrated nanostructured gas sensor for indoor safety monitoring; the limit of detection of the sensor toward H_2 was 10 ppb. Through the integration of a solar cell and a lithium-ion battery, the device attained self-power functionality in low-light interior environments. The sensor signals were measured and transferred to a smartphone in real time via a wireless Bluetooth connection. The acquired four-dimensional data, corresponding to four types of gas sensors, were condensed into two-dimensional data through principal component analysis (PCA), and the support vector machine (SVM) algorithm was subsequently utilized for gas discrimination, providing a decision space for the automatic classification of unknown gases. Chen et al. [72] reported a SnO_2 nanotube sensor array for H_2 and benzene identification using learning vector quantization (LVQ). The mixed ambient was created using three gases: 5000 ppm of H_2, 25 ppm of NO_2, and 5 ppm of benzene (using N_2 as carrier gas). The limit of detection for the SnO_2 sensor is 10 ppm of H_2, and the signal-to-noise ratio is 2. The e-nose, equipped with a readout circuit, transmits sensing data via a Bluetooth unit to a mobile phone, facilitating the development of high-performance smart sensor networks. Pd-SnO_2 nanoflowers and Pd-WO_3 microparticle-based sensor arrays were reported by Li et al. [73]. The operating temperature of this sensor is 300°C, and the corresponding power consumption is 17 mW. The identification of the gases was done using p-neighbors back-propagation neural network (*p*N-BPNN) model over the H_2 concentration range 15–500 ppm among the seven other gases.

Indeed, power consumption is also a key feature for the long-term monitoring of H_2. Hence, Chen et al. [72] and Li et al. [73] focused on low-powered e-nose for indoor monitoring, whereas Song et al. [71] integrated a solar cell and a Li-battery to make the device self-powered. Li et al. [73] reduced the power consumption to 17 mW for an operation temperature of 300°C by reducing the size of the sensors; the reported dimension is 110^2 $\mu m^2/2$ (/2 represents the number of sensors, Pd-SnO_2 and Pd-WO_3). Chen et al. [72] reduced power consumption by doing the measurement in the duty cycle mode rather than the continuous mode. Similarly, a duty-cycling mode with nano-heater-based MOS gas sensor approach is used to address the high-power consumption by Kim et al. [74]. To heat the sensor to 250°C, the duty-cycling mode drastically cuts the power usage to 160 μW, with a reduction of up to 90% compared to the constant power consumption mode. Additionally, a convolutional neural network (CNN) algorithm converts the air, SO_2, CO, H_2, and NO_2 responses into 2D spectrogram images and analyzes them in real time by identifying the gas within the sliding time window. The above-discussed four reports are chemi-resistive-based e-nose/sensor arrays for indoor gas monitoring applications [74].

Another example of sensor arrays is the chemFET-based ones, which were fabricated with metal-decorated carbon nanotubes by Star et al. [75]. In this case, the signals were analyzed using PCA and partial least squares regression (PLS Regression) to measure not only H_2 but also NO_2^-, NH_3, CO, and H_2S. Another relevant example of a chemFET-based approach is the aerial H_2 sensing probe installed in a microdrone that

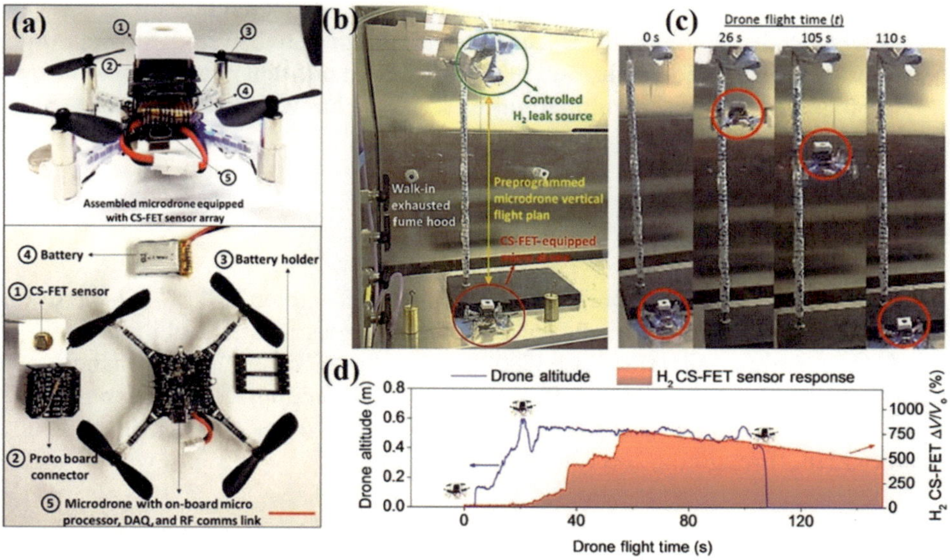

Figure 5.
Aerial chemical sensing probes. (a) Image showing a microdrone equipped with a Ni-Pd-based H_2-sensitive CS-FET (top) and an unassembled drone showing the individual components (bottom). DAQ, data acquisition. (b) Image depicting the experimental setup used for the aerial hydrogen sensing experiment. (c & d) Image capture of the CS-FET-equipped microdrone in motion at different times with corresponding altimetry and generated H_2 sensor response, respectively (Reused with permission from the reference [76] © 2025 American Association for the Advancement of Science).

included, beyond the chemFET sensors, a microprocessor, data acquisition unit, RF comms link, and a battery (**Figure 5**) [76]. The reported chemFET-based H_2 sensor array is capable of operating at room temperature. The reported sensing elements are PdAu, NiPd, and Ni for H_2S, H_2, and NO_2, respectively.

5. Conclusion

Being a promising carbon-free energy carrier and the lightest element in the periodic table with a superior reducing nature, monitoring H_2 is still a complex task that has generated a quest for reliable sensing systems. This monitoring is critical to deal with safety issues associated with the use of this gas.

Therefore, in situ monitoring of H_2 leakage is needed, and the electrical and optomechanical sensors discussed in this chapter can be promising tools for this end. This chapter overviews these two approaches capable of sensing H_2 in different scenarios. These sensors can be miniaturized and introduced in different sensor arrays, providing effective monitoring tools with low-power consumption. Another key aspect is related to the synthesis and deposition of the sensing elements. The advances in this field can provide innovative solutions for the enhancement of the sensing features of these materials and for the adoption of emerging materials with H_2 sensing capabilities.

The future directions of these two sensing approaches will focus on improving the current features of the sensors. This includes the enhancement of the sensing performance, the integration of the sensors in sensing systems and monitoring tools, and the adoption of innovative methods for data analysis. These advances will certainly impact the availability of analytical information related to H_2. The hope is that stable, portable, and reliable H_2 sensors with integrated data transferring units and (semi)quantitative identification in real environments will become possible by bringing together the latest combinations of scientific knowledge and technological advancement.

Acknowledgements

The authors thank the support from the projects "H2 Green Valley" (C645551860-00000064), "PRODUTECH R3: Production Technologies Cluster Mobilizing Agenda for Reindustrialization" (C645808870-00000067), and "ATE—Aliança para a Transição Energética" (C644914747-00000023), supported by Component C5—Capitalisation and Business Innovation, under the Portuguese Resilience and Recovery Plan, through the NextGenerationEU Fund.

Appendices and nomenclature

g or kg	gram or kilogram
m/mm/cm/nm	meter/millimeter/centimeter/nanometer
K/°C	kelvin/degree celsius
mJ/kJ	millijoule/kilojoule
S	second
%	percentage
ppm/ppb	parts-per-million/parts-per-billion
mV	millivolt
μA	microampere
μW	microwatt
O_2^-, O^- and O^{2-}	superoxide, atomic oxygen, and peroxide

Author details

Nancy Anna Anasthasiya Anthuvan[1], Karuppiah Veerathangam[2], João Piteira[1] and Hugo M. Oliveira[1]*

1 International Iberian Nanotechnology Laboratory (INL), Braga, Portugal

2 Institute for Energy Research, School of Chemistry and Chemical Engineering, Jiangsu University, Zhenjiang, China

*Address all correspondence to: hugo.oliveira@inl.int

References

[1] Wen JX, Hecht ES, Mevel R. Recent advances in combustion science related to hydrogen safety. Progress in Energy and Combustion Science. 2025;**107**:101202. DOI: 10.1016/j.pecs.2024.101202

[2] Nancy AAA, Prabhu E, Jayaraman V, et al. Wide range H2 sensors for process streams, based on multilayered Pd thin films dispersed with Ag nanoparticles and chromium as buffer layer. International Journal of Hydrogen Energy. 2020;**45**:22195-22203. DOI: 10.1016/j.ijhydene.2020.06.010

[3] Liu RX, Zhang X, Zhu XS, et al. Optical fiber hydrogen sensor based on the EVA/Pd coated hollow fiber. IEEE Photonics Journal. 2022;**14**:6825806. DOI: 10.1109/JPHOT.2022.3168009

[4] Dong R, Yang M, Zuo Y, et al. Conducting polymers-based gas sensors: Principles, materials, and applications. Sensors. 2025;**25**:1-33. DOI: 10.3390/s25092724

[5] Neri G. First fifty years of chemoresistive gas sensors. Chem. 2015;**3**:1-20. DOI: 10.3390/chemosensors3010001

[6] Majhi SM, Mirzaei A, Kim HW, et al. Recent advances in energy-saving chemiresistive gas sensors: A review. Nano Energy. 2021;**79**:105369. DOI: 10.1016/j.nanoen.2020.105369

[7] Kim YJ, Lee S, Choi S, et al. Highly durable chemoresistive micropatterned PdAu hydrogen sensors: Performance and mechanism. ACS Sensors. 2024;**9**:5363-5373. DOI: 10.1021/acssensors.4c01599

[8] Beatriceveena TV, Gnanasekar KI. Highly sensitive and wide range hydrogen sensor based on thin films of AgInO2 by pulsed laser deposition. Sensors and Actuators, B: Chemical. 2023;**383**:133599. DOI: 10.1016/j.snb.2023.133599

[9] Qian J, Hua Z, Zhi Z, et al. Dual-gate field-effect transistor ultrahighly sensitive hydrogen sensors enabled by an ultrahigh inherent gain. ACS Applied Electronic Materials. 2025;**7**:4938-4945. DOI: 10.1021/acsaelm.5c00381

[10] Sharma B, Kim JS. MEMS based highly sensitive dual FET gas sensor using graphene decorated Pd-Ag alloy nanoparticles for H2 detection. Scientific Reports. 2018;**8**:5902. DOI: 10.1038/s41598-018-24324-z

[11] Sun A, Yu H, Zhou Y, et al. Pd gated AlGaN/GaN high electron mobility transistor for ppb level hydrogen gas detection. International Journal of Hydrogen Energy. 2022;**47**:17494-17503. DOI: 10.1016/j.ijhydene.2022.03.185

[12] Huang Z, Yang W, Zhang Y, et al. Miniaturized electrochemical gas sensor with a functional nanocomposite and thin ionic liquid Interface for highly sensitive and rapid detection of hydrogen. Analytical Chemistry. 2024;**96**:17960-17968. DOI: 10.1021/acs.analchem.4c02561

[13] Lee EG, Jung SW, Jo YE, et al. Electrochemical hydrogen sensor assembly for monitoring high-concentration hydrogen. Physica Status Solidi A: Applications and Materials Science. 2022;**219**:2100782. DOI: 10.1002/pssa.202100782

[14] Tang Y, He J, Gao X, et al. Continuous amperometric hydrogen gas sensing in ionic liquids. Analyst. 2018;**143**:4136-4146. DOI: 10.1039/c8an00577j

[15] Liu Y, Zhou R, Xu X, et al. Fiber-optic FPI-based hydrogen sensor with polymer/air cascaded cavity and PDMS/Pt/WO3 layered structure. Optics Communication. 2025;**586**:131893. DOI: 10.1016/j.optcom.2025.131893

[16] Zhang X, Li X, Zhang X, et al. Fiber optics-mechanics coupling sensor for high-performance hydrogen detection. Photonic Sensors. 2025;**15**:250314. DOI: 10.1007/s13320-025-0742-8

[17] Liu Y, Zhou R, Xu X, et al. Sensitivity enhanced fiber-optic hydrogen sensor based on Vernier effect of polymer-air cascaded FPI. IEEE Sensors Journal. 2025;**25**:9626-9632. DOI: 10.1109/JSEN.2025.3529581

[18] McKeown SJ, Wang X, Yu X, et al. Realization of palladium-based optomechanical cantilever hydrogen sensor. Microsystems and Nanoengineering. 2017;**3**:1-6. DOI: 10.1038/micronano.2016.87

[19] Li H, Li Y, Wang K, et al. Ultra-high sensitive micro-chemo-mechanical hydrogen sensor integrated by palladium-based driver and high-performance piezoresistor. International Journal of Hydrogen Energy. 2021;**46**:1434-1445. DOI: 10.1016/j.ijhydene.2020.10.013

[20] Khan D, Li H, Gajula D, et al. H2Detection using plasmonically generated surface photoacoustic waves in Pd nanoparticle-deposited GaN microcantilevers. ACS Sensors. 2020;**5**:3124-3132. DOI: 10.1021/acssensors.0c01181

[21] Zhong J, Lu S, Liu S, et al. Fiber-tip Fabry-Pérot interferometer with a graphene-Au-Pd cantilever for trace hydrogen sensing. Lab on a Chip. 2023;**23**:3518-3526. DOI: 10.1039/d2lc01149b

[22] Xiong C, Zhou J, Liao C, et al. Fiber-tip polymer microcantilever for fast and highly sensitive hydrogen measurement. ACS Applied Materials and Interfaces. 2020;**12**:33163-33172. DOI: 10.1021/acsami.0c06179

[23] Zhang X, Zhang X, Li X, et al. The nanophotonic machinal cavity and its hydrogen sensing application. Sensors and Actuators, B: Chemical. 2022;**367**:132095. DOI: 10.1016/j.snb.2022.132095

[24] Zhang P, Xiao Y, Zhang J, et al. Highly sensitive gas sensing platforms based on field effect transistor-a review. Analytica Chimica Acta. 2021;**1172**:338575. DOI: 10.1016/j.aca.2021.338575

[25] Korotcenkov G, Do HS, Stetter JR. Review of electrochemical hydrogen sensors. Chemical Reviews. 2009;**109**:1402-1433. DOI: 10.1021/cr800339k

[26] Zhang Y, Peng H, Qian X, et al. Recent advancements in optical fiber hydrogen sensors. Sensors and Actuators, B: Chemical. 2017;**244**:393-416. DOI: 10.1016/j.snb.2017.01.004

[27] Liu YY, Li Z, Liang Y, et al. Recent advances in nanomaterial-enabled chemiresistive hydrogen sensors. Chemical Communications. 2024;**60**:14497-14520. DOI: 10.1039/d4cc05430j

[28] Hinojo A, Lujan E, Colominas S, et al. Optimization of 3D printing conditions for BaCe0.6Zr0.3Y0.1O3-δ in the construction of amperometric high-temperature H2 sensors. Talanta Open. 2025;**11**:100454. DOI: 10.1016/j.talo.2025.100454

[29] Wetchakun K, Samerjai T, Tamaekong N, et al. Semiconducting metal oxides as sensors for

environmentally hazardous gases. Sensors and Actuators, B: Chemical. 2011;**160**:580-591. DOI: 10.1016/j.snb.2011.08.032

[30] Toksha B, Gupta P, Rahaman M. Hydrogen sensing with palladium-based materials: Mechanisms, challenges, and opportunities. Chemistry, an Asian Journal. 2024;**19**:202400127. DOI: 10.1002/asia.202400127

[31] Ding J, Zheng Z, Song Z, et al. Recent progress in chemiresistive gas sensors with 1D nanostructured sensing materials: Insights into the structure-morphology-performance relationship. Materials and Design. 2023;**234**:112360. DOI: 10.1016/j.matdes.2023.112360

[32] Nguyet TT, Thanh Le DT, Van Duy N, et al. A sigh-performance hydrogen gas sensor based on Ag/Pd nanoparticle-functionalized ZnO nanoplates. RSC Advances. 2023;**13**:13017-13029. DOI: 10.1039/d3ra01436c

[33] Chen W, Zhou Q, Wan F, et al. Gas sensing properties and mechanism of Nano-SnO2-based sensor for hydrogen and carbon monoxide. Journal of Nanomaterials. 2012;**2012**:1-9. DOI: 10.1155/2012/612420

[34] Chen C, Chen W, Liu Q, et al. Electrospinning of Pd-In2O3Nanofibers for high-performance room temperature hydrogen sensors. ACS Applied Nano Materials. 2022;**5**:12646-12655. DOI: 10.1021/acsanm.2c02436

[35] Tasyurek LB, Isik E, Isik I, et al. Enhancing the performance of TiO2 nanotube-based hydrogen sensors through crystal structure and metal electrode. International Journal of Hydrogen Energy. 2024;**54**:678-690. DOI: 10.1016/j.ijhydene.2023.08.202

[36] Mobtakeri S, Habashyani S, Gür E. Highly responsive Pd-decorated MoO3Nanowall H2Gas sensors obtained from In-situ-controlled thermal oxidation of sputtered MoS2Films. ACS Applied Materials and Interfaces. 2022;**14**:25741-25752. DOI: 10.1021/acsami.2c04804

[37] Wang YT, Whang WT, Chen CH. Hollow V2O5 nanoassemblies for high-performance room-temperature hydrogen sensors. ACS Applied Materials and Interfaces. 2015;**7**:8480-8487. DOI: 10.1021/am509182s

[38] Horprathum M, Srichaiyaperk T, Samransuksamer B, et al. Ultrasensitive hydrogen sensor based on Pt-decorated WO3 nanorods prepared by glancing-angle dc magnetron sputtering. ACS Applied Materials and Interfaces. 2014;**6**:22051-22060. DOI: 10.1021/am505127g

[39] Badie C, Mirzaei A, Lee JH, et al. Selective detection of H2 gas in gas mixtures using NiO-shelled Pd-decorated ZnO nanowires. Advanced Materials Technologies. 2024;**9**:2302081. DOI: 10.1002/admt.202302081

[40] Güldüren ME, İskenderoğlu D, Güney H, et al. Structural, optical, and H2 gas sensing analyses of Cr doped CuO thin films grown by ultrasonic spray pyrolysis. International Journal of Hydrogen Energy. 2023;**48**:20804-20814. DOI: 10.1016/j.ijhydene.2023.03.057

[41] Zhang S, Chang X, Zhou L, et al. Stabilizing single-atom Pt on Fe2O3 nanosheets by constructing oxygen vacancies for ultrafast H2 sensing. ACS Sensors. 2024;**9**:2101-2109. DOI: 10.1021/acssensors.4c00162

[42] Askar P, Kanzhigitova D, Ospanova A, et al. 1 ppm-detectable hydrogen gas sensor based on nanostructured polyaniline. Scientific Reports. 2024;**14**:26984. DOI: 10.1038/s41598-024-77083-5

[43] Al-Mashat L, Debiemme-Chouvy C, Borensztajn S, et al. Electropolymerized polypyrrole nanowires for hydrogen gas sensing. Journal of Physical Chemistry C. 2012;**116**:13388-13394. DOI: 10.1021/jp3015854

[44] Yadhukrishnan KV, Jose SP, Vasu V, et al. Fabrication of polythiophene/graphitic carbon nitride IDE sensors for exceptional room temperature hydrogen sensitivity. International Journal of Hydrogen Energy. 2024;**93**:1088-1099. DOI: 10.1016/j.ijhydene.2024.10.073

[45] Verma G, Kadiri H, Gokarna A, et al. Flexible hybrid nanostructured chemiresistive sensing platform for low-temperature trace hydrogen detection. Sensors and Actuators, B: Chemical. 2025;**422**:136621. DOI: 10.1016/j.snb.2024.136621

[46] Baig N, Kammakakam I, Falath W, et al. Nanomaterials: A review of synthesis methods, properties, recent progress, and challenges. Materials Advances. 2021;**2**:1821-1871. DOI: 10.1039/d0ma00807a

[47] Abid N, Khan AM, Shujait S, et al. Synthesis of nanomaterials using various top-down and bottom-up approaches, influencing factors, advantages, and disadvantages: A review. Advances in Colloid and Interface Science. 2022;**300**:102597. DOI: 10.1016/j.cis.2021.102597

[48] Li Z, Mi B, Ma X, et al. Review of thin-film resistor sensors: Exploring materials, classification, and preparation techniques. Chemical Engineering Journal. 2023;**477**:147029. DOI: 10.1016/j.cej.2023.147029

[49] Kyokunzire P, Zaraket J, Fierro V, et al. Recent developments in the use of activated carbon-based materials for gas sensing applications. Journal of Environmental Chemical Engineering. 2024;**12**:113702. DOI: 10.1016/j.jece.2024.113702

[50] Hummers WS, Offeman RE. Preparation of graphitic oxide. Journal of the American Chemical Society. 1958;**80**:1339. DOI: 10.1021/ja01539a017

[51] Pei S, Cheng HM. The reduction of graphene oxide. Carbon N Y. 2012;**50**:3210-3228. DOI: 10.1016/j.carbon.2011.11.010

[52] Parvez K, Wu ZS, Li R, et al. Exfoliation of graphite into graphene in aqueous solutions of inorganic salts. Journal of the American Chemical Society. 2014;**136**:6083-6091. DOI: 10.1021/ja5017156

[53] Shi Q, Tokarska K, Ta HQ, et al. Substrate developments for the chemical vapor deposition synthesis of graphene. Advanced Materials Interfaces. 2020;**7**:1902024. DOI: 10.1002/admi.201902024

[54] Çelebi C, Yanik C, Demirkol AG, et al. The effect of a SiC cap on the growth of epitaxial graphene on SiC in ultra high vacuum. Carbon N Y. 2012;**50**:3026-3031. DOI: 10.1016/j.carbon.2012.02.088

[55] Movaghgharnezhad S, Kang P. Laser-induced graphene: Synthesis advances, structural tailoring, enhanced properties, and sensing applications. Journal of Materials Chemistry C. 2024;**12**:6718-6742. DOI: 10.1039/d3tc04677j

[56] Teymouri A, Haji Hossein A, Khoshnazar R, et al. A review on carbon nanofiber production and application in cementitious mixtures. Journal of Building Engineering. 2024;**84**:108519. DOI: 10.1016/j.jobe.2024.108519

[57] Naguib M, Kurtoglu M, Presser V, Jun L, Niu J, Heon M, et al.

Two-dimensional nanocrystals produced by exfoliation of Ti3AlC2. Advanced Materials. 2011;**23**:4248-4253. DOI: 10.1002/adma.201102306

[58] Bhargava Reddy MS, Kailasa S, Marupalli BCG, et al. A family of 2D-MXenes: Synthesis, properties, and gas sensing applications. ACS Sensors. 2022;**7**:2132-2163. DOI: 10.1021/acssensors.2c01046

[59] Sun W, Shah SA, Chen Y, et al. Electrochemical etching of Ti2AlC to Ti2CT: X (MXene) in low-concentration hydrochloric acid solution. Journal of Materials Chemistry A. 2017;**5**:21663-21668. DOI: 10.1039/c7ta05574a

[60] Li M, Lu J, Luo K, et al. Element replacement approach by reaction with Lewis acidic molten salts to synthesize nanolaminated MAX phases and MXenes. Journal of the American Chemical Society. 2019;**141**:4730-4737. DOI: 10.1021/jacs.9b00574

[61] Husmann S, Naguib M, Shim H, et al. Ionic liquid-based synthesis of MXene. Chemical Communications. 2020;**56**:11082-11085. DOI: 10.1039/D0CC03189E

[62] Xu C, Wang L, Liu Z, et al. Large-area high-quality 2D ultrathin Mo2C superconducting crystals. Nature Materials. 2015;**14**:1135-1141. DOI: 10.1038/nmat4374

[63] Zou Y, Wang Q, Xiang C, et al. Doping composite of polyaniline and reduced graphene oxide with palladium nanoparticles for. International Journal of Hydrogen Energy. 2016;**41**:5396-5404. DOI: 10.1016/j.ijhydene.2016.02.023

[64] Shahzamani Z, Ranjbar M, Comini E, et al. Palladium thin films on microfiber filtration paper as flexible substrate and its hydrogen gas sensing mechanism. International Journal of Hydrogen Energy. 2019;**44**:17185-17194. DOI: 10.1016/j.ijhydene.2019.04.277

[65] Naa A, Kampara RK, Rai PK, et al. Highly sensitive graphene oxide functionalized ZnO nanowires for ammonia vapour detection at ambient temperature. Sensors and Actuators, B: Chemical. 2018;**255**:1064-1071. DOI: 10.1016/j.snb.2017.08.148

[66] Kondalkar VV, Park J, Lee K. MEMS hydrogen gas sensor for in-situ monitoring of hydrogen gas in transformer oil. Sensors and Actuators, B: Chemical. 2021;**326**:128989. DOI: 10.1016/j.snb.2020.128989

[67] Choudhary M, Shrivastav A, Sinha AK, et al. Emerging nanomaterials for hydrogen sensing: Mechanisms and prospects. International Journal of Hydrogen Energy. 2024;**77**:557-574. DOI: 10.1016/j.ijhydene.2024.06.149

[68] Rohit KS, Hussain S, et al. Size dependent dual functionality of CeO2 quantum dots: A correlation among parameters for hydrogen gas sensor and pollutant remediation. Chemosphere. 2024;**364**:142959. DOI: 10.1016/j.chemosphere.2024.142959

[69] Gao M, Cho M, Han HJ, et al. Palladium-decorated silicon nanomesh fabricated by nanosphere lithography for high performance. Room Temperature Hydrogen Sensing. Small. 2018;**14**:1703691. DOI: 10.1002/smll.201703691

[70] Katoch A, Kim JH, Kwon YJ, et al. Bifunctional sensing mechanism of SnO2-ZnO composite nanofibers for drastically enhancing the sensing behavior in H2 gas. ACS Applied Materials and Interfaces. 2015;**7**:11351-11358. DOI: 10.1021/acsami.5b01817

[71] Song Z, Ye W, Chen Z, et al. Wireless self-powered high-performance integrated nanostructured-gas-sensor network for future smart homes. ACS Nano. 2021;**15**:7659-7667. DOI: 10.1021/acsnano.1c01256

[72] Chen J, Chen Z, Boussaid F, et al. Ultra-low-power smart electronic nose system based on three-dimensional tin oxide nanotube arrays. ACS Nano. 2018;**12**:6079-6088. DOI: 10.1021/acsnano.8b02371

[73] Li Z, Yu J, Dong D, et al. E-nose based on a high-integrated and low-power metal oxide gas sensor array. Sensors and Actuators, B: Chemical. 2023;**380**:133289. DOI: 10.1016/j.snb.2023.133289

[74] Kim T, Kim Y, Cho W, et al. Ultralow-power single-sensor-based E-nose system powered by duty cycling and deep learning for real-time gas identification. ACS Sensors. 2024;**9**:3557-3572. DOI: 10.1021/acssensors.4c00471

[75] Star A, Joshi V, Skarupo S, et al. Gas sensor array based on metal-decorated carbon nanotubes. The Journal of Physical Chemistry. B. 2006;**110**:21014-21020. DOI: 10.1021/jp064371z

[76] Fahad HM, Shiraki H, Amani M, et al. Room temperature multiplexed gas sensing using chemical-sensitive 3.5-nm-thin silicon transistors. Science Advances. 2017;**3**:1602557. DOI: 10.1126/sciadv.1602557

Chapter 8

Enhancing AI Data Center Resiliency through Metal Hydride Powered Backup Systems

Jalaal A. Hayes

Abstract

As computational technology rapidly advances, the demand for electric power to support massive data flows is increasing, posing a challenge to existing electrical grids. Policymakers and industry leaders are working to modernize these systems, seeking cleaner and more energy-efficient methods to sustainably power data centers in our communities and lower the price of electricity demand overall. In addition to wind and solar, hydrogen has been demonstrated to be a promising candidate for enhancing energy security and resilience, capable of serving as both a primary and backup power source for renewable systems such as solar and wind. Metal hydrides are gaining attention for their unique properties, which make them excellent energy storage solutions compared to compressed and liquefied hydrogen storage methods. This chapter examines various metal hydrides and their potential role in supporting AI-focused data centers. The key factors, including activation behavior, thermal stability, electrochemical performance, cycle stability, and environmental and sustainability aspects of metal hydrides used with data centers, are discussed in detail. Overall, metal hydrides are concluded to be a promising solution if the technology advances within the scope of AI data centers, along with the suggested research and future directions for metal alloy improvements.

Keywords: data centers, hydrogen, metal hydrides, backup power, renewable energy

1. Introduction

In today's world, artificial intelligence (AI) and related technologies have increased in popularity. With new software and the Internet of Things (IoT) on the way, there will be more demand for the current and next generation data centers throughout the world. Although this is excellent in the computer engineering industry, this increase will require a significant amount of power, especially for the data centers that will serve this computing power. Currently, the data centers across the United States are considered small, with a power requirement range of 1–3 MW total power [1], running only cloud-based computing servers and possibly some telecommunication servers.

However, the emerging data centers in the United States and even around the world (recently called hyperscale data centers) will need a power range of 10–100 MW, which is ten to a hundred times the power currently used, respectively. **Figure 1** shows the difference between the small data centers and the hyperscale data centers for perspective [1–3].

Based on this information, electrical power will be pivotal in the resilience of data centers as advancements are made in the contemporary world. Power Usage Effectiveness (PUE) is regarded as a vital metric for the development of sophisticated and efficient data centers, particularly for artificial intelligence and other technological innovations [2]. PUE can be quantified through the consideration of the following factors: (1) type of data center; (2) energy source; (3) geographic region; and (4) operational efficiency.

By utilizing these four variables alongside current historical data of the data center, an accurate PUE can be determined. For instance, according to Supermicro.com, PUE is defined as the ratio of total data center power to the power consumed by IT and computing equipment. This ratio is instrumental in assessing the energy consumption per location, thereby enabling the implementation of cooling systems and other mechanisms to enhance overall efficiency [4]. For example, for an emerging data center, the PUE has to be around 1.5 or lower to have an efficient yet significant operating center for the AI industry.

Currently, fossil fuels are used to power small and large data centers. Several studies are exploring nuclear power to help sustain and improve the energy required for critical infrastructure for data centers and beyond [4, 5]. Although these developments are significant, concerns about the electricity demands of the current infrastructure raise environmental and community benefit issues. Regarding environmental concerns, 13.8 billion kilograms of e-waste have already been generated since 2022, and this number is expected to grow as technology advances, according to the United Nations [6]. This data raises worries for communities near emerging hyperscale data centers across the United States and beyond, mainly due to emissions from fossil fuel-based power sources used by these centers [7]. Furthermore, there is concern about the impact of electricity demand on existing grids. Some research suggests that grid modernization is necessary as AI data centers are built, due to the power demands previously mentioned [1–5].

To ensure the sustainability of hyperscale data centers and enhance overall data center resilience, novel, and innovative approaches are being developed to address current concerns. These approaches include the utilization of solar, wind, and hydrogen-powered systems (**Figures 2** and **3**).

Figure 1.
Current Data Center (left) and Emerging Data Center (right) for the future AI-data-centered application.

Figure 2.
Solar farms with wind turbines. These are potential renewables for data center resilience.

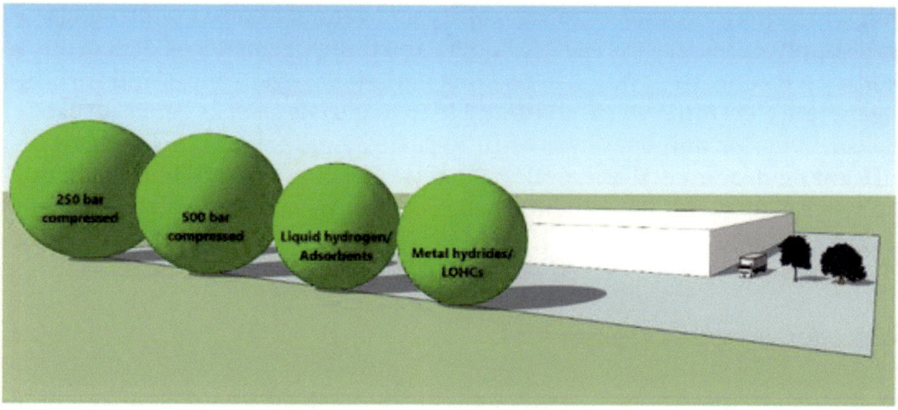

Figure 3.
Different types of hydrogen storage systems considered by large scale data centers [8, 9].

1.1 Solar as power for data center resilience

According to research by the Solar Energy Industries Association (SEIA), solar power has been explored for enhancing data center resilience [10]. According to current data, solar power for data centers can be feasible when paired with energy storage, such as batteries and other types of long-duration storage, due to the intermittent nature of solar energy. As a solution, companies like Google and Cassava Technologies in South Africa have been investing in large-scale solar farms for hyperscale data centers [11–14]. For instance, Google announced plans to build "energy parks" across the United States to host data centers on their developing microgrids. This will ensure electricity demands are met without affecting the surrounding communities' current energy utility prices [11]. Additionally, Cassava Technologies and Distributed Power Africa are currently building a 12 MW solar photovoltaic (PV) plant in South

Africa [13]. This will be one of the first models of a solar microgrid and data center infrastructure on the continent. To support these projects, universities and companies are utilizing HOMER-modeled hybrid software to observe and manage power flow within the energy system and the designed microgrid before construction, as noted in a 2023 study [14].

1.2 Wind as power for data center resilience

According to industry experts, wind power is considered a highly suitable option for providing silence in data centers due to its low greenhouse gas emissions throughout its lifecycle. Compared to the emission of 100 grams of CO_2 per hour, wind power only emits between 10 and 50 grams of CO_2 per kilowatt-hour, and these emissions can be further reduced through carbon sequestration and other technological methods [15]. Given this information, western China has considered using wind turbines to power its 100 MW data center, thanks to the region's windy climate [16]. Additionally, modeling studies in the United States suggest that a similar strategy could be adopted in other windy regions, such as parts of Texas and Kansas.

While wind power presents considerable opportunities, there are inherent challenges associated with the integration of these technologies. Similar to solar power, wind energy is subject to variability and cannot assure a continuous energy supply necessary for the 24/7 operational requirements of data centers. An additional concern pertains to residual emissions; over time, wind turbines may generate electronic waste attributable to materials such as concrete used in the construction and operation of fully-scaled wind farms. This issue raises significant environmental and policy considerations within offshore communities. Consequently, the subsequent discussion will focus on the advancement of hydrogen-powered technologies [15, 16].

2. Hydrogen powered system

In 2019, the United States Department of Energy held a workshop and reported that a hydrogen-powered system for data center applications is feasible [17]. To make this hydrogen-powered system feasible for the data center market, understanding the different hydrogen storage technologies is very important. Each major storage approach operates through different mechanisms, which can affect safety, overall performance, and ultimately the suitability for AI applications requiring long-duration storage. In this chapter, we will explore the three main types of storage: physical storage, cryogenic liquid storage, and material storage for hydrogen [17, 18].

2.1 Compressed gas storage

Compressed gas storage keeps hydrogen in its molecular form (H2) under pressure in specialized vessels, as shown in **Figure 4**. The four types of vessels have their own advantages and disadvantages when it comes to storage with high pressures ranging from 350 to 700 bar. These pressures balance the gravimetric storage densities of 4–6 wt% hydrogen needed for practical applications in various markets, according to the United States Department of Energy. In addition,

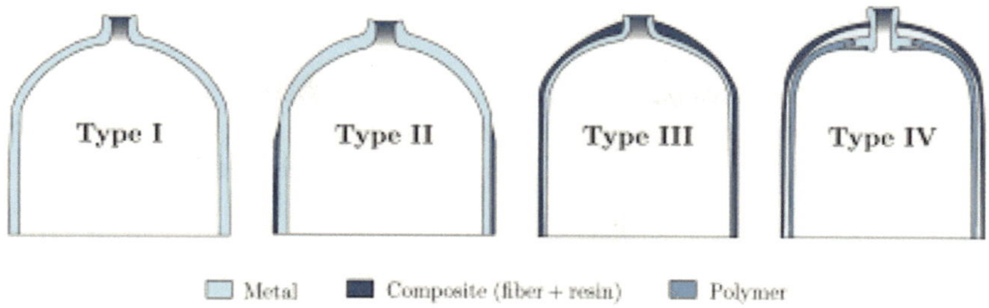

Figure 4.
Four types of vessels for compressed high-pressure hydrogen storage.

hydrogen can provide rapid charging and discharging through valve systems and pipelines due to its molecular structure remaining unchanged. This offers a considerable advantage for industrial infrastructure for hydrogen handling and applications due to its availability and complete reversibility without any degradation of the molecular hydrogen [19, 20]. For a clean-powered hyperscale data center, this could be an ideal storage unit to start with.

There are some drawbacks to this storage technology. Due to its high-pressure containment in the vessels, this can cause various safety hazards in handling hydrogen over time on a large scale. Because of the connotation with hydrogen being very explosive, it is critical to make sure that hydrogen is handled with precise measures. In addition, there is a 10–15% decrease in energy content due to the energy required for compression [19, 20]. This reduces the overall efficiency of the hydrogen-powered system and is based on the energy content of the allocated space for the storage capacity of the compression vessel [19, 20].

2.2 Liquid cryogenic storage

Liquid cryogenic storage (or liquid hydrogen for short) embraces the phase transition of hydrogen from its gaseous state to a liquid at −253 degrees Celsius, which is a cryogenic temperature. The liquid state of hydrogen provides 2.5 times more storage density than compressed gas at 700 bar [21]. This advantage will allow for compact storage installations for potential applications in data center cooling systems [21].

Although the possibilities are promising, there are significant limitations to liquid hydrogen. One such limitation is the constant energy requirement to maintain cryogenic temperatures. Additionally, continuous leak detection and monitoring for boil-off are necessary when hydrogen expands above cryogenic temperatures to prevent gradual energy loss [21, 22]. These limitations make this storage technology less ideal for backup applications in current and future data center resiliency [21, 22].

2.3 Material-based storage

Unlike compressed and cryogenic hydrogen storage, materials-based storage affects the chemical makeup of molecular hydrogen by using pressure, temperature, and chemical reactions to absorb and release hydrogen. In materials-based storage, two main mechanisms are identified: physisorption and chemisorption. After analyzing these mechanisms, the focus will shift to metal hydride storage and its potential influence on data center resiliency [23–31].

2.4 Physiosorption and its systems

Physiosorption (or physically bound) is a process of storing the hydrogen gas into the pores of the hydrogen storage materials [23–29]. This type of storage process involves binding hydrogen to the compound through Van der Waals forces, allowing for rapid absorption and release from the system. As a result, reversibility and faster kinetics can serve as an excellent advantage for the physically bound process [23–29]. Current hydrogen storage materials: metal organic frameworks (MOFs) and carbon nanotubes (CNTs) will be discussed in more detail in the succeeding section.

2.5 Metal: Organic frameworks

Metal organic frameworks (MOFs) are crystalline materials that have micro pores that absorb gases, particularly hydrogen. The discovery of this storage material was made by Yaghi and his colleagues in 1995 [25, 29]. The structure of MOFs include organic polymers, which are interconnected with metal linkers. **Figure 5** presents examples of some of the MOFs that have been used and can be used for potential hydrogen storage [29]. Other than its primary characteristics, its strengths and weaknesses will be further discussed in terms of use as a hydrogen storage material for the future [23–29].

The major advantage of MOFs is the high surface area that adsorbs a large amount of hydrogen on the surface of the material. As an example, MOF-5 and Cu-BTC produce a higher storage capacity than other carbon materials at 0.6 MPa [29]. The storage capacity reflects the large surface area that is accompanied with that MOF. Another distinct advantage is the high porosity of the hydrogen materials. For example, many metal-organic frameworks, such as MOF-5, have high pore diameters for hydrogen to be stored according to Hirscher and his group members [25, 29].

However, one of MOFs shortcomings is storage at low temperatures (i.e. cryogen temperature) [29–31]. Based on the existing research literature, hydrogen adsorption and desorption is mostly conducted between temperature levels of 20–120 K, which is lower than room temperature (298 K [29–31]. Another significant shortcomings is the low weight percent of hydrogen that is released from MOFs. Based on findings from other research investigations, the average weight percent that is released from metal organic frameworks is 1.4–2.5 wt% which does not meet the existing DOE standard which is 6–9 wt% for hydrogen release and desorption [29–31].

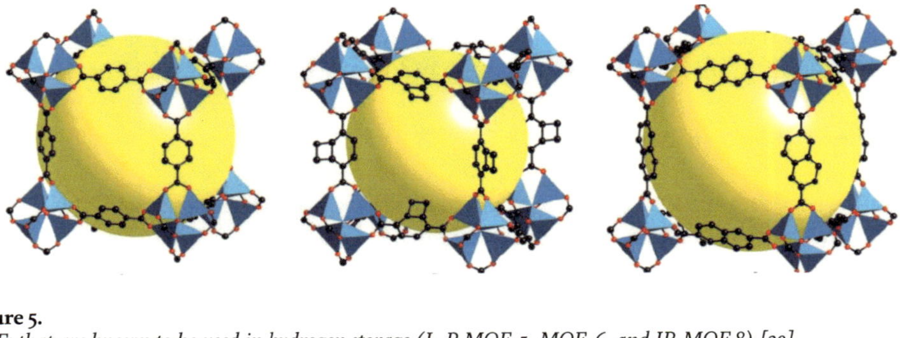

Figure 5.
MOFs that are known to be used in hydrogen storage (L-R MOF-5, MOF-6, and IR-MOF 8) [29].

DOI: http://dx.doi.org/10.5772/intechopen.1012860

2.6 Carbon nanotubes

Carbon nanotubes (CNTs) are carbon molecules that are microporous with nanotubular structures [30, 31]. There are four types of carbon nanotubes, which are: single-walled carbon nanotubes, double-walled carbon nanotubes, multi-walled carbon nanotubes, and coiled carbon nanotubes [20–32]. Single-walled (SWNTs) are nanotubes that only contain one wall of pores on their surface. Double-walled carbon nanotubes (DWCNTs) consist of two concentric walls that are thickened on the nanotubes for increased porosity for gases. Multi-walled (MWCNTs) are similar to DWCNTs. However, they have more concentrically attached walls and are very thick inside the tubular structure. Coiled carbon nanotubes (CCNTs) are quite unique. The structure of coiled carbon nanotubes consist of cylindrically-shaped walls [30, 31]. **Figure 6** present the structures of classified carbon nanotubes. Iijima and colleagues discovered those carbon molecules in 1991 [30–33]. That discovery was very important because of its potential use for hydrogen storage. Consequently, strengths and weaknesses for the current research will be discussed further in this section [30–33].

The strengths of carbon nanotubes are similar to those of the metal-organic frameworks discussed previously regarding hydrogen energy [30]. One strength is the low density of the material. Another strength is the large surface area on the CNTs [30, 31]. For example, Gayathri and colleagues demonstrated that the ring-coiled structure (known as Y-configuration) of a sample could absorb up to 6.8 wt% hydrogen due to its large surface area [30].

A weakness or shortcoming of the latter hydrogen material is the need for transitional metals to increase adsorption. Wu and his research group used Pd to increase the adsorption for hydrogen from 4 wt% to about 4.5 wt%, which is 0.5 wt% closer to DOE's goal for hydrogen storage [31]. Another shortcoming of the CNTs is the need for activating agents due to the high stability of the compound [29–31]. For example, to increase the adsorption with the Pd, Wu and his research group used potassium hydroxide, KOH, in order to activate/ destabilize for proper experiments [29–31].

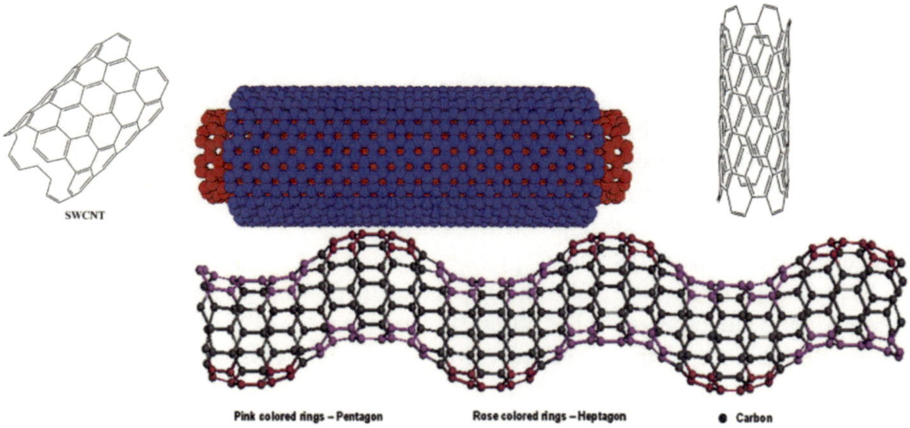

Figure 6.
Different types of carbon nanotubes analyzed for hydrogen storage [29].

2.7 Chemisorption and its systems

Chemically bound storage or chemisorption is a process in which hydrogen is adsorbed and desorbed within the storage material. Over the years, numerous research investigations into hydrogen storage materials have revealed various practical methods for storing hydrogen chemically [22, 32–36].

3. Metal hydrides

Metal hydride systems present several benefits for data center applications, particularly in backup power systems. A primary advantage is their operation at considerably lower pressures compared to other hydrogen storage techniques, thereby mitigating high-pressure safety concerns associated with the technology. Additionally, these systems operate predominantly at ambient temperatures during hydrogen storage. In the absence of backup power operation, there is no requirement for continuous heating and cooling systems, enabling hydrogen to be stored over extended periods without degradation [22, 32–36].

While these benefits are significant, there are some drawbacks. The primary concern is the kinetics of the materials currently used. According to Celestine et al., the rate of hydrogen absorption and desorption in these materials can be slow compared to compressed or liquid hydrogen. This is due to complex interactions between the metal and hydrogen, which will be discussed in the next section. Another disadvantage is the cost of metal hydrides, as they are scarce globally due to the reliance on rare-earth metals. However, ongoing research is exploring alternative metal hydrides that are more abundant and readily available for potential hydrogen storage solutions in the market. Because of this, metal hydrides remain a backup power option, particularly for the emerging data center industry worldwide [22, 32–36].

3.1 Foundation of metal hydrides

Metal hydride storage focuses on the ability of metallic compounds to absorb and desorb hydrogen atoms into their crystalline structure. This is different from the physisorption process because the absorbed hydrogen occupies the interstitial sites within the crystal structure. This creates a solid-state hydrogen storage solution that can store hydrogen at energy densities higher than liquid hydrogen [32, 33].

The thermodynamics of charging and discharging hydrogen from a metal follow well-established principles based on pressure-composition isotherms (PCIs). **Figure 7** illustrates what a PCI typically looks like during hydriding (charging) and dehydriding (discharging) of hydrogen from a metal. The straight line, known as a thermodynamic plateau, represents a two-phase region where the metal hydride and the metal coexist. This plateau also corresponds to flat voltage profiles, making metal hydrides ideal for applications that require stable power delivery, such as backup power for data centers [22, 32].

According to various research studies, the most stable and commercially adopted metal hydrides belong to the AB5 family, which includes LaNi5 and other rare-earth nickel metal hydrides. The crystal structure of these materials has large interstitial cavities that can hold hydrogen atoms while maintaining its lattice structure through absorption and desorption cycles. For example, for LaNi5, the lanthanum atoms provide the primary hydrogen storage site. The nickel atoms facilitate the movement of hydrogen atoms in and out of the lattice structure via the diffusion process [34–36].

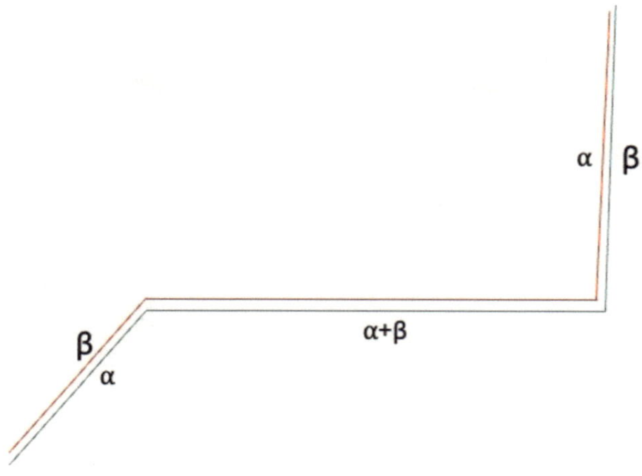

Figure 7.
Pressure-composition isotherm curve in relation to hydrogen storage.

Due to the commercial success of those metal classes, various other metal alloys are being developed to improve the hydrogen storage materials market for different applications. **Figure 8** shows the different alloys designed as hydrogen storage systems [37]. These compositional changes demonstrate the impressive flexibility of metal hydride chemistry. The kinetics of hydrogen absorption and desorption in metal hydrides involve complex, reaction-controlled, and solid-state diffusion processes. Having established an understanding of metal hydrides, the discussion can now move to the proposed use of metal hydrides for backup power in AI data centers. The chemistry of hydrogen absorption and desorption inherent to this unique class of materials

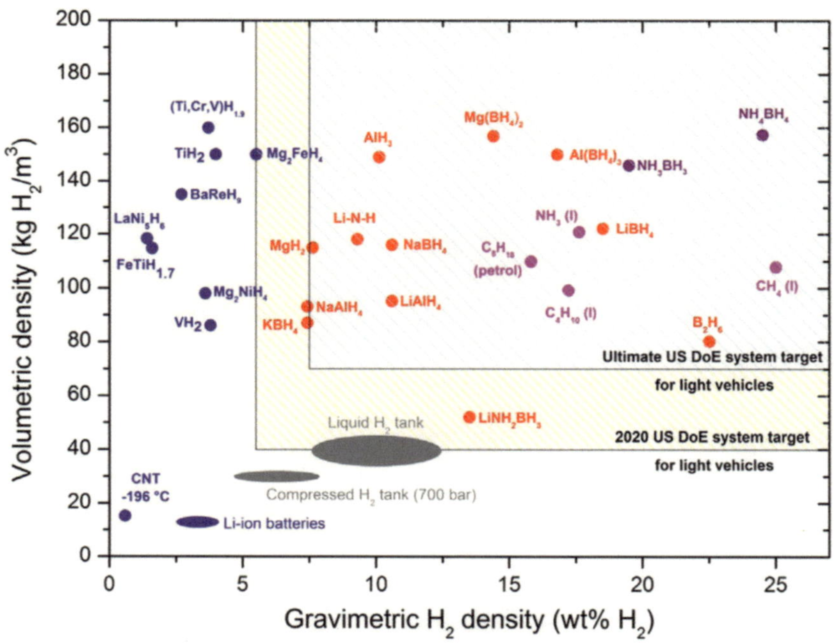

Figure 8.
Graph of metal hydrides considered for hydrogen storage for the US DOE targets for 2020 [37].

offers natural advantages that align very well with the characteristics of AI workloads, and the reliability needs crucial for critical computing infrastructure.

4. Applications for data center

For data center infrastructure and environments, several key chemical factors are crucial for optimal performance of the metal hydride alloys [37]. Throughout this section, we will discuss these factors and provide comprehensive examples based on current literature. Each factor is currently being refined through ongoing research and by modifying different alloy compositions.

4.1 Activation behavior

Activation behavior is truly key for metal hydrides to be fully optimal for data center applications. Freshly synthesized metal alloys require an initial activation process to reach full hydrogen storage capacity. This process involved the fracturing of alloy particles during the first volume expansion of hydrogenation, creating a fresh open surface area [21, 38]. **Figure 9** shows the process of hydrogenation and dehydrogenation on the metal lattice structure [39].

4.2 Plateau pressure characteristics

The plateau pressure characteristics are crucial for understanding the cycling duration of metal hydrides in data center applications. The plateau pressure can be identified in PCIs. **Figure 8** shows the basic concept of a PCI curve and the multiphase within the plateau pressure. For optimal performance, the hydrogen equilibrium should remain stable throughout the temperature range for both hydride

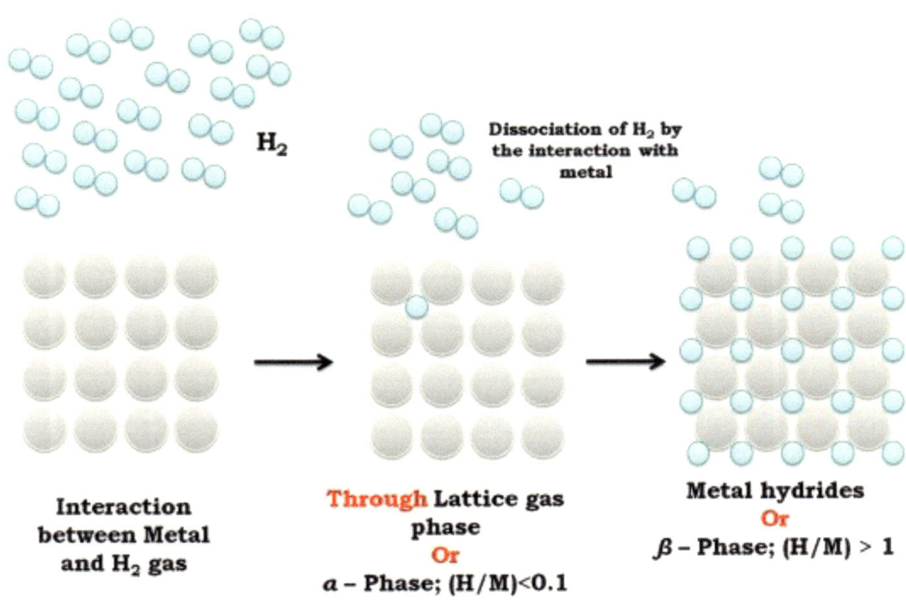

Figure 9.
Process of hydrogenation and dehydrogenation [39].

and dehydride forms of the metal alloy materials. This ensures that there is sufficient driving force for rapid charge and discharge for applications. If there is any hysteresis (a gap between the PCIs of charging and discharging), then the kinetics may be slower than predicted. In this case, catalysts with the metal hydride can be added to improve the process [19, 20, 23–32].

4.3 Thermal stability

Thermal stability consists of whether the metal hydride alloys can handle various temperatures within the data center environment. These can be the cooling facilities or equipment heat loads that the computing hardware can give off. Overall, the decomposition temperature of the metal hydride must exceed the maximum expected operating temperature by a sufficient margin to prevent hydrogen loss during operation. For example, lithium borohydride, LiBH4, a metal alloy, has a decomposition temperature of 450 degrees Celsius for the material degrades [Source]. If the data center is operating at 27 degrees Celsius, then metal hydride is safe due to its decomposition temperature being higher than 27 degrees (400 degrees) [31, 40].

4.4 Electrochemical performance

The interfacial chemistry between the metal hydride electrode and the alkaline electrolyte is crucial for the performance of the electrochemical cell for data center applications. The reason is due to the interface governs the kinetics of hydrogen absorption and desorption as well as the long-term stability for the cell before it becomes oxidized [34].

Figure 10 shows the electrochemical cell with the metal hydride. The electrode reaction in metal hydride systems involves a reversible formation of hydrogen at the metal hydride surface. During discharge, absorbed hydrogen atoms react with hydroxide ions to form water and release electrons. The reaction mechanism differs fundamentally from lithium-ion battery systems, providing various advantages for structural stability and overall electrochemical cell life [34].

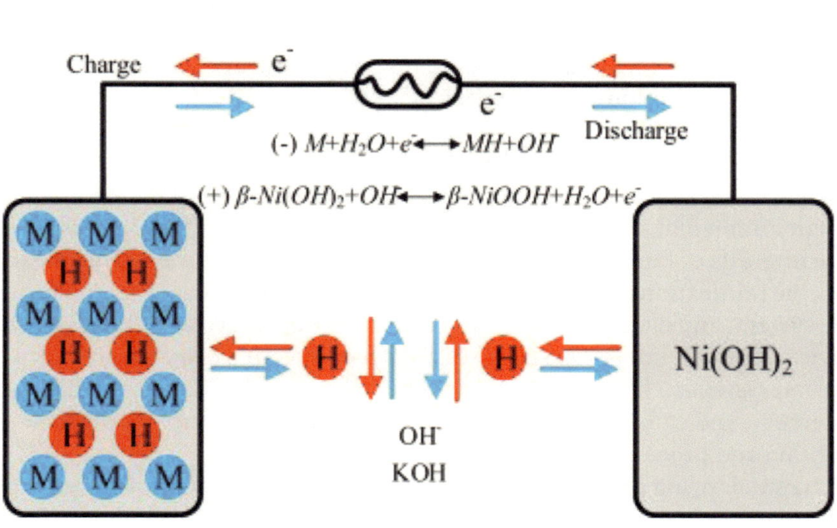

Figure 10.
Nickel metal hydride electrochemical cell [41].

4.5 Cycle stability

Cycle stability for the metal hydride systems is based on the mechanical stability of the metal hydride particles under repeated volume expansion and contraction of the lattice structure during hydrogen absorption and desorption processes. The change in volume within the lattice structure associated with hydrogen absorption can reach 25–30% in some metal hydride systems creating potential mechanical stresses that affect particles and electrical contact [36]. The proposed solution will be to prepare advanced alloys compositions and electrode techniques to minimize degradation over time [22, 36].

With this solution and the characteristics of various metal hydrides such as rare-earth metals, metal hydride systems are shown to have very low discharge rates due to thermodynamic stability of absorbing and desorbing hydrogen overtime compared to current lithium-ion battery systems in the market. Due to this characteristic, this ensures that backup power systems based on metal hydride systems can remain at full capacity even after extended times without cycling [22, 36].

4.6 Thermal management and safety characteristics

Metal hydrides are very unique systems compared to other energy storage systems that are currently in the market regarding thermal management. The thermodynamics of hydrogen absorption and desorption can create opportunities and advantages for integrated thermal management design for backup power systems while maintaining storage performance. This is due to the enthalpy values for typical metal hydrides. According to various studies, metal hydrides enthalpy (or heats of formation) values ranges from 25 to 35 kJ/mol. [Source]. This represents a significant thermal energy source that can be captured and utilized for other applications. For example, waste heat from IT computing can be used to assist in optimizing discharging performance of metal hydride systems thus creating a thermal coupling between the metal hydride storage and primary data center loads [40, 42–46].

For safety characteristics, metal hydride has shown to be major advantage compared to alternative storage technologies. The reason is because the hydrogen is stored in a chemically bound state at low pressures, eliminating major hazards compared to compressed high pressure hydrogen. Also, metal hydrides are considered less toxic compared to current lithium-ion systems [45, 46].

4.7 Environmental and sustainability impact

For environmental and sustainability impact, the metal hydrides aligns well with the interests of building sustainable data centers for the future. The materials used for the fabrication of metal hydrides such as rare-earth metals, complex metal hydride system, and elemental hydride systems are fully recyclable, biodegradable, and environmentally sounds based on established processes respectively compared to lithium-ion systems. These materials as well as materials that do not contain such as lead, mercury, and cadmium also eliminates environmental concerns unless it contains mining and processing [8, 9].

For manufacturing processes, the metal hydride has well-understood environmental impacts and guidelines from the conventional metallurgical techniques. Based on the current cost structure for energy-service life, metal hydride storage systems are comparable to current energy storage methods and can become cheaper over time due to the extended service life of the metal hydride. In addition, the carbon footprint for maintaining metal hydride systems is considered relatively low compared to other methods due to their low safety profile and minimal environmental impact [8, 9].

5. Overall integration with data center infrastructure

For consideration of integrating metal hydride backup power systems in the data centers, there needs to be successful chemical and operational compatibility factors. The characteristics of metal hydride storage systems have advantages for enhanced integration and allows modular design for specific requirements. The modularity of metal hydride systems can be scalable deployment which accommodates the various designs of data center across the world [47, 48]. However, the difference between the traditional battery storage profiles and hydrogen storage profiles are very crucial for monitoring and control systems. For batteries, the monitoring profiles are just voltage and current while metal hydrides have to observe voltage, current, and hydrogen pressure, temperature, and at times electrolyte conditions. These extra parameters are crucial in providing overall system power and resiliency for the data center. For the thermal management, the thermodynamic conditions of metal hydrides can be leveraged to complement existing HVAC systems and other heating cooling systems to assist in reducing overall facility energy consumption [18, 47, 48].

5.1 Economic analysis and cost optimization

The economic viability of metal hydride backup power systems for data center applications demonstrates significant advantages when evaluated through total cost of ownership rather than initial capital expenditure alone. Recent techno economic analyses indicate that metal hydride systems can achieve competitive storage costs of $0.38-0.45/kWh, making them cost-competitive with compressed gas storage systems, while industrial-grade metal hydride systems demonstrate exceptional cycle life reaching 1500 cycles compared to consumer applications. The inherent safety characteristics and wide temperature operational range of nickel-metal hydride systems, combined with their double energy density compared to lead-acid batteries, provide substantial value propositions for data center operators facing increasing demands for reliable, long-duration backup power [4, 18].

The maintenance and operational cost advantages of metal hydride systems become particularly compelling when considered over extended deployment periods typical of critical data center infrastructure. Metal hydride batteries can achieve up to 500 charge cycles at 100% depth of discharge with specific power capabilities up to 1000 W/kg, while exhibiting significantly reduced thermal runaway risks with peak temperatures of only 80°C compared to lithium-ion systems reaching 240°C under abuse conditions. The recyclability of metal hydride materials provides additional economic benefits through end-of-life value recovery, as the rare earth elements, nickel, and other metals retain significant market value that can offset initial system investments. Cost reduction opportunities continue to emerge through advanced alloy development and manufacturing scale improvements, while the integration potential with data center thermal management systems creates synergistic value that can enhance overall facility efficiency and reduce operational expenses [4, 18].

6. Future developments and research directions

Though there are some companies that are focused on metal hydride systems in the markets, there are still some developments emerging in this class of storage. The field of metal hydrides will continue to evolve rapidly based on the advances of material science and the demand of sustainable energy storage systems. For example, the computational chemistry techniques are being revolutionized to look at the discovery of energy material compositions specifically for metal hydrides [48–50].

A novel direction of nanostructured metal hydrides continues to show promise for enhance performance of other types of hydrogen storage systems such as the metal organic frameworks. Advanced synthesis of mechanical alloying, chemical vapor deposition, and electrochemical deposition enables precise control over morphology, surface chemistry, and particle size which can enable even better metal hydride storage systems in the near future [48–50].

Hybrid storage developments and studies continue to emerge such as PV-hydrogen systems and wind-hydrogen storage system for the advancement of the overall sustainability and renewable energy markets. This will also be key for modern hyperscale data center that will emerge in the near future [48–50].

7. Conclusion

Metal hydride energy storage technology represents a compelling solution for enhancing data center resilience in the era of artificial intelligence. The unique chemistry of hydrogen absorption and desorption in crystalline metal lattices provides inherent advantages that align exceptionally well with the demanding requirements of AI workloads and data center operations.

The fundamental chemical properties of metal hydrides—including their flat thermodynamic plateaus, excellent cycle stability, and robust thermal characteristics—translate directly into superior performance for backup power applications. Unlike alternative energy storage technologies that may struggle with the dynamic power profiles and stringent reliability requirements of AI applications, metal hydride systems thrive under these demanding conditions.

The safety and environmental characteristics of metal hydride systems address growing concerns about the sustainability and safety of data center operations. The absence of toxic materials, the inherent safety of the hydrogen storage mechanism, and the excellent recyclability of system components align with corporate sustainability goals while maintaining the performance standards required for mission-critical applications.

Perhaps most importantly, the chemical stability and longevity of metal hydride systems provide data center operators with backup power infrastructure that can reliably support their operations for decades rather than years. This extended service life reduces both operational complexity and total cost of ownership while providing the stability needed to support long-term strategic planning.

The continued evolution of metal hydride chemistry, driven by advances in materials science and computational discovery techniques, promises further enhancements in performance and cost-effectiveness. As AI workloads continue to grow in complexity and importance, metal hydride backup power systems offer a chemical foundation for data center resilience that can evolve and adapt to meet future challenges.

The integration of metal hydride technology into data center infrastructure represents more than just a technological upgrade—it constitutes a strategic investment in chemical energy storage that can provide reliable, sustainable, and cost-effective backup power for the intelligent future. As data center operators face increasing pressure to enhance both reliability and sustainability, metal hydride systems offer a proven chemical solution that addresses both imperatives simultaneously.

Author details

Jalaal A. Hayes
Chemistry, Lincoln University of PA, Lincoln University, PA, USA

*Address all correspondence to: jalaalhayes@eviince.com

References

[1] International Energy Agency. Energy and AI Analysis [Internet]. 2025. Available from: https://iea.blob.core.windows.net/assets/601eaec9-ba91-4623-819b-4ded331ec9e8/EnergyandAI.pdf [Accessed: August 01, 2025]

[2] Lei N, Masanet E. Climate and technology-specific PUE and WUE estimations for U.S. data centers using a hybrid statistical and thermodynamics-based approach. DOI: 10.1016/j.resconrec.2022.106323

[3] Brady G, Kapur N, Summers J, Thompson HM. A case study and critical assessment in calculating power outage usage effectiveness for a data centre. Energy Conversion Management. 2013;**76**:155-161. DOI: 10.1016/j.enconman.2013.07.035

[4] Wang Y, Han Y, Shen J, Wang B, Li X, Zhang Y, et al. Data center integrated energy systems: Generalization, approaches, methods, techniques, and future perspectives. The Innovation Energy. 2024;**1**:100014. DOI: 10.59717/j.xinn-energy.2024.100014

[5] Amin SMM, Hossain N, Lipu MSH, Urooj S. Development of a PV/battery micro-grid for a data center in Bangladesh: Resilience and sustainability analysis. Sustainability. 2023;**15**:15691. DOI: 10.3390/su152215691

[6] The Global E-Waste Statistics Partnership. Global E-Waste Monitor 2024 [Internet]. 2024. Available from: https://api.globalewaste.org/publications/file/297/Global-E-waste-Monitor-2024.pdf [Accessed: August 01, 2025]

[7] Diahovchenko IM, Huseynov M, Karpenko A, Trypolska G. Dynamic microgrids for distribution-system resiliency. Electric Power Systems Research. 2023;**221**:109417. DOI: 10.1016/j.epsr.2023.109417

[8] Solar Energy Association. Why America's AI Leaders are Pumping Billions into Solar+ Storage. 2025. Available from: https://seia.org/blog/ai-leaders-are-pumping-billions-into-solar-storage/ [Accessed: April 15, 2025]

[9] United States Department of Energy. First Made-in-USA Solar Plant Set to Power Google Data Centers, Boosting Local Economy and American Manufacturing. 2025. Available from: https://www.energy.gov/policy/articles/first-made-usa-solar-plant-set-power-google-data-centers-boosting-local-economy-and [Accessed: July 26, 2025]

[10] Agarwal A, Sun J, Noghabi S, Iyengar S, Badam A, Chandra R, et al. Redesigning data centers for renewable energy. In: Proceedings of the Twentieth ACM Workshop on Hot Topics in Networks. Virtual Event, United Kingdom. Association for Computing Machinery. Nov 2021. pp. 45-52. DOI: 10.1145/3484266.3487394

[11] Cassava Technologies. Africa Data Centers Turns to Solar Solutions in South Africa and Nigeria. 2021. Available from: https://www.cassavatechnologies.com/africa-data-centres-turns-to-solar-solutions/ [Accessed: August 04, 2025]

[12] Liang Y, Li P, Su W, Li W, Xu W. Development of green data center by configuring photovoltaic power generation and compressed air energy storage systems. Energy. 2024;**292**:130516. DOI: 10.1016/j.energy.2024.130516

[13] Gao V, Zeng Z, Liu X, Kumar PR. The answer is blowing in the wind: Analysis of powering internet data centers with wind energy. In: IEEE Annual Joint Conference: INFOCOMM, IEEE Computer and Communications Societies; 14-19 April 2013; Turin, Italy. pp. 520-524. Klopci. DOI: 10.1109/INFCOM.2013.6566827

[14] Zhang Y, Li H, Wang S. Decarbonizing data centers through regional bits mitigation: A comprehensive assessment of China's "eastern data, western computing" initiative and its global implications. Applied Energy. 2025;**392**:126020. DOI: 10.1016/j.apenergy.2025.126020

[15] Kurtz J, Ma Z, Saur G, Wrubel JA, Mount R, Hammond S. Analysis of hydrogen infrastructure for the feasibility, economics, and sustainability of a fuel cell powered data center. Sustainable Energy Technologies and Assessments. 2023;**57**:103257. DOI: 10.1016/j.seta.2023.103357

[16] Xie Y, Cui Y, Wu D, Zeng Y, Sun L. Economic analysis of hydrogen powered data centers. International Journal of Hydrogen Energy. 2021;**46**:27841-27850. DOI: 10.1016/j.ijhydene.2021.06.048

[17] Klopcic N, Grimmer I, Winkler F, Sartory M, Trattner A. A review on metal hydride materials for hydrogen storage. International Journal for Hydrogen Energy. 2023;**72**:108456. DOI: 10.1016/j.est.2023.108456

[18] Nemukula E, Mtshali CB, Nemagwele F. Metal hydride for sustainable hydrogen storage: A review. International Journal of Energy Research. 2025:630225. DOI: 10.1155/er/6300225

[19] Kapischke J, Berger A, Fieg G. Measurement of pressure-composition isotherms of high temperature and low temperature metal hydrides. Experiential Thermal and Fluid Science. 1998;**18**:70-81. DOI: 10.1016/S0894-1777(98)10007-9

[20] Wolf RJ, Yao JM, Pierotti RA, Kunc JA. Pressure-composition isotherms for palladium hydride: Monte Carlo calculations. Physical Review B. 1993;**48**:12415-12426. DOI: 10.1103/PhysRevB.48.12415

[21] Liang G, Huot J, Schulz R. Hydrogen storage properties of mechanically alloys LaNi5-based materials. Journal of Alloys and Compounds. 2001;**320**:133-139. DOI: 10.1016/S0925-8388(01)00929-X

[22] Zhu S, Zhang X, Li H, Wang Z, Wu X. Long-term H2 absorption/desorption properties of an AB5- type LaNi4.75MnO.25. Material Science and Engineering B. 2020;**262**:114777. DOI: 10.1016/j.mseb.2020.114777

[23] Suh MP, Park HJ, Prasad TK, Lim DW. Hydrogen storage in metal-organic frameworks. Chemical Reviews. 2012;**112**:782-735. DOI: 10.1021/cr200274s

[24] Murray LJ, Dinca M, Long JR. Hydrogen storage metal-organic frameworks. Chemical Society Reviews. 2009;**38**:1294-1314. DOI: 10.1039/B802256A

[25] Rosi NL. Design, Synthesis, and Control of Metrics, Functionality, and Interpenetration in Metal-Organic Frameworks and Their Application in Hydrogen Storage [Dissertation]. Omar M. Yaghi: University of Michigan; 2003

[26] Hu YH, Zhang L. Hydrogen storage in metal organic frameworks. Advanced Materials. 2010;**22**:E117-E130. DOI: 10.1002/adma.200902096

[27] Langmi HW, Ren J, North BC, Mathe MK. Hydrogen storage in

metal organic frameworks: A review. Electrochimica Acta. 2014;**128**:368-392. DOI: 10.1016/j.electacta.2013.10.190

[28] Kassaoui MEL et al. Enhancement of hydrogen storage of metal organic framework-5 by substitution (Zn, Cd, Mg) and decoration (Li, Be, Na). International Journal of Hydrogen Energy. 2021;**46**:26426-26436. DOI: 10.1016/j.ijhydene.2021.05.107

[29] Furukawa H, Cordova KE, O'Keefe M, Yaghi OM. The chemistry and applications of metal-organic frameworks. Science. 2013;**341**:1230444. DOI: 10.1126/science.1230444

[30] Liu Y, Pan H, Gao M, Li R, Wang Q. Intrinsic/extrinsic degradation of Ti-V-based hydrogen storage alloys upon cycling. Journal of Physical Chemistry C. 2008;**112**:16682-16690. DOI: 10.1021/jp8005052

[31] Lamb KE, Webb CJ. Core-shell, critical-temperature-suppressed V alloy-Pd alloy hydrides for hydrogen storage - A technical evaluation. Molecules. 2023;**28**:3024. DOI: 10.3390/molecules28073024

[32] Goto K, Hirata T, Yamamoto I, Nakao W. Suitability evaluation of LaNi5 as hydrogen-storage-alloy actuator by in-situ displacement measurement during hydrogen pressure change. Molecules. 2019;**24**:2420. DOI: 10.3390/molecules24132420

[33] Strozi RB, Ivanisenko J, Koudriachova N, Huot J. Effect of HPT on first hydrogenation of LaNi5 metal hydride. Energies. 2021;**14**:6710. DOI: 10.3390/en14206710

[34] Shen Y, Zhang F, Ma H, Zhou Y, Wang Y, Wang X, et al. Electrochemical and kinetic performance of low-cobalt and cobalt-free rare-earth AB5-type hydrogen storage alloys. Materials. 2025;**18**:3317. DOI: 10.3390/ma18143317

[35] Zurek M, Zajac W. Investigation of the hydrogen absorption by the LaNi5 and LaNi4.75Pb0.25 alloys using a statistical physics model. Hydrogen. 2022;**3**:361-378. DOI: 10.3390/hydrogen3030022

[36] Zhang X, Li J, Zhao Y, Zhang G, He X, Bai J, et al. Cycle decay mechanism of AB5-type hydrogen storage alloy for metal hydride hydrogen compressor. Journal of Rare Earths. 2025. DOI: 10.1016/j.jre.2025.04.026

[37] Afzal M, Sharma P. Design of a large scale metal hydride based hydrogen storage reactor: Simulation and heat transfer optimization. International Journal for Hydrogen Energy. 2018;**43**:13356-13372. DOI: 10.1016/j.ijhydene.2018.05.084

[38] Tong L, Hu X, Yuan C, Yang F, Li Y. Thermal management of metal hydride hydrogen storage tank coupled to PEM fuel cells. Case Studies in Thermal Engineering. 2023;**43**:102812. DOI: 10.1016/j.csite.2023.102812

[39] Kukkapalli VK, Kim S, Thomas SA. Thermal management of techniques in metal hydrides for hydrogen storage: A review. Energies. 2023;**16**:3444. DOI: 10.3390/en16083444

[40] Couturier K, Joppich F, Wörner A, Tamme R. Tank design for on-board hydrogen storage in metal hydrides. In: Proceedings of the ASME 2008 2nd International Conference on Energy Sustainability Collocated with the Heat Transfer, Fluids Engineering, and 3rd Energy Nanotechnology Conferences; 10-14 August 2008; Jacksonville, Florida, USA. ASME; 2008. pp. 517-524. DOI: 10.1115/ES2008-54031

[41] Mehrpooya M, Zhalehrajabi E. Thermal management in hydrogen storage tanks using metal hydrides and phase change materials. Industrial and Engineering Chemistry Research. 2024;**63**:21860-21874. DOI: 10.1021/acs.iecr.4c03638

[42] Zhu J, Lin X, Lv L, Li M, Luo Q, Kudiiarov VN, et al. The relationship between thermal management methods and hydrogen storage performance of the metal hydride tank. Journal of Materials Science and Technology. 2024;**203**:66-77. DOI: 10.1016/j.jmst.2024.03.018

[43] Ambrosino A, Chabane D, Hissel D, Djerdir A, Elkedim O, Sorrentino M, et al. Thermal management of metal hydride hydrogen tank using lattice structures. Journal of Energy Storage. 2025;**131**:117402. DOI: 10.1016/j.est.2025.117402

[44] Shahbazbegia V, Shafie-khah M, Laaksonen H, Strbac G, Ameli H. Power-to-hydrogen resiliency framework. Applied Energy. 2023;**348**:121429. DOI: 10.1016/j.apenergy.2023.121429

[45] Yu J, Li T, Pan J. Optimizing hydrogen integrated micro CHP systems with demand response for sustainable energy management. Journal of Building Engineering. 2025;**99**:111625. DOI: 10.1016/j.jobe.2024.111625

[46] Verma A, Joshi K. MH-PCTpro: A machine learning model for rapid prediction of pressure-composition-temperature (PCT) isotherms. iScience. 2025;**28**:112251. DOI: 10.1016/j.isci.2025.112251

[47] Tong L, Yuan C, Yang T, Yuan Y. Thermal management of metal hydride hydrogen storage tank coupled with proton exchange membrane fuel cells. Case Studies in Thermal Engineering. 2023;**43**:102812. DOI: 10.1016/j.csite.2023.102812

[48] Moller KT, Jensen TR, Akiba E, Li H. Hydrogen- A sustainable energy carrier. Progress in Natural Science: Materials International. 2017;**27**:34-40. DOI: 10.1016/j.pnsc.2016.12.014

[49] Yuvaraj AR, Jayarama A, Sharma D, Nagakar SS, Duttagupta SP, Pinto R. Role of metal-organic framework in hydrogen gas storage: A critical review. International Journal of Hydrogen Energy. 2024;**59**:1434-1458. DOI: 10.1016/j.ijhydene.2024.02.060

[50] Wang S, Xie Y, Guerrero JM. Market batteries and their characteristics. In: Farhad S, Gupta RK, Yasin G, Nguyen TA, editors. Nanotechnology for Battery Recycling, Remanufacturing, and Reusing. DOI: 10.1016/B978-0-323-91134-4.00010-8

Chapter 9

Hydrogen Storage: Challenges, Technologies, and Potential Applications

Renata Pereira Lopes Moreira, Noemí Cristina Silva de Souza and Marcela de Oliveira Brahim Cortez

Abstract

Hydrogen is widely considered the fuel of the future due to its superior properties compared to other fuels, such as gasoline, especially in terms of energy density by mass. Its combustion produces only water vapor, making it a highly attractive alternative for mitigating climate change. Based on the amount of CO_2 emitted during the production process, hydrogen is classified by color, such as green, blue, or gray. However, significant challenges remain regarding its storage and transportation. In this context, the use of chemical carriers has emerged as a promising alternative. In recent years, public policies around the world have increasingly supported the development of hydrogen technologies. As a result, the global use of hydrogen is expected to grow substantially in the coming decades. This chapter discusses the main storage methods and technological applications of hydrogen, with a focus on the most innovative perspectives for its large-scale use.

Keywords: hydrogen storage, elementary hydrides, intermetallic hydrides, complex hydrides, liquid organic hydrogen carriers

1. Introduction

Hydrogen is the most abundant element in the universe and exhibits singular chemistry, which explains why it is not assigned to a specific group in the Periodic Table. It takes part in a wide range of organic and inorganic compounds. Hydrogen (H) takes part in a wide range of organic and inorganic compounds. In organic chemistry, hydrogen is present in hydrocarbons (e.g., methane, CH_4), alcohols (e.g., ethanol, C_2H_5OH), carboxylic acids (e.g., acetic acid, CH_3COOH), and in vital biomolecules, such as proteins, carbohydrates, lipids, and nucleic acids (DNA and RNA), where it plays a crucial role in stabilizing molecular structures through hydrogen bonding.

In inorganic chemistry, hydrogen occurs in numerous substances, including water (H_2O), hydrochloric acid (HCl), sulfuric acid (H_2SO_4), nitric acid (HNO_3), and metal

 IntechOpen

hydrides (e.g., LiH). Its ability to form covalent bonds with a wide variety of elements, along with its participation in intermolecular interactions, makes hydrogen a fundamental component in both biological systems and numerous industrial and technological applications.

Hydrogen naturally exists as a diatomic molecule, H_2, in which two hydrogen atoms are covalently bonded. This homonuclear diatomic molecule is stabilized by the equal sharing of electrons between atoms of identical electronegativity. One of the key features of hydrogen is its versatility. It can be produced from multiple sources, including water electrolysis, biomass, natural gas reforming, and other renewable or nonrenewable processes. Another important characteristic is its high gravimetric energy density. Moreover, its combustion generates only water vapor, producing no harmful emissions that contribute to climate change, thereby positioning hydrogen as a promising clean energy carrier for a sustainable future.

Global demand for pure hydrogen has grown substantially over the past decades, rising from less than 20 million tons (Mt) in 1975 to over 70 Mt. in 2018 [1]. According to the International Energy Agency (IEA), demand reached approximately 97 Mt. in 2023, marking an increase of around 2.5% compared to the previous year [2]. It is estimated that hydrogen energy will meet approximately 11% of the global energy demand by 2025, increasing to 34% by 2050 [3].

The use of hydrogen as a fuel has been referenced at three key historical moments: During the oil crisis of the 1970s; in the 1990s and 2000s, driven by growing concerns over climate change and peak oil scenarios; and more recently, in the context of the global energy transition [1]. This renewed interest is largely due to hydrogen's key characteristics, such as its high energy density, along with recent advances in its production, storage, and transportation.

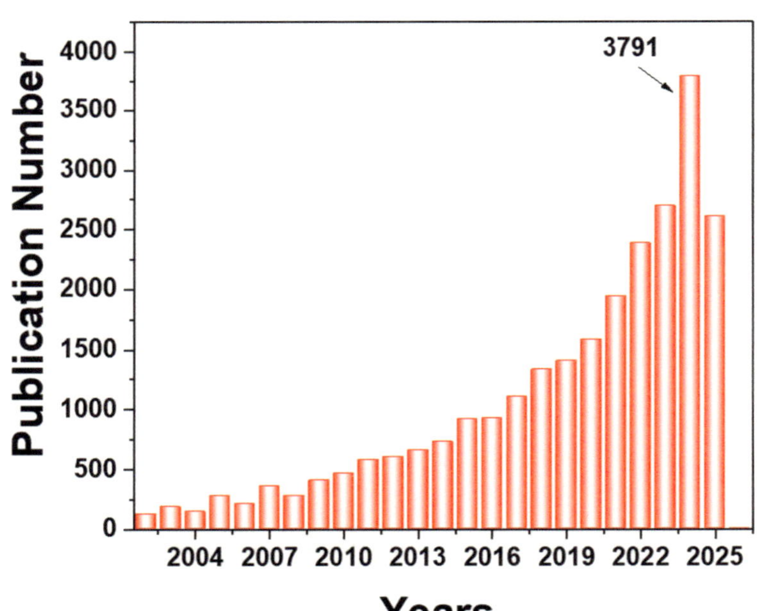

Figure 1.
Annual number of publications on hydrogen, storage, and applications retrieved from the Web of Science database.

A search in the Web of Science database using the keywords hydrogen, storage, and applications resulted in the graph shown below. A progressive increase in the number of publications over the years can be observed, reflecting the growing interest of the scientific community in this field. In 2024 alone, more than 3700 papers were published on the topic (**Figure 1**).

According to the review by Cevahir Tarhan and Mehmet ali Çil, it is estimated that by 2040, 35% of the vehicle fleet in Europe will be powered by hydrogen. Excluding the United States, there are currently 401 hydrogen refueling stations in operation worldwide, with another 159 planned for construction. In the United States, there are currently 46 stations in operation [3].

2. Hydrogen: Properties, safety, and regulations

Some properties of hydrogen (H_2) are presented in **Table 1**. It can be observed that the density of gaseous H_2 is approximately one-third that of natural gas, while, in its liquid form, it is about one-sixth. Its boiling point is roughly 90°C lower than that of Liquefied Petroleum Gas (LPG). The lower density and boiling point require more critical and stringent storage conditions. However, in terms of gravimetric energy density, hydrogen has a value approximately three times higher than that of gasoline, making it highly efficient per unit of mass. Hydrogen is also highly explosive, which demands careful handling.

The laminar flame velocity, the speed at which a flame front propagates under stable, nonturbulent conditions, is about three times greater than that of methane [4]. Additionally, the autoignition temperature, defined as the lowest temperature at which a substance spontaneously ignites without an external ignition source such as a spark or flame, is another important characteristic of hydrogen.

Several accidents have occurred in different parts of the world, such as the Hindenburg airship disaster in 1937, which resulted in 39 deaths. The explosion at the hydrogen plant in Tessenderlo, Belgium, in 1942 caused approximately 190 fatalities. The incident at the hydrogen facility in Norco, Louisiana, USA, in 1988 also had serious consequences. These accidents prompted significant advancements in safety standards, including the development of key guidelines by the International Organization for Standardization (ISO). Notably, the ISO/TR 15916, a technical report published

Property	Value
Color, odor	Colorless, odorless
Toxicity	None, simple asphyxiant
Phase	Gas
Density	Gas: 0.089 g L^{-1}, liquid: 0.07 g cm^3
Melting and boiling point	−259.14 °C, −252.87°C
Gravimetric energy density	142 MJ kg^{-1}
Laminar flame velocity	3.06 m s^{-1}
Auto ignition temperature	585°C

Table 1.
Properties of hydrogen gas. Adapted from (Schlapbach and Züttel, 2001) [4].

in 2015, provides safety recommendations for the storage, handling [5]. Additionally, ISO 16110, published in 2007, outlines the requirements for hydrogen production through various methods, emphasizing process control and operational safety to minimize risks [6].

ISO/TR 15916:2015 establishes safety guidelines for the storage, handling, and use of hydrogen, addressing aspects such as the physical properties of the gas, associated risks, adequate ventilation, leak detection, and ignition prevention. ISO 16110:2007, on the other hand, deals with the requirements for hydrogen production through various methods, emphasizing process controls, operational safety, and measures to minimize risks during gas generation.

3. Hydrogen color classification

Hydrogen can be classified using a color-based system that reflects the different production methods, commonly referred to as the "hydrogen rainbow" [1]. This nomenclature does not relate to the physical or chemical properties of the element, but rather to the environmental impact and energy source used in its production (**Figure 2**).

Green hydrogen, for instance, is produced through water electrolysis powered by renewable energy sources, such as solar or wind. This method emits no carbon dioxide (CO_2) during the production process, making it a sustainable and environmentally responsible alternative to conventional hydrogen production. However, it requires a substantial amount of electrical energy, which remains a major challenge for its large-scale economic and technological viability. Purple hydrogen, on the other hand, is generated using electricity from nuclear power plants to drive the electrolysis process.

In contrast, gray hydrogen, also referred to as brown or black hydrogen depending on the feedstock, is produced *via* the reforming of fossil fuels, primarily natural gas (steam methane reforming) and coal (gasification), processes that release substantial amounts of carbon dioxide (CO_2) into the atmosphere. To mitigate the environmental burden associated with these conventional pathways, blue hydrogen employs the same fossil-based production routes but integrates carbon capture and storage (CCS) technologies, which significantly curb greenhouse gas emissions by sequestering the CO_2 generated during the process. Turquoise hydrogen, on the other hand, is obtained through methane pyrolysis, a thermochemical process that decomposes natural gas into hydrogen and solid carbon, without generating CO_2 as a direct byproduct.

Despite recent progress, further advancements are still needed, as green hydrogen, produced *via* electrolysis with efficiencies ranging from 55–80%, remains expensive, costing between US$ 2.28 and US$ 7.39 per kilogram [7]. In contrast, gray

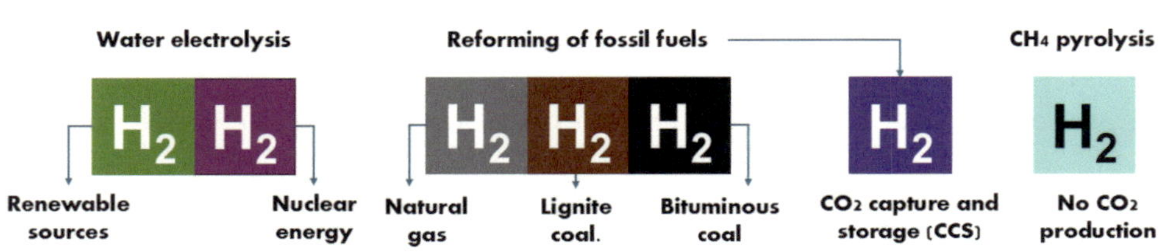

Figure 2.
Color-coded classification of hydrogen types.

hydrogen, which emits approximately 8.5 kg of CO_2 per kg of hydrogen produced, is significantly cheaper, at just US$ 0.67 to US$ 1.31 per kilogram.

4. Hydrogen storage

Given the advantages of hydrogen, one major concern is its storage, due to the inherent risks. Moreover, because hydrogen has a low energy density per unit volume, large storage tanks are required, which poses a significant limitation, along with the danger of explosion. Hydrogen can be stored in different ways, including (1) physical methods such as compression and liquefaction; (2) adsorption onto porous materials; and (3) in the form of chemical carriers, as shown in **Figure 3**.

4.1 Physical storage

Physical storage refers to methods that store hydrogen in its molecular form without altering its chemical composition. These methods typically involve compressing hydrogen gas at high pressures or liquefying it at very low temperatures, enabling efficient storage and transport. Physical storage is a critical aspect for the practical use of hydrogen as an energy carrier.

4.1.1 Liquefaction

The liquefaction process involves converting hydrogen into its liquid state. This requires cryogenic storage, which demands extremely low temperatures, around 20 K [9]. Under these conditions, the density of liquid hydrogen at 20 K and atmospheric pressure is 71.28 kg/m^3, which is significantly higher than that of gaseous hydrogen at 300 K (0.0816 kg/m^3) [10]. However, liquefied hydrogen is recommended only for short-term storage, since its low boiling point may lead to evaporation, reducing storage efficiency [3]. Additional drawbacks include the requirement for extremely low temperatures and the need for storage tanks made of thick insulating materials to

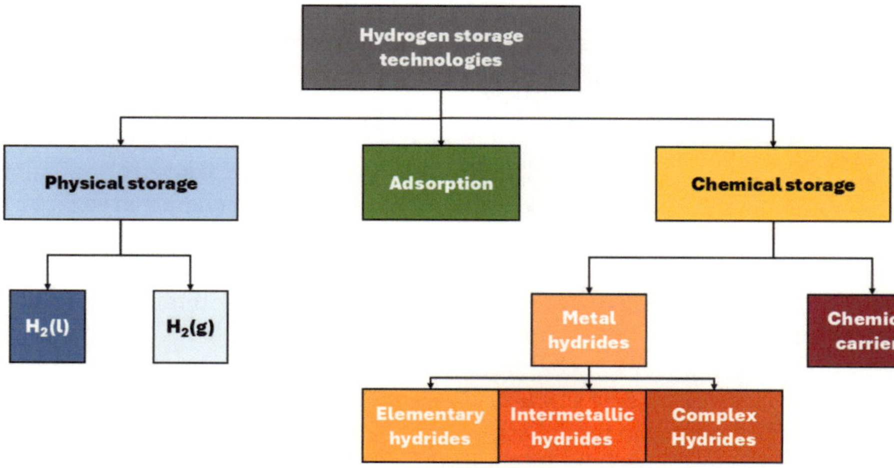

Figure 3.
Hydrogen storage technologies. Adapted from the Ref. [8].

prevent hydrogen loss [9]. Spherical shapes minimize the surface area relative to the stored volume, which reduces thermal losses, a critical factor in cryogenic storage, such as that of liquid hydrogen.

4.1.2 Gasification

Currently, the most used hydrogen storage methods involve gasification and liquefaction processes. Gasification consists of compressing hydrogen to store it in its gaseous state, typically at high pressures ranging from 350 to 700 bar under ambient temperature [11]. Increasing the pressure enhances energy density, as it allows more hydrogen to be stored within a given volume. To ensure efficient storage, both gravimetric and volumetric properties must remain low [12]. Storage tanks must exhibit high mechanical strength due to the applied pressure, chemical inertness to hydrogen to prevent material degradation, and long-term durability. This process offers advantages such as fast and reliable refueling, as well as immediate usability due to its physical state. However, it also has drawbacks: The storage tanks have a limited service life, posing safety risks, including potential leaks and explosions caused by material degradation and wear-related defects [13].

4.2 Adsorption

Hydrogen storage by adsorption involves retaining H_2 in porous solid materials through physical adsorption. This method employs materials with a high specific surface area, such as porous activated carbon, zeolites, covalent organic frameworks (COFs), and metal-organic frameworks (MOFs), to achieve high storage density. The process relies on van der Waals forces between the porous surfaces and hydrogen molecules [14]. Due to the reversibility of physical adsorption, this method has low energy consumption during regeneration [15]. The storage capacity is influenced by the material's structure and porosity. However, in carbon-based materials, it has been observed that a chemical adsorption process also occurs, involving hydrogen dissociation at specific sites within the carbon structure, such as edges or small distinct regions [16].

Hydrogen storage by adsorption into porous materials offers several advantages. Among them, the low energy consumption during regeneration stands out, since physical adsorption is a reversible process that allows hydrogen to be released with minimal energy input. Additionally, it is considered a safer alternative, as hydrogen is stored in molecular form and at lower pressures compared to compressed or liquefied systems, thereby reducing the risk of explosions. Another positive aspect is the wide range of porous materials that can be employed. Although physical adsorption is not efficient at room temperature, it can operate effectively under moderate cryogenic conditions, such as 77 K (the boiling point of liquid nitrogen), which is technically feasible for certain applications. Adsorbent materials also generally exhibit good cyclic stability, allowing them to be reused over multiple cycles without significant loss of performance.

On the other hand, the process has limitations, such as the need to operate at low temperatures to ensure adequate storage capacity. Moreover, the volumetric hydrogen density tends to be lower than that of other methods, such as metal hydrides or compressed hydrogen, which may limit its application in systems that require high energy density per unit volume.

4.3 Chemical storage

A particularly promising avenue lies in the use of chemical hydrogen storage materials, which offer high hydrogen density and the potential for safe, compact storage. These substances incorporate hydrogen in their structure and can release it under controlled and safe conditions. Among the various chemical hydrogen storage options, metal hydrides are particularly noteworthy and can be classified into elemental hydrides, intermetallic hydrides, and complex hydrides. Some information about the different types of storage is presented in **Table 2**.

4.3.1 Metal hydrides

The hydride ion (H^-) consists of one proton and two electrons. It is highly reactive and unstable in its free state due to its low electron affinity and high charge density. It acts as a strong reducing agent and reacts exothermically with protons to form H_2 ($\Delta H = -1676$ kJ/mol) [17]. Hydride ions are commonly found in elemental, intermetallic, and complex hydrides.

4.3.1.1 Elementary hydrides

Elementary hydrides include compounds such as MgH_2, which contains 7.6 wt% of hydrogen and offers a volumetric hydrogen storage density of 110 kg/m^3 [18]. However, it requires high temperatures for hydrogen desorption due to its thermal stability [19, 20]. AlH_3, on the other hand, is a kinetically stable metal hydride with a favorable volumetric hydrogen density of 149 g H_2/L, nearly twice that of cryogenic liquid hydrogen [21]. A key advantage of AlH_3 is its low decomposition temperature, typically below 100°C, which makes it attractive for low-temperature hydrogen release. Nevertheless, the regeneration of both MgH_2 and AlH_3 remains challenging, limiting their practical applications. While AlH_3 is a well-known hydride, its use has been largely restricted to solid rocket propellants rather than hydrogen storage for energy systems. The decomposition of MgH_2 and AlH_3 proceeds according to Eqs. (1) and (2), respectively, leading to the release of H_2 and the formation of metal powder.

$$MgH_2\,(s) \rightarrow Mg\,(s) + H_2\,(g) \qquad \Delta H \approx +75\,kJ\,/\,mol\,H_2 \tag{1}$$

$$AlH_3\,(s) \rightarrow Al\,(s) + 3/2\,H_2\,(g) \qquad \Delta H \approx +10\,kJ\,/\,mol\,H_2 \tag{2}$$

However, the regeneration of these materials is quite complex, which limits their practical application. Although AlH_3 is a well-known metal hydride, its use has been mainly focused on solid rocket propellants rather than as a hydrogen storage material for energy applications.

4.3.1.2 Intermetallic hydrides

Metal alloys are solid solutions formed by two or more metals, in which the composition can vary continuously between the constituent elements, as represented by

Hydrogen storage performance	Physical storage		Adsorption	Chemical storage			
	Gasification	Liquefaction		Elementary hydrides	Intermetallic hydrides	Complex hydrides	Chemical carriers
Volumetric hydrogen storage density (kg H_2/m^3)	7–27	70	20–50	110–150	60–120	80–140	50–70
Gravimetric storage capacity (wt%)	Not applicable	Not applicable	Not applicable	1–7	1–2	5–18	4–18
Operating conditions	Requires high pressures (350–700 bar) RT*.	Requires cryogenic storage (~20 K).	Operates under moderate cryogenic conditions (~77 K).	Requires high temperatures for H_2 release due to high thermal stability.	Operates at moderate temperature and pressure.	Hydrogen can be released *via* hydrolysis (catalytic process) and thermolysis (100–300°C).	H_2 release from formic acid occurs *via* catalytic decomposition.
Notes	Requires tanks with high mechanical strength, chemical resistance, and durability.	High energy cost for liquefaction.	Lower volumetric density than hydrides or compressed H_2, limiting high-energy-density applications.	Complex regeneration limits practical application.	Limited by low gravimetric capacity.	Regeneration of reaction by-products is complex and limits practical use.	Produces CO_2 as a byproduct.

**RT = Room temperature.*

Table 2.
Hydrogen storage technologies.

$A_{1-x}B_x$ ($0 < x < 1$), commonly found among similar metals such as the alkali metals [17]. In contrast, intermetallic compounds have a fixed stoichiometry and an ordered crystal structure, with defined formulas like AB_2 (e.g., YFe_2) or AB_5 (e.g., $LaNi_5$). The formation of alloys or intermetallic imparts new physical and chemical properties to the resulting materials. In applications related to hydrogen storage, intermetallic compounds such as YFe_2 are particularly relevant because they can absorb up to five hydrogen atoms per formula unit, while pure iron does not form stable hydrides. This demonstrates how the combination of metals in intermetallic structures can modify their behavior toward hydrogen, enabling advanced applications in energy technologies.

Intermetallic compounds (IMCs) are often recognized as formers of interstitial hydrides, capable of stable hydrogen absorption and desorption under moderate conditions. During the absorption process, hydrogen atoms occupy interstitial sites within the crystal structure, typically octahedral and tetrahedral positions coordinated by six and four metal atoms, respectively. This arrangement allows the metallic lattice to maintain its structural integrity while accommodating hydrogen [22]. IMCs include various stoichiometric classes, such as AB (e.g., TiNi, TiFe, ZrNi), A_2B (e.g., Zr_2Fe, Ti_2Ni, Mg_2Ni), AB_2 (e.g., ZrV_2, $ZrMn_2$), and AB_5 (e.g., $LaNi_5$) [23]. Element A, typically a hydride-forming metal, combines with element B, which may be either a transition or non-transition metal, to form intermetallic hydrides. Under normal conditions, these hydrides tend to have low hydrogen content, yet they are of interest due to their stability and reversibility.

Alloys derived from $LaNi_5$ exhibit excellent performance in hydrogen storage, with fast and reversible absorption and desorption, operating at moderate pressures (~2 bar) at room temperature. The volumetric hydrogen density in $LaNi_5H_{6.5}$ is comparable to that of gaseous hydrogen at 1800 bar, but with safe release at low pressure. Although the hydrogen content by mass is below 2% due to the heavy elements involved, these alloys are widely used in NiMH batteries, achieving electrochemical capacities of up to 330 mAh/g [4].

4.3.1.3 Complex hydrides

As in elementary hydrides, hydrogen in complex hydrides is covalently bonded to central atoms within complex anions, such as $[AlH_4]^-$ or $[BH_4]^-$ [23]. This behavior contrasts with interstitial hydrides, where hydrogen atoms occupy interstitial sites within a metallic lattice without forming well-defined covalent bonds with specific atoms.

$NaBH_4$ has a high gravimetric hydrogen storage density (10.5 wt%) and can release H_2 through hydrolysis under mild temperature and atmospheric pressure conditions, as described by Eq. (3). However, the hydrogen evolution reaction from $NaBH_4$ exhibits slow kinetics, limiting its practical efficiency. To overcome this, metals such as platinum, cobalt, and nickel are commonly used as catalysts to accelerate the hydrolysis reaction of NaBH4 [24–27].

$$NaBH_4 + 2H_2O \rightarrow NaBO_2 + 4H_2 \tag{3}$$

Ammonia borane (NH_3BH_3) is a chemical compound containing 19.6 wt% hydrogen, exhibiting low toxicity and solubility in water at room temperature. These properties facilitate its hydrolysis and subsequent hydrogen release, as described by Eq. (4). The use of suitable catalysts is also necessary [28].

$$NH_3BH_3 + 4H_2O \rightarrow NH_4B(OH)_4 + 3H_2 \tag{4}$$

Although the use of complex hydrides is highly promising, the regeneration of metaborate, a byproduct of the reaction, remains a challenge to be overcome.

4.3.1.4 Chemical carriers

Liquid organic hydrogen carriers (LOHCs) are compounds that allow reversible hydrogen storage and release in the liquid state, both in their hydrogenated and dehydrogenated forms [29]. First mentioned in the 1980s, they stand out for their high gravimetric and volumetric hydrogen densities, as well as their operational safety and compatibility with existing transportation infrastructure. Early research on liquid organic hydrides for hydrogen storage focused primarily on cycloalkanes, such as cyclohexane, methylcyclohexane, and decalin [30].

Catalysis plays a crucial role in the underlying reactions of LOHC systems. Both the hydrogenation of the unloaded LOHC (H_0-LOHC) and the dehydrogenation of the loaded LOHC (H_n-LOHC) are typically catalyzed processes [31]. Heterogeneous catalysis is particularly beneficial for such storage systems, as it eliminates the need to separate the catalyst from the reaction mixture.

One LOHC that has gained significant attention in recent years is formic acid, particularly due to its renewable origin from biomass [32, 33]. This compound undergoes dehydrogenation, releasing hydrogen in the gas phase. With a hydrogen content of 4.4 wt%, low toxicity, and a liquid state at room temperature, formic acid stands out as a promising material for safe hydrogen transport and storage. The dehydrogenation reaction occurs as described in Eq. (5).

$$\mathrm{HCOOH} \rightarrow CO_2 + \mathrm{H}_2 \tag{5}$$

5. Technological applications of hydrogen

5.1 Aerospace sector

Hydrogen is a highly efficient fuel, with increasing use in aerospace, aviation, and energy applications. Its growing adoption is driven by significant energy and environmental advantages over traditional liquid hydrocarbon fuels [34]. This potential was recognized early in the aerospace sector. The first hydrogen/oxygen rocket engines, RL-10 (68 kN) and J-2 (1020 kN), were developed in the United States in the 1960s, though research began as early as 1944 under the Department of Energy [35]. In the same decade, the Soviet Union also explored the use of hydrogen/oxygen propellants for upper rocket stages, initiating the development of the 11D56 (73.5 kN) and 11D57 (392 kN) engines in 1962 for future upgrades to the N1-L3 rocket system.

5.2 Transportation sector

Hydrogen technology has advanced significantly across various sectors. In the maritime industry, this progress is exemplified by hydrogen-powered submarines, such as Germany's Type 214 [7]. In the automotive sector, the GM Electrovan, developed in 1966, was the first functional hydrogen fuel cell vehicle prototype, although it never reached commercial production. More recently, Toyota introduced the Mirai, a fuel cell-powered vehicle, in 2016. The first hydrogen-powered bus was launched in

1993 by Ballard Power Systems, a Canadian pioneer in fuel cell technology. Developed in collaboration with British Columbia Transit, this vehicle marked the world's first hydrogen fuel cell bus. Gazda-Grzywacz et al. conducted a life cycle assessment (LCA) comparing diesel, electric, and hydrogen busses in a major metropolitan area, focusing on hydrogen produced from coke oven gas, a steel industry byproduct [36]. The study found that hydrogen busses had lower environmental impacts than both electric and diesel busses across most categories, with notable reductions in ozone depletion (−91.4%) and global warming potential (−29.9%) compared to diesel busses.

In the transport sector, hydrogen poses safety risks due to its low ignition energy, wide flammability range, high leak potential, and the use of high-pressure tanks. In the event of a collision, these factors can significantly increase the risk of fire or explosion, underscoring the need for rigorous safety design and impact assessment [37].

5.2.1 Cars

The Toyota Mirai is a mid-size hydrogen fuel cell vehicle produced by Toyota and holds the distinction of being the first fuel cell car mass-produced and sold commercially worldwide. Hydrogen is stored as compressed gas at 700 bar (about 70 MPa) in advanced Type IV tanks featuring an internal polymer liner reinforced with carbon fiber for enhanced mechanical strength. The first-generation Mirai accommodates roughly 5 kg of hydrogen, while the second-generation increased capacity to approximately 5.6 kg, held in two tanks totaling about 122.4 liters. The driving range varies between 500 and 650 kilometers per fill under the Worldwide Harmonized Light Vehicles Test (WLTP) cycle, with refueling times ranging from 3 to 5 minutes [38].

Launched in 2018, the Hyundai Nexo is a hydrogen fuel cell-powered crossover sport utility vehicle unveiled at the Consumer Electronics Show (CES) in January of that year. It succeeded the Hyundai Tucson Fuel Cell Electric Vehicle (FCEV) and serves as Hyundai's flagship eco-friendly model. Hydrogen storage uses Type IV tanks at 700 bar pressure, lined internally with polymer and carbon fiber for safety and durability. The total hydrogen capacity is around 6.3 kg, distributed across three tanks with a combined volume near 156 liters, strategically placed throughout the chassis. Its WLTP-estimated driving range is approximately 609 kilometers, with refueling times between 3 and 5 minutes [39].

Honda's Clarity lineup focused on alternative fuel vehicles, beginning in 2008 with the Honda Fuel Cell eXperimental (FCX) Clarity, the company's first hydrogen fuel cell model. In 2017, the range expanded to include three variants: The Honda Clarity Fuel Cell, successor to the FCX Clarity; the Honda Clarity Electric, a battery-electric vehicle; and the Honda Clarity Plug-in Hybrid. Production ended in August 2021, though U.S. leases for the fuel cell model continued through 2022. This vehicle stored hydrogen at 700 bar in robust tanks and offered an EPA-estimated range of about 589 kilometers, with refueling taking 3–5 minutes [40].

5.2.2 Busses

The Solaris Urbino 18 hydrogen bus, developed with advanced Polish technology, earned the 2025 Bus of the Year award due to its engineering that enables a range of up to 600 kilometers on a single refuel with zero emissions. Its combination of high range, quiet operation, and low environmental impact makes it a sustainable and efficient option for modern public transportation. Solaris, a leading European manufacturer based in Poland, has delivered over 260 hydrogen busses [41].

The Hyundai Elec City FCEV is an urban bus powered by a fuel cell system rated at 180 kW, featuring a hydrogen tank capacity of 875 liters and a high-performance 78.4 kWh battery. It offers a driving range of up to 550 kilometers per charge. The Elec City was exported to Saudi Arabia in 2020 and Austria in 2022. By September 2024, total sales surpassed 1000 units [42].

The Wright StreetDeck, a double-decker bus model built by British company Wrightbus, has recently seen versions powered by hydrogen fuel cells. The StreetDeck Hydrogen is among the first commercially available double-decker hydrogen busses in the United Kingdom.

5.3 Ammonia production

Due to its essential role in fertilizer production, ammonia synthesis stands out as one of the most important chemical processes of the past century [43]. In 1908, Fritz Haber developed a method to synthesize ammonia from N_2 and H_2 under high-pressure and temperature (Eq. 6). Carl Bosch later industrialized the process in 1911, leading to what became known as the Haber-Bosch process. Both were later awarded the Nobel Prize in Chemistry.

$$N_2 + 3H_2 \rightarrow 2NH_3 + 46.1\,\text{kJ mol}^{-1} \tag{6}$$

Currently, about 90% of the hydrogen produced is used for ammonia synthesis, petroleum refining, and methanol production, although this proportion is expected to decrease in the coming years [3]. A large share of this hydrogen is directed toward ammonia synthesis, of which over 96% is currently produced through the Haber-Bosch process [44]. This process relies heavily on fossil fuels (natural gas, oil, and coal) as feedstock for hydrogen production *via* steam reforming. As a result, it accounts for approximately 1.2% of global anthropogenic CO_2 emissions.

To reduce the environmental impact of ammonia production, decarbonization efforts are increasingly focused on replacing fossil-based hydrogen with green hydrogen, produced *via* electrolysis powered by renewable energy sources. In this context, the growing demand for low-carbon solutions is driving the expansion of the green ammonia market. Although conventional hydrogen remains the dominant feedstock due to its lower cost, green and blue hydrogen are gaining traction despite their higher production costs. Similarly, green ammonia is emerging as a sustainable alternative, though still produced in smaller volumes and at a higher cost compared to its conventional counterpart [45].

5.3.1 Green ammonia production

On April 8, 2024, Yara International officially inaugurated its renewable hydrogen plant in Herøya, Norway, marking a significant step toward decarbonizing fertilizer production. The facility now produces green hydrogen and ammonia and has already delivered the first tons of fertilizers made with this renewable ammonia. Hydrogen is generated through water electrolysis using energy sourced exclusively from renewable sources, replacing natural gas as the primary feedstock. For storage, the hydrogen is kept in compressed gaseous form in high-pressure tanks, typically between 30 and

80 bar, with cylindrical or spherical tanks designed to ensure operational safety and stability [46].

Meanwhile, CF Industries has already started producing green ammonia at its Donaldsonville, Louisiana, facility. In 2024, the company installed a 20 MW electrolyzer, initiating green ammonia production from hydrogen generated through water electrolysis powered by renewable energy [47].

5.4 Additional industrial applications

In the metallurgical sector, hydrogen plays a key role in direct reduced iron processes. In this method, hydrogen replaces coke as the reducing agent to convert iron oxides into metallic iron, enabling steel production with significantly lower carbon dioxide emissions [48].

Beyond steelmaking, although still less established, hydrogen is being explored as a fuel in cement kilns, which currently rely heavily on coal and other fossil fuels. Despite challenges in scaling and achieving the high temperatures required, hydrogen-based cement production is a promising research area [49].

Another major application of hydrogen lies in the petrochemical industry, particularly in methanol synthesis. Under high temperature and pressure, hydrogen reacts with carbon monoxide (CO) in the presence of suitable catalysts to produce methanol, a key building block for various industrial chemicals [50].

6. Conclusions

Hydrogen has emerged as one of the most promising energy alternatives today. Never has there been such a high level of production and investment in its development, with a wide range of technologies being tested across the entire hydrogen value chain, from production and storage to transportation and application. Climate change, combined with sustainability-oriented policies, has accelerated the advancement of hydrogen-based solutions, leading to projections of a significant increase in hydrogen consumption over the coming decades. However, despite the progress achieved, important challenges remain—particularly in improving efficiency, reducing costs, and ensuring safety throughout all stages of the hydrogen cycle.

Acknowledgements

The authors express their gratitude to the Coordenação de Aperfeiçoamento de Pessoal de Nível Superior - Brazil (CAPES), the Conselho Nacional de Desenvolvimento Científico e Tecnológico (CNPq, 312400/2021-7, 405828/2022-5, 305649/2021-3, and 407799/2022-2), and the Fundação de Amparo à Pesquisa do Estado de Minas Gerais (FAPEMIG, RED-00144-22, APQ-01275-18, and APQ-03572-23) for their financial support.

Conflict of interest

The authors declare no conflict of interest.

Declaration of use of AI-assisted technologies

During the development of this proposal, ChatGPT was used exclusively for grammatical review. All ideas and technical content are the sole responsibility of the authors.

Author details

Renata Pereira Lopes Moreira*, Noemí Cristina Silva de Souza and Marcela de Oliveira Brahim Cortez
Department of Chemistry, Universidade Federal de Viçosa (UFV), Viçosa, Brazil

*Address all correspondence to: renata.plopes@ufv.br

References

[1] Noussan M, Raimondi PP, Scita R, Hafner M. The role of green and blue hydrogen in the energy transition—A technological and geopolitical perspective. Sustainability. 2021;**13**(1):298. DOI: 10.3390/su13010298(2021)

[2] International Energy Agency. Global Hydrogen Review 2024. Paris: IEA; 2024. Available from: https://www.iea.org/reports/global-hydrogen-review-2024 [Accessed: July 1, 2025]

[3] Tarhan C, Çil MA. A study on hydrogen, the clean energy of the future: Hydrogen storage methods. Journal of Energy Storage. 2021;**40**:102676. DOI: 10.1016/j.est.2021.102676

[4] Mazloomi K, Gomes C. Hydrogen as an energy carrier: Prospects and challenges. Renewable and Sustainable Energy Reviews. 2012;**16**(5):3024-3033. DOI: 10.1016/j.rser.2012.02.028

[5] ISO. ISO/TR 15916. Basic Considerations for the Safety of Hydrogen Systems. Geneva: International Organization for Standardization; 2015

[6] ISO. ISO 16110. Hydrogen Generators Using Fuel Processing Technologies — Part 1: General Safety. Geneva: International Organization for Standardization; 2006

[7] Hossain Bhuiyan MM, Siddique Z. Hydrogen as an alternative fuel: A comprehensive review of challenges and opportunities in production, storage, and transportation. International Journal of Hydrogen Energy. 2025;**102**(10):1026-1044. DOI: 10.1016/j.ijhydene.2025.01.033

[8] Andersson J, Grönkvist S. Large-scale storage of hydrogen. International Journal of Hydrogen Energy. **44**(23):11901-11919. DOI: 10.1016/j.ijhydene.2019.03.063

[9] Kannaiyan K, Lekshmi GS, Ramakrishna S, Kang M, Kumaravel V. Perspectives for the green hydrogen energy-based economy. Energy. 2023;**284**:129358. DOI: 10.1016/j.energy.2023.129358

[10] Wang Z, Mérida W. Thermal performance of cylindrical and spherical liquid hydrogen tanks. International Journal of Hydrogen Energy. 2024;**53**:667-683. DOI: 10.1016/j.ijhydene.2023.11.287

[11] Ma N, Zhao W, Wang W, Li X, Zhou H. Large scale of green hydrogen storage: Opportunities and challenges. International Journal of Hydrogen Energy. 2024;**50**(Part B):379-396. DOI: 10.1016/j.ijhydene.2023.09.021

[12] Sikiru S, Oladosu TL, Amosa TI, Olutoki JO, Ansari MNM, Abioye KJ, et al. Hydrogen-powered horizons: Transformative technologies in clean energy generation, distribution, and storage for sustainable innovation. International Journal of Hydrogen Energy. 2024;**56**(22):1152-1182. DOI: 10.1016/j.ijhydene.2023.12.186

[13] Massaro MC, Biga R, Kolisnichenko A, Marocco P, Monteverde AHA, Santarelli M. Potential and technical challenges of on-board hydrogen storage technologies coupled with fuel cell systems for aircraft electrification. Journal of Power Sources. 2023;**555**(30):232397. DOI: 10.1016/j.jpowsour.2022.232397

[14] Huang X, Jin S, Yu M, Li Y, Li M, Chen J. Numerical studies of a new device for a cryo-adsorption hydrogen

storage system. International Journal of Hydrogen Energy. 2024;**82**:1051-1059. DOI: 10.1016/j.ijhydene.2024.08.058

[15] Wu JK, Chen S, Yu M, Zhang XJ, Jiang L. Techno-economic analysis on low-temperature and high-pressure cryo-adsorption hydrogen storage. Fuel. 2025;**381**:133532. DOI: 10.1016/j.fuel.2024.133532

[16] Xu WC, Takahashi K, Matsuo Y, Hattori Y, Kumagai M, Ishiyama S, et al. Investigation of hydrogen storage capacity of various carbon materials. International Journal of Hydrogen Energy. 2007;**32**:2504-2512. DOI: 10.1016/j.ijhydene.2006.11.012

[17] Matar SF. Intermetallic hydrides: A review with ab initio aspects. Progress in Solid State Chemistry. 2010;**38**(1-4):1-37. DOI: 10.1016/j.progsolidstchem.2010.08.00

[18] Evard E, Gabis I, Yartys VA. Kinetics of hydrogen evolution from MgH_2: Experimental studies, mechanism and modelling. International Journal of Hydrogen Energy. 2010;**35**:9060-9069. DOI: 10.1016/j.ijhydene.2010.05.092

[19] Jiang Y, Si N, Jiang W, Wang Z, Zhang H. FeNiCu-based composite catalyst for hydrogen storage in MgH2. Chemical Engineering Journal. 2024;**499**. DOI: 10.1016/j.cej.2024.156449

[20] Zanberi A, Yusnizam NY, Ismail M. Catalytic effect of TiH2 on the dehydrogenation properties of AlH3. Materials Letters. 2025;**395**:138696. DOI: 10.1016/j.matlet.2025.138696

[21] Jiang W, Wang H, Zhu M. AlH3 as a hydrogen storage material: Recent advances, prospects and challenges. Rare Metals. 2021;**40**:3337-3356. DOI: 10.1007/s12598-021-01769-2

[22] Sato T, Saitoh H, Utsumi R, Ito J, Nakahira Y, Obana K, et al. Hydrogen absorption reactions of hydrogen storage alloy LaNi5 under high pressure. Molecules. 2023;**28**:1256. DOI: 10.3390/molecules28031256

[23] Bhaskar A, Muduli RC, Kale P. Prediction of hydrogen storage in metal hydrides and complex hydrides: A supervised machine learning approach. International Journal of Hydrogen Energy. 2025;**98**:1212-1225. DOI: 10.1016/j.ijhydene.2024.12.121

[24] Junior IM, Sperandio GH, Lopes RP. Efficient hydrogen evolution from NaBH4 using bimetallic nanoparticles (Ni–Co) supported on recycled Zn–C battery electrolyte paste. International Journal of Hydrogen Energy. 2024;**53**:1323-1331. DOI: 10.1016/j.ijhydene.2023.11.362

[25] Sperandio G, Junior IM, Bernardo E, Moreira R. Graphene oxide from graphite of spent batteries as support of nanocatalysts for fuel hydrogen production. PRO. 2023;**11**:3250. DOI: 10.3390/pr11113250

[26] Sperandio GH, de Carvalho JP, de Jesus CBR, Junior IM, de Oliveira KLA, Puiatti GA, et al. Hydrogen evolution from NaBH4 using novel Ni/Pt nanoparticles decorated on a niobium-based composite. International Journal of Hydrogen Energy. 2024;**83**:774-783. DOI: 10.1016/j.ijhydene.2024.07.022

[27] Bousada GM, Nogueira da Silva V, Fernandes de Souza B, de Oliveira RS, Machado Junior I, da Cunha CHF, et al. Niobic acid as a support for microheterogeneous nanocatalysis of sodium borohydride hydrolysis under mild conditions. RSC Advances. 2024;**14**:19459-19471. DOI: 10.1039/d4ra01879f

[28] Wang C, Tuninetti J, Wang Z, Zhang C, Ciganda R, Salmon L, et al. Hydrolysis of ammonia-borane over Ni/ZIF-8 nanocatalyst: High efficiency, mechanism, and controlled hydrogen release. Journal of the American Chemical Society. 2017;**139**:11610-11615. DOI: 10.1021/jacs.7b06859

[29] Makaryan IA, Sedov IV, Maksimov AL. Hydrogen storage using liquid organic carriers. International Journal of Hydrogen Energy. 2020;**45**(46):24902-24916. DOI: 10.1016/j.ijhydene.2020.04.274

[30] He T, Pei Q, Chen P. Liquid organic hydrogen carriers. Journal of Energy Chemistry. 2015;**24**(5):587-594. DOI: 10.1016/j.jechem.2015.08.007

[31] Niermann M, Beckendorff A, Kaltschmitt M, Bonhoff K. Liquid organic hydrogen carrier (LOHC) – Assessment based on chemical and economic properties. International Journal of Hydrogen Energy. 2019;**44**(13):6631-6654. DOI: 10.1016/j.ijhydene.2019.01.199

[32] Liu Y, Fu F, Salmon L, Espuche B, Moya S, Berlande M, et al. Facile MOF support improvement in synergy with light acceleration for efficient nanoalloy-catalyzed H_2 production from formic acid. ACS Applied Materials & Interfaces. 2023;**15**:23343-23352. DOI: 10.1021/acsami.3c03684

[33] Huang W, Xu F, Li DS, Astruc D, Liu X. "On–off" switch for H_2 and O_2 generation from HCOOH resp. H_2O_2. Carbon Energy. 2023;**5**:e269. DOI: 10.1002/cey2.269

[34] Levikhin AA. Boryaev: Physical properties and thermodynamic characteristics of hydrogen. Heliyon. 2024;**10**:e36414. DOI: 10.1016/j.heliyon.2024.e36414

[35] Galeev AG. Review of engineering solutions applicable in tests of liquid rocket engines and propulsion systems employing hydrogen as a fuel and relevant safety assurance aspects. International Journal of Hydrogen Energy. 2017;**42**:25037-25047. DOI: 10.1016/j.ijhydene.2017.06.242

[36] Gazda-Grzywacz M, Grzywacz P, Burmistrz P. Environmental benefits of hydrogen-powered buses: A case study of coke oven gas. Energies. 2024;**17**(20):5155. DOI: 10.3390/en17205155

[37] Sapnken FE, Posso F, Kibong MT, Tamba JG. The potential of green hydrogen fuel as an alternative in Cameroon's road transport sector. International Journal of Hydrogen Energy. 2024;**49**:433-449. DOI: 10.1016/j.ijhydene.2023.08.339

[38] Toyota Motor Corporation. The New Toyota Mirai: Advanced Hydrogen Fuel Cell Technology for a Cleaner Future. Brussels, Belgium: Toyota Europe Newsroom; 2020. Available from: https://newsroom.toyota.eu/the-new-toyota-mirai/ [Accessed: July 22, 2025]

[39] Hyundai Motor Company. The Innovative Clean Mobility Technology – Hydrogen Tank: The Safe, Durable Energy Storage System. Seoul, South Korea: Hyundai Motor Company. NEXO Specifications; 2025. Available from: https://www.hyundai.com/kr/en/eco/nexo/specifications [Accessed: July 22, 2025]

[40] Honda Motor Co., Ltd. Honda Clarity Fuel Cell: Pioneering Alternative Fuel Vehicle Technology. Honda Motor Co., Ltd. Honda News: Fuel Cell Vehicle Update; s.d. Available from: https://global.honda/newsroom/news/2017/4170912eng.html [Accessed: July 22, 2025]

[41] Solaris Bus & Coach. Solaris Urbino 18 Hydrogen Wins Bus of the Year 2025. Solaris Bus & Coach. Solaris Urbino 18 Hydrogen named Bus of the Year 2025; 2025. Available from: https://www.solarisbus.com/news/solaris-urbino-18-hydrogen-wins-bus-of-the-year-2025 [Accessed: July 17, 2025]

[42] Hyundai Motor Company. Elec City FCEV: Specifications and global expansion. Available from: https://www.hyundai.com/worldwide/en/eco/elec-city-fcev/highlights [Accessed: July 17, 2025]

[43] Humphreys J, Lan R, Tao S. Development and recent progress on ammonia synthesis catalysts for Haber–Bosch process. Advanced Energy and Sustainability Research. 2021;**2**:2000043

[44] Spatolisano E, Figini D. Absorption-enhanced Haber-Bosch for small-scale green NH_3 production. A feasibility study. Energy Conversion and Management. 2025;**337**:119904. DOI: 10.1016/j.enconman.2025.119904

[45] Wang H, Lin N, Arzumanyan M. The market for low-carbon-intensity ammonia. Gases. 2024;**4**(3):224-235

[46] Yara International ASA. Yara opens renewable hydrogen plant: "A major milestone". Yara International ASA; 2024. Available from: https://www.yara.com/corporate-releases/yara-opens-renewable-hydrogen-plant-a-major-milestone/ [Accessed: July 22, 2025]

[47] CF Industries Holdings, Inc. CF Industries and POET to Demonstrate the Use of Low-Carbon Ammonia in Corn Production to Reduce Carbon Intensity of Ethanol. CF Industries Holdings, Inc; 2024. Available from: https://www.cfindustries.com/newsroom/2024/cf-poet-low-carbon-ammonia [Accessed: July 22, 2025]

[48] Vergara D, Fernández-Arias P, Lampropoulos G, Antón-Sancho Á. Hydrogen revolution in Europe: Bibliometric review of industrial hydrogen applications for a sustainable future. Energies (Basel). 2024;**17**:3658. DOI: 10.3390/en17153658

[49] Behrendt F. The future of industrial hydrogen: Renewable sources and applications for the next 15 years. Clean Energy. 2025;**9**(1):3-8. DOI: 10.1093/ce/zkae103

[50] Naseem K, Qin F, Khalid F, Suo G, Zahra T, Chen Z, et al. Essential parts of hydrogen economy: Hydrogen production, storage, transportation and application. Renewable and Sustainable Energy Reviews. 2025;**210**:115196. DOI: 10.1016/j.rser.2024.115196

Chapter 10

Investigating the Use of Hydrogen Technologies in Today's Road Vehicles

Yasin Karagoz and Saban Pusat

Abstract

In recent years, there has been an increasing demand for clean and environmentally friendly vehicles, driven by stringent emission standards and growing environmental awareness. In this context, electric vehicles, plug-in hybrid technologies and hydrogen-fueled vehicles are of particular significance. The prevalence of technologically advanced issues in electric vehicles remains a subject of ongoing concern, particularly with regard to the challenges associated with charging and the effective range of these vehicles. Despite the evident decrease in emissions, plug-in hybrid electric vehicles (PHEVs) still possess conventional internal combustion engines, thus indicating that they may not be a wholly environmentally sustainable solution. In this context, hydrogen-powered vehicles are of particular significance. It is evident that both fueling processes can be executed expeditiously, and vehicles can be operated with disparate technologies. In the domain of hydrogen technologies, the most prevalent methods are the PEM fuel cell and the hydrogen-powered internal combustion engine. It is important to note that both methods have their respective advantages and disadvantages. In terms of pollutants, hydrogen facilitates the attainment of a clean and environmentally friendly result.

Keywords: hydrogen, PEM fuel-cell, hydrogen powered vehicles, NOx, emissions

1. Introduction

The global warming problem, growing concerns, and the increasingly stringent emission standards in recent years have led researchers continuing to search for new solutions [1]. This capability is maintained in electric vehicles, plug-in hybrid technologies, and hydrogen-fueled vehicles [2]. Electric vehicles continue to face challenges such as range improvements [3]. Plug-in hybrid electric vehicles, on the other hand, still rely on conventional internal combustion engines, and although emissions are reduced, they are not a completely environmentally friendly solution. This potential is highlighted by hydrogen-fueled vehicles. Hydrogen-fueled vehicles can utilize both internal combustion engines and fuel cell technologies [4].

In internal combustion engine technologies, high thermal efficiencies can be achieved with techniques such as hydrogen jet [5]. Alternatively, existing engines can be operated with 100% hydrogen fuel using three different methods: the old technology of injecting hydrogen through the manifold *via* carburetion, the port injection method of injecting hydrogen into the port *via* an electronic control unit, or the direct injection method of injecting hydrogen directly into the cylinder [6]. Fuel cell technology, as is well known, includes various technologies such as alkaline, PEM, and solid oxide fuel cells [7]. Among these, PEM fuel cell technologies are the most suitable for use in vehicles [8]. PEM fuel cell technology is the most suitable fuel cell type when considering compactness, vehicle application, efficiency, and safety. It offers both higher efficiency and greater productivity. PEM fuel cells can be easily used in vehicles due to their high efficiency, compactness, and practicality. They are widely used in land vehicles today.

In the use of hydrogen in internal combustion engines, the most common methods used today are multi-point port hydrogen injection and direct cylinder hydrogen injection [9]. Among these, direct cylinder hydrogen injection prevents engine power loss [10]. On the other hand, combustion problems such as backfire, pre-ignition, and knock are also prevented [11]. However, high-pressure direct injection hydrogen injectors are produced by only a few companies. On the other hand, the method of sending hydrogen to the combustion chamber through a carburetor is not currently used because it is outdated technology. On the other hand, port injection is another method used [12]. Multi-point hydrogen injection with the assistance of the ECU through a port gas injector is a practical method. However, due to the very low density of hydrogen in the port hydrogen injection method, it replaces the intake air and causes a power loss of up to 15% in the engine [13]. On the other hand, if engine calibration, injection timing, and strategies are not adjusted correctly, problems such as backfire and preignition can occur [14]. This may be due to the fact that hydrogen has a relatively low ignition energy despite its high ignition temperature [15]. As is well known, strict safety precautions must be taken in hydrogen-fueled vehicles. It is anticipated that no problems will be encountered if the fuel system complies with the relevant standards. Seamless 316 series stainless steel pipes, stainless steel equipment, and special, leak-proof gas connections are used in hydrogen vehicle applications. If these requirements are followed, hydrogen can be safely used as a fuel in vehicles.

Some studies on hydrogen-fueled vehicles in the literature are summarized below.

In their study, Pramuanjaroenkij, et al. [16] investigated vehicles powered by PEM fuel cells and concluded that they could easily replace conventional internal combustion engine vehicles in the future.

In their study, Cunanan et al. [17] investigated environmentally friendly solutions for heavy-duty vehicles. Their results concluded that using fuel cells in heavy-duty vehicles, which are generally used for transportation, is much more appropriate.

In their study, Thomas et al. [18] compared battery-powered and fuel-cell vehicles using LCA analysis. Their results demonstrated that fuel-cell vehicles with a service life of 160 km or more are significantly more environmentally friendly than battery-powered vehicles in terms of LCA.

As is known, no pollutant emissions are observed in PEM fuel cell vehicles. However, gasoline-fueled engines experience emissions such as CO, CO_2, HC, and NOx. While carbon-derived pollutants such as CO, CO_2, and HC are not observed in hydrogen-fueled internal combustion engine systems, high NOx emissions are a disadvantage of hydrogen-fueled engines. In the study, fuel consumption and energy consumption were comprehensively examined using vehicle model data obtained experimentally from the vehicle, and emissions were excluded from the scope of this study. In this study, a sample passenger vehicle model was developed and fuel consumption values were obtained using Newtonian equations, taking into account vehicle resistance. In this context, a conventional vehicle was compared with both a fuel cell-powered vehicle and an internal combustion engine-powered vehicle. Consequently, fuel consumption and emission values were compared for current vehicle technologies.

2. Mathematical model

Within the scope of the study, a sample gasoline-powered passenger car was selected. Following this, diagnostic data was collected from the sample vehicle to obtain an engine map. Engine maps were generated using the obtained data. A one-dimensional engine model was then developed using AVL Boost software. The model was then revised to obtain data from the hydrogen-fueled internal combustion engine. The data were then compared with the vehicle model. In the final stage, a vehicle model for PEM FC was developed, and the outputs obtained from the model developed with the hydrogen-fueled ICE and the gasoline-fueled ICE were compared.

2.1 Vehicle model

In this study, a vehicle model was developed in MATLAB for a conventional vehicle. The model was compared with a gasoline-only passenger car, a PEM fuel cell-powered passenger car, and a hydrogen internal combustion engine passenger car. In this context, both equivalent fuel consumption and emissions were compared. A schematic view of the model is shown in **Figure 1**.

In **Figure 2**, the same vehicle model was developed for a hydrogen PEM fuel cell. Here, the electric motor and drive selections were also made. Thus, the hydrogen fuel cell model was also examined in terms of fuel consumption and emissions as an alternative to the internal combustion engine.

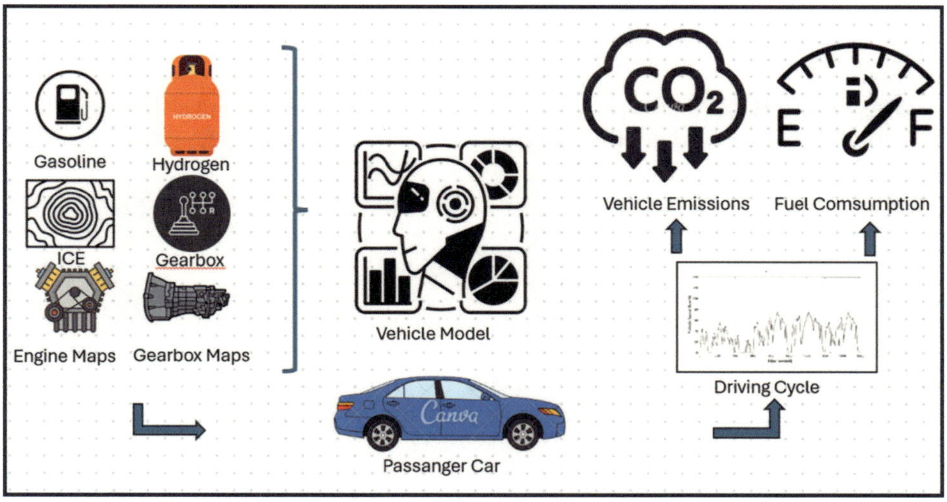

Figure 1.
Schematic representation of a hydrogen-fueled motor vehicle or gasoline-powered motor vehicle model.

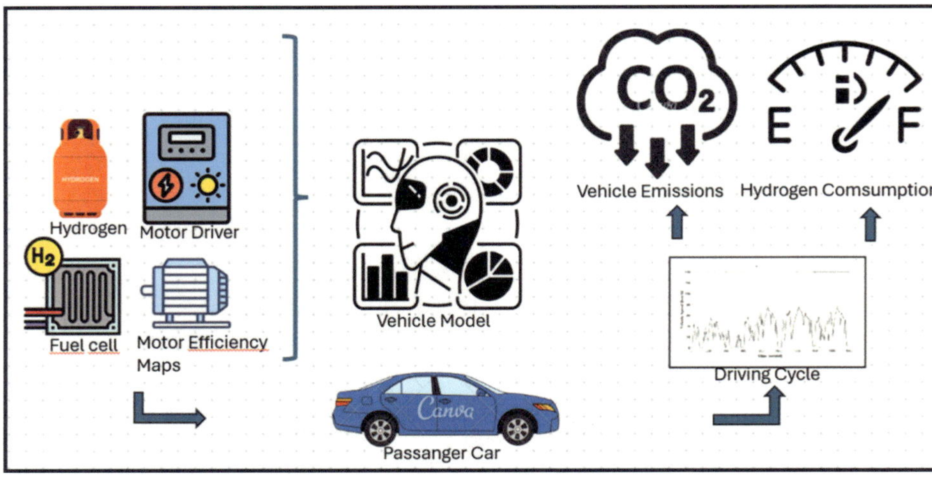

Figure 2.
Schematic representation of the hydrogen fuel cell vehicle model.

2.2 Driving resistance

In the vehicle model, total running resistance is first calculated by taking into account road resistance forces. By accounting for rolling resistance, aerodynamic resistance, acceleration resistance, and slope resistance, total running resistance and the instantaneous power and torque required for this resistance are calculated. Initially, total running resistance is given in Eq. (1).

$$W_T = W_R + W_A + W_{ACC} + W_S \tag{1}$$

Here, WT represents the total road resistance, WR represents the rolling resistance, WA represents the aerodynamic resistance, WACC represents the acceleration resistance, and WS represents the slope resistance in N. The calculation of WR rolling resistance is given in Eq. (2).

$$W_R = C_R mgcos\theta \tag{2}$$

Here, CR is the rolling resistance coefficient, m is the vehicle mass in kg, g is the gravitational acceleration in m/s2, and Θ is the road slope angle. The vehicle aerodynamic resistance is given in Eq. (3).

$$W_A = \frac{1}{2}\rho AC_D V^2 \tag{3}$$

Here, ρ is the density of air in kg/m3, A is the projected area of the vehicle in m^2, CD is the aerodynamic resistance coefficient, and V is the vehicle's cruising speed in m/s. The acceleration resistance of the vehicle is given in Eq. (4).

$$W_{ACC} = \alpha mga \tag{4}$$

Here, α is the acceleration resistance correction coefficient for rotating masses, m is the vehicle mass in kg, and a is the acceleration value in m/s^2. The slope-dependent resistance of the vehicle is given in Eq. (5).

$$W_S = mgsin\theta \tag{5}$$

Here, m is the vehicle mass in kg, g is the gravitational acceleration in m/s2, and ϴ is the road slope angle.

After calculating the vehicle's total road resistance, the instantaneous power requirement of the vehicle was calculated as given in Eq. (6).

$$N_e = W_T x V \tag{6}$$

Here, WT is the total travel resistance of the vehicle in N units, V is the travel speed of the vehicle in m/s.

2.3 Gasoline fueled ICE model

Here, data was first collected from the sample vehicle engine *via* the OBD-2 port using a diagnostic device. Using Car Scanner Pro software, engine and vehicle data were recorded *via* the diagnostic device over the CANBUS protocol while the vehicle was in motion. Fuel consumption values were obtained at different power and engine speed values. Subsequently, a detailed engine map was created using the obtained data, as seen in **Figure 3**, using a MATLAB ANN model. Engine speed, demanded engine power, MAP, fuel quantity, EGT, and accelerator pedal position were entered into the model as inputs. Fuel consumption and engine operation data were then obtained using the MATLAB ANN model. Data were obtained from 1,892,421 operating points using the ANN. The Levenberg-Marquardt training method was then used. 70% of the data was used for training, 15% for validation, and 15% for testing. Regression values remained at 86% and above, demonstrating sufficient model accuracy. Thus, detailed engine maps could be created for the gasoline engine. Instantaneous fuel consumption and CO2 emissions data were obtained. From the obtained data, fuel consumption and CO2 emissions data were obtained in three dimensions (based on speed and engine power).

2.4 Hydrogen-fueled ICE model

A mathematical model was constructed using AVL Boost software, using data from the four-stroke, four-cylinder, gasoline-fueled, naturally aspirated spark

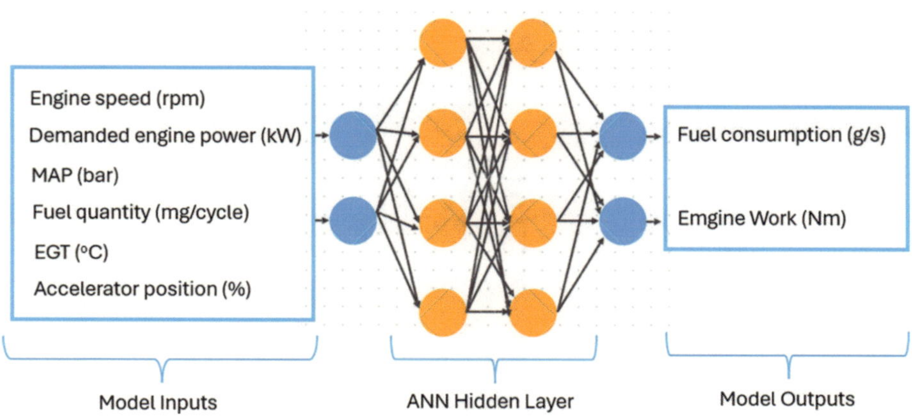

Figure 3.
Creating the motor map with the ANN structure.

ignition engine on the sample vehicle. The mathematical model was used to generate engine maps that accurately matched fuel consumption with 96% accuracy or higher. **Figure 4** shows the basic parameters required for the model. Data such as cylinder diameter, stroke length, cylinder volume, crank and connecting rod lengths, air filter volume, intake manifold volume, ignition timing, and so on were entered into the model. The model was then run using a simulation using hydrogen injection via port injection. Thus, fuel consumption and emissions data at various points were obtained by operating the engine with hydrogen fuel.

2.5 PEM FC model

As can be seen in **Figure 5**, a thermal efficiency model was established for the PEM fuel cell currently used in vehicles. Catalogs from hydrogen fuel cell manufacturers were used to create this model. However, due to the electrochemical reaction in PEM fuel cells, efficiency maps vary depending on instantaneous power and current requirements, particularly temperature. Fuel cell efficiencies are also affected by

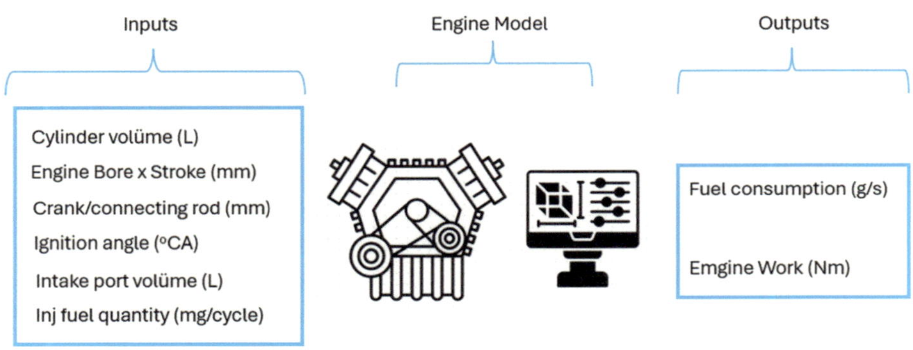

Figure 4.
Required parameters for one-dimensional engine model.

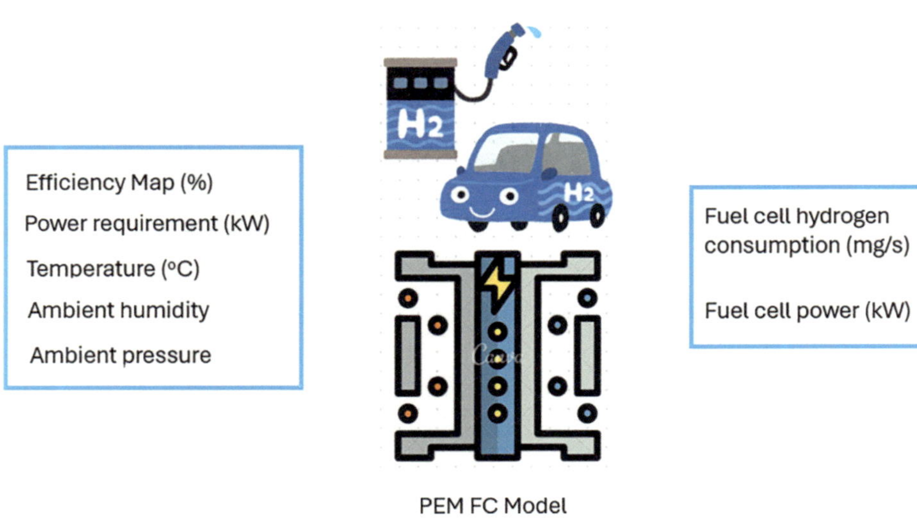

Figure 5.
PEM fuel cell model.

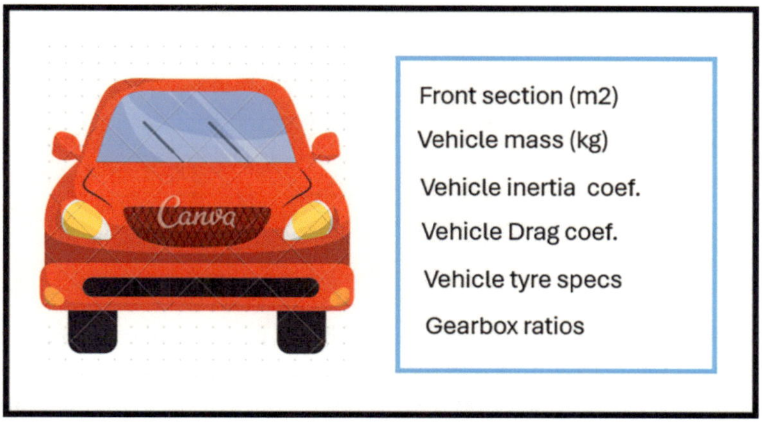

Figure 6.
Sample vehicle inputs.

factors such as ambient pressure, humidity, and temperature. Therefore, a mathematical model was constructed to include correction coefficients, and the actual efficiency values were determined using these correction coefficients.

2.6 EM and driver model

A model was constructed considering the efficiency maps for the electric motor and drive. This included utilizing the efficiency maps based on the speed and torque of the electric motor. Three-dimensional efficiency maps for both the drive and the electric motor were obtained in the form of look-up tables and incorporated into the model. This resulted in more realistic maps and models.

2.7 Sample vehicle

Figure 6 shows the basic parameters required for the selected example vehicle. A passenger car was selected as the example vehicle. Vehicle parameters such as front section area, vehicle mass, vehicle inertia coefficient, vehicle drag coefficient, vehicle tire specifications, and gearbox ratios were entered into the model. This allowed for a comparison of fuel consumption for a modern passenger car with hydrogen fuel cell, hydrogen-fueled internal combustion engine, and gasoline-fueled engines.

2.8 Driving cycle

The vehicle's cruise cycle, designated FTP 75 (Federal Test Procedure), was used. The cruise cycle characteristics are summarized in **Table 1**. Furthermore, FTP 75 is a US cruise cycle derived from FTP 72. By inputting the speed-time profile into the model, the vehicle's fuel consumption per 100 km was obtained using a mathematical model.

3. Results and discussion

In this study, vehicle data was first read from a sample vehicle via the CANBUS protocol, enabling the vehicle's engine and transmission data to be read. Subsequently,

Specification	Value
Time	1877 s
Distance traveled by vehicle	17.77 km
Average speed	34.12 km/h
Maximum speed	91.25 km/h

Table 1.
FTP 75 Driving cyle specifications [19].

a model was developed using Artificial Neural Networks to generate an engine map. A one-dimensional engine analysis was then performed, and a mathematical model was established using AVL Boost software. The model then provided fuel consumption data for the engine operating on 100% hydrogen fuel. The resulting model allowed the creation of an engine map.

Subsequently, a vehicle model was constructed using MATLAB software to obtain fuel consumption data for a sample passenger car. Subsequently, a fuel cell vehicle model was constructed. Fuel cell fuel consumption and efficiency maps based on electrical power requirements were used. Efficiency maps for the electric motor and drive were also included. Finally, the vehicle was simulated as if it were driving in a cruise cycle, designated FTP 75 (Federal Test Procedure), and equivalent fuel consumption and energy consumption results were compared.

Figure 7 shows the model results for a gasoline-fueled spark-ignition engine-powered vehicle, a hydrogen-fueled spark-ignition engine-powered vehicle, and a PEM hydrogen fuel cell- and electric motor-powered vehicle. Fuel consumption for the example passenger vehicle is shown according to the FTP 75 cruising cycle. All fuel consumptions were calculated using the MATLAB equivalent gasoline fuel consumption model, taking into account the lower heating values. Subsequently, the results were extrapolated to 100 km, and the equivalent gasoline fuel consumption per 100 km was calculated. From the obtained results, the gasoline-fueled spark-ignition

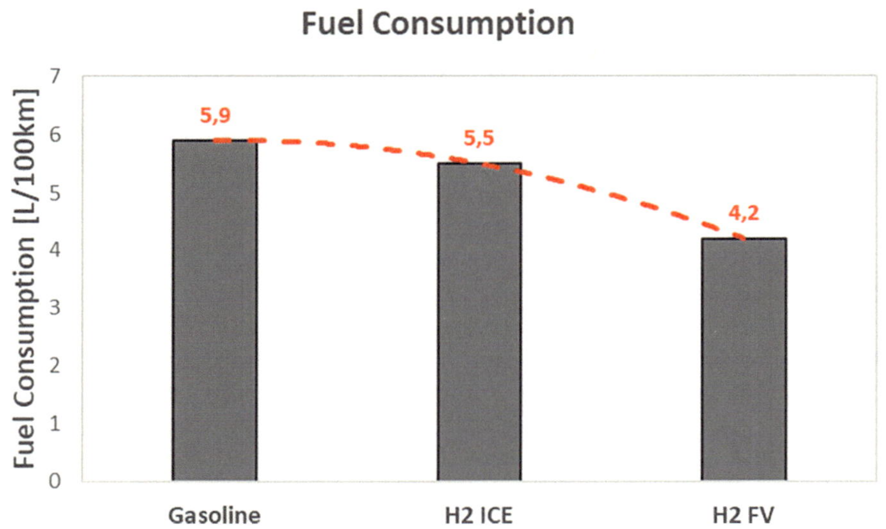

Figure 7.
Fuel consumption values.

engine consumes 5.9 L/100 km, the hydrogen-fueled spark-ignition engine consumes 5.5 L/100 km, and the PEM hydrogen fuel cell-powered vehicle consumes 4.2 L/100 km equivalent. The simulation results obtained with the gasoline-fueled internal combustion engine vehicle show a 6.8% improvement compared to the gasoline-fueled internal combustion engine vehicle.

On the other hand, it was concluded that the PEM hydrogen fuel cell-electric motor-driven system achieved a 28.8% improvement compared to a gasoline-fueled internal combustion engine vehicle. Considering the results obtained, hydrogen's high combustion efficiency and increased engine thermal efficiency, a decrease in fuel consumption compared to gasoline is expected when hydrogen is used in internal combustion engines. Furthermore, due to the PEM fuel cell's high energy-efficient electricity generation through electrochemical reactions, a decrease in equivalent fuel consumption is also expected.

Figure 8 shows the energy consumption results for a gasoline-fueled spark-ignition engine-powered vehicle, a hydrogen-fueled spark-ignition engine-powered vehicle, and a PEM hydrogen fuel cell- and electric motor-powered vehicle, calculated using the MATLAB model. For a sample passenger car, the energy consumption is shown for the FTP 75 cruising cycle. All energy consumptions were calculated using the MATLAB model, taking into account the lower heating values of the vehicle. The results were then extrapolated to 100 km, and the equivalent gasoline fuel consumption per 100 km was calculated. From the obtained results, the gasoline-fueled spark-ignition engine consumed 194.7 kJ/100 km, the hydrogen-fueled spark-ignition engine consumed 181.5 kJ/100 km, and the PEM hydrogen fuel cell-powered vehicle consumed 138.6 kJ/100 km.

Energy simulation results obtained with a gasoline-fueled internal combustion engine vehicle indicate a 6.8% improvement compared to a gasoline-fueled internal combustion engine vehicle. Conversely, a 28.8% improvement in energy consumption was observed in the PEM hydrogen fuel cell-electric motor driven system compared to a gasoline-fueled internal combustion engine vehicle. Considering hydrogen's high in-cylinder temperature and pressure, low thermal losses, high

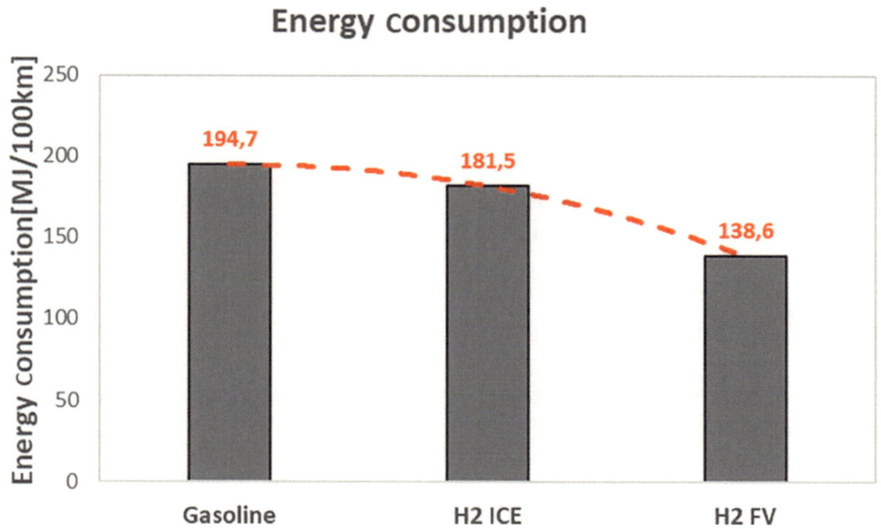

Figure 8.
Energy consumption values.

diffusion coefficient, and narrow flame quenching distance, the results indicate that the use of hydrogen in internal combustion engines is expected to increase thermal efficiency. Therefore, the use of hydrogen as a fuel in internal combustion engines is an expected outcome. Furthermore, considering the high energy efficiency of PEM hydrogen fuel cells together, a decrease in energy consumption is an expected result.

4. Conclusions

In this study, engine and transmission data were obtained from a sample vehicle via a CAN BUS diagnostic device, and a gasoline engine map was generated using Artificial Neural Networks. Subsequently, after verifying the spark ignition engine model with existing engine data using AVL Boost software, engine map data was obtained by performing hydrogen combustion. Subsequently, a mathematical model for the vehicle's fuel consumption was developed using MATLAB. A vehicle model was also developed for the PEM hydrogen fuel cell. Finally, equivalent gasoline fuel consumption and energy consumption values for the FTP 75 USA cruise cycle were obtained and compared per 100 km. The results are summarized below.

- A gasoline-fueled spark-ignition engine-powered vehicle consumes 5.9 L/100 km, a hydrogen-fueled spark-ignition engine-powered vehicle consumes the equivalent of 5.5 L/100 km, and a PEM hydrogen fuel cell-powered vehicle consumes the equivalent of 4.2 L/100 km. Simulation results show a 6.8% improvement in fuel mileage for a hydrogen-fueled internal combustion engine vehicle compared to a gasoline-fueled internal combustion engine vehicle. Conversely, a PEM hydrogen fuel cell-electric motor-powered system provides a 28.8% improvement in fuel mileage compared to a gasoline-fueled internal combustion engine vehicle.

- A gasoline-fueled spark-ignition engine consumes 194.7 kJ/100 km, a hydrogen-fueled spark-ignition engine consumes 181.5 kJ/100 km, and a PEM hydrogen fuel cell-powered vehicle consumes 138.6 kJ/100 km. The results indicate that hydrogen's combustion properties, such as high in-cylinder temperature and pressure, low thermal losses, high diffusion coefficient, and narrow flame quenching distance, will increase combustion and thermal efficiency, thereby reducing energy consumption compared to a gasoline engine. Furthermore, considering the high energy efficiency of PEM hydrogen fuel cells, which generate electricity through electrochemical reactions, it is expected that energy consumption will decrease with the use of PEM fuel cells.

Acknowledgements

We would like to thank Erin Motor company for their equipment and device support during the work.

Conflict of interest

The authors declare no conflict of interest.

Author details

Yasin Karagoz* and Saban Pusat
Yildiz Technical University, Istanbul, Turkey

*Address all correspondence to: yasin.karagoz@yildiz.edu.tr

References

[1] Hurren T, Durbin TD, Johnson KC, Karavalakis G. The impacts of improving heavy-duty internal combustion engine technology on reducing NOx emissions inventories going into the future. Science of the Total Environment. 2025;**986**:179781

[2] Wen C, Li J, Wang B, Lu G, Xu H. Expertise-guided NOx emission modeling of hybrid vehicle engines via peak-valley-enhanced Gaussian process regression. Energy. 2025;**322**:135166

[3] Tetik T, Karagoz Y. Remarkable results of energy consumption and CO2 emissions for gasoline and electric powered vehicle. Science Progress. 2025;**108**(1):00368504241305897

[4] Turoń K. Hydrogen-powered vehicles in urban transport systems–current state and development. Transportation Research Procedia. 2020;**45**:835-841

[5] Caramia G, Amirante R, De Palma P. Unsteady RANS simulations of under-expanded hydrogen jets for internal combustion engines. International Journal of Hydrogen Energy. 2024;**96**:849-859

[6] Musy F, Ortiz R, Ortiz I, Ortiz A. Hydrogen-fuelled internal combustion engines: Direct injection versus port-fuel injection. International Journal of Hydrogen Energy. 2025;**137**:925-938

[7] Calise F, Cappiello FL, Cimmino L, Cutolo L, Vicidomini M. Thermoeconomic comparison of alkaline, solid oxide and anion exchange membrane Electrolyzers for power-to-gas applications. Renewable and Sustainable Energy Reviews. 2025;**224**:116115

[8] Aghakhani A, Haque N, Saccani C, Pellegrini M, Guzzini A. Direct carbon footprint of hydrogen generation via PEM and alkaline electrolysers using various electrical energy sources and considering cell characteristics. International Journal of Hydrogen Energy. 2023;**48**(77):30170-30190

[9] Lu Y, Que J, Xia Y, Li X, Jiang Q, Feng L. A comparative study of the effects of EGR on combustion and emission characteristics of port fuel injection and late direct injection in hydrogen internal combustion engine. Applied Energy. 2024;**375**:123830

[10] Zhang Z, Song Z, Wang Y, Liu C, Mao C, Yin Z, et al. A comprehensive review of hydrogen production technology and the performance, combustion, and emissions of hydrogen-fueled internal combustion engines. Journal of the Energy Institute. 2025;**122**:102229

[11] Park H, Lee J, Oh S, Kim C, Park Y, Yoo D. Strategies of hydrogen-air mixture formation for suppressing backfire, pre-ignition, and knock in full-load operations of hydrogen heavy-duty engines. Energy. 2025;**326**:136343

[12] Lu Y, Que J, Xia Y, Li X, Jiang Q, Feng L. A comparative study of the effects of EGR on combustion and emission characteristics of port fuel injection and late direct injection in hydrogen internal combustion engine. Applied Energy. 2024;**375**:123830

[13] Zhang Z, Song Z, Wang Y, Liu C, Mao C, Yin Z, et al. A comprehensive review of hydrogen production technology and the performance, combustion, and emissions of hydrogen-fueled internal combustion

engines. Journal of the Energy Institute. 2025;**122**:102229

[14] Dhyani V, Subramanian KA. Control of backfire and NOx emission reduction in a hydrogen fueled multi-cylinder spark ignition engine using cooled EGR and water injection strategies. International Journal of Hydrogen Energy. 2019;**44**(12):6287-6298

[15] Wang Q, Shen J, Li X, Lyu D, Li L, Han D. Combustion and emissions of a spark-ignition ammonia-hydrogen engine at lean combustion conditions. International Journal of Hydrogen Energy. 2025;**136**:261-274

[16] Pramuanjaroenkij A, Kakaç S. The fuel cell electric vehicles: The highlight review. International Journal of Hydrogen Energy. 2023;**48**(25):9401-9425

[17] Cunanan C, Tran MK, Lee Y, Kwok S, Leung V, Fowler M. A review of heavy-duty vehicle powertrain technologies: Diesel engine vehicles, battery electric vehicles, and hydrogen fuel cell electric vehicles. Clean Technologies. 2021;**3**(2):474-489

[18] Thomas CE. Fuel cell and battery electric vehicles compared. International Journal of Hydrogen Energy. 2009;**34**(15):6005-6020

[19] DieselNet. Emission Test Cycles-Summary of worldwide engine and vehicle test cycles. Available from: http://www.dieselnet.com/standards/cycles/ftp75.html

Chapter 11

Water Electrolysis Overview: A Progressive Origination of Renewable Hydrogen Energy

Relebohile Mokete

Abstract

There is necessity to shift from the exhaustion of green-house gases emitting sources to carbon-neutral energy means. Water electrolysis (WE) is governed by thermodynamics and kinetics that are involved during the hydrogen and oxygen evolution reactions. Among various WE techniques, the polymer electrolyte membrane (PEM) WE, referred to as PEMWE, is the newest technology that involves green H_2 generation *via* electrolysis. The concept of electrolysis incorporates water splitting under the influence of electrical potential to produce hydrogen and oxygen gases from the cathode and anode chambers, respectively. During water splitting, water forms oxygen and hydrogen ions or protons (H^+) in the anode. The protons move from the anode to the cathode through a PEM, whereas the electrons (e^-) travel through the electron path (wire) to combine with protons thus producing hydrogen gas (H_2). Many factors contribute toward the H_2 production using the PEMWE and such include the power source, the electrodes and their catalysts as well as the PEM, its modification, and overall properties. Modified PEMs tend to show improved properties and research has to focus more on finding affordable and effective materials that can improve the PEMWE. Hydrogen storage and transport are crucial. Thus far, the solid-state storage of hydrogen has proven to be effective and the hydrides exhibit high storage capacity. Emphasis on the green H_2 research and application is important since it promotes the zero-emission atmosphere.

Keywords: water electrolysis, hydrogen, renewable energy, polymer electrolyte membrane (PEM), electrolyzer

1. Introduction

There is currently an apparent shift from the exhaustion of green-house gases emitting sources to carbon-neutral energy means [1]. This is due to the fact that the high energy demand globally has led to the exploitation of resources, including the polluting fuels that exert pressure on the environment thus leaving it in an unnatural state [1, 2]. Today, the research efforts focus on finding solutions of utilizing green energy alternatives. In particular, renewable energy has been one promising energy form to outperform the carbon emitting energy forms [1–3]. Among various examples, hydrogen energy is a typical renewable energy form that is gaining research interest [2]. Why hydrogen? Hydrogen is among the fuels that fulfill the energy needs without harming the environment and its key roles as an energy carrier are evident in energy storage, chemical processes as well as transportation [4, 5]. Also, hydrogen is affordable with higher efficiency as opposed to conventional fossil fuels [4, 6]. As the Earth's most abundant element, hydrogen exists as part of the water compound and it is also bound to either gaseous or liquid hydrocarbons [4, 7].

The hydrogen production methods include hydrocarbon pyrolysis, hydrocarbon reforming, and biomass processing [8]. Some various renewable sources of hydrogen include wind, sun, and the Earth's sub-surface wherein the hydrogen gas can be obtained *via* processes like water electrolysis and thermochemical cycle [8, 9]. In comparison with other chemical fuels, hydrogen's chemical energy per unit mass is three times larger [4]. Therefore, the development of hydrogen production systems and technology is driven by the escalating clean and efficient energy systems' demand [4]. The commercialization of the fuel cell technology for hydrogen production is based on low cost and efficient fuel systems.

1.1 Water electrolysis (WE)

A typical technique for hydrogen generation is water electrolysis (WE) [4]. WE is regarded as a promising alternative for hydrogen production at low cost and zero carbon footprints that is applicable in both low-scale and large-scale hydrogen production [10]. The electrolyzer is a unit or reactor in which electrolysis occurs [8]. The electrolyzer types include alkaline electrolyzers that have both the cathode and anode with a ceramic diaphragm separation that is immersed in the alkaline electrolyte [8, 11]. The water electrolyzers utilize a solid polymer electrolyte referred to as a membrane (PEM), which is embedded between the anode and cathode chambers [10, 11].

1.1.1 Polymer electrolyte membrane water electrolysis (PEMWE)

The PEMWE is the newest and advanced hydrogen generation technology to date which incorporates the PEM within an electrolyzer [8, 12, 13]. Briefly, the concept of PEMWE involves water splitting under the influence of electrical potential to produce hydrogen and oxygen gases from the cathode and anode chambers, respectively [14]. During water splitting, water forms oxygen and hydrogen ions or protons (H^+) in the anode [15]. The protons move from the anode to the cathode through a Proton Exchange Membrane (PEM), whereas the electrons (e^-) travel through the external electron path (wire) to combine with protons in the cathode thus producing hydrogen gas (H_2). This WE process is represented respectively by Eqs. (1)–(3) [15].

$$\textit{Cathode reaction}: 2H^+ + 2e^- \rightarrow H_2 \tag{1}$$

$$\textit{Anode reaction}: H_2O \rightarrow 1/2O_2 + 2H^+ + 2e^- \tag{2}$$

$$\textit{Overall reaction}: H_2O \rightarrow H_2 + 1/2O_2 \tag{3}$$

The schematic diagram of a PEM electrolyzer is shown in **Figure 1** [8, 14].

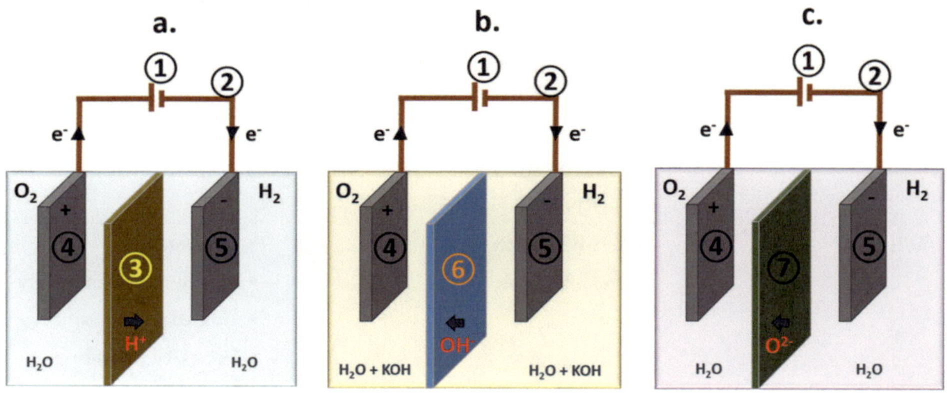

Figure 1.
Representation of WE that includes (a) PEMWE, (b) AWE, and (c) SOWE. ① Power source, ② electrons' path, ③ PEM and ④ anode and ⑤ cathode electrodes, ⑥ diaphragm, and ⑦ electrolyte.

The overall PEMWE system (**Figure 1**) can be is subdivided into the power source (1), electrons' path or current medium (2) and the electrolyzer (3) [16, 17]. As mentioned elsewhere in the text, the electrochemical cells such as electrolyzers (and fuel cells) comprise of the two electrodes (cathode and anode), a separating electrolyte (PEM) that serves as a passage for ions resulting from the redox half reactions and the ion-transport medium which is usually a liquid (water) [16–18]. In order for the PEMWE system to be functional, all its components in **Figure 1** have to be connected and in good condition for smooth operation [8]. The absence or dysfunction of one component may lead to the system failure [19]. Generally, Research has continued to focus on the development of the PEMWE system in an effort to achieve the best efficiency of hydrogen generation [20]. Various materials have been explored for making effective electrodes, power source and PEM in order to achieve the best performance [16]. The PEMWE production system requires the power source in order to initiate the water splitting process [20]. **Table 1** summarizes the PEMWE components, whereas **Table 2** shows the advantages and disadvantages of various types of WE units [8, 21, 22].

Category	Component		Function
Power supply	Direct current (e.g., PV cell)		Power source to electrolyzer
Electron path	Wire (e.g., Cu)		Electron passage or current medium
Electrolyzer	Electrodes:	Anode (e.g., IrO_2 catalyst)	OER
		Cathode (e.g., Pt catalyst)	HER
	Gas diffusion layer (MEA)	Anode (e.g., Sintered titanium)	O_2 and H_2 gases' diffusion
		Cathode (e.g., Toray carbon paper (TGP-H-120))	
Electrolyte	PEM		Proton passage
Ions medium	Water		Dissolved ions medium

Table 1.
PEMWE components.

Electrolysis type	Advantages	Disadvantages
PEMWE	• High performance (efficiencies, current density) • Good partial load range • Low gas crossover • High pressure H_2 production	• Costly materials • Prone to corrosion • Short lifetime
AWE	• Relatively affordable • Prolonged stability	• Gas cross over • Minimum current density • Limited partial load range • Decreased pressure (operational) • Insufficient purity of generated gases
SOE	• Reduced Ohmic losses	• Low current density • Accelerated degradation • Shorter system lifetime • Electrode contamination • High operation temperature

Table 2.
Comparison of the PEMWE, AWE and SOWE.

1.1.2 Alkaline water electrolysis (AWE)

The commercialization of alkaline electrolysis in order to produce H_2 is evident [21]. The concept of alkaline water electrolysis involves the use of the liquid alkaline electrolyte within which the two electrodes operate [23]. Due to the affordability of nickel (Ni), Ni-associated catalysts are commonly utilized in AWE [21]. The principle of AWE is according to Eqs. (4)–(6).

$$\textit{Cathode reaction}: 2H_2O + 2e^- \rightarrow H_2 + 2OH^- \quad (4)$$

$$\textit{Anode reaction}: 2OH^- \rightarrow \frac{1}{2}O_2 + H_2O + 2e^- \quad (5)$$

$$\textit{Overall reaction}: H_2O \rightarrow H_2 + \frac{1}{2}O_2 \quad (6)$$

In AWE, the lifetime of cell stack is deemed very long and the stability is prolonged [21, 24]. Although deemed relatively cost-effective in terms of the capital cost, the limitation of AWE include less ability to function under increased current densities, decreased operational pressure, insufficient purity of the gas products and low operational pressure (**Table 2**) [24–26]. Kuleshov et al. [26] synthesized the hydrogel of zirconium hydroxide within a porous diaphragm consisting of TiO_2 ceramic hydrophillic filler. As the hydrogel was incorporated into the membrane, there was significant voltage increase to 2 V when the current was 300 mA cm^{-2}.

The incorporation of the TiO_2 filler showed a voltage decrease to 1.83 V at 300 mA cm^{-2}. The hydroxyl ion transfer was successfully exhibited in the diaphragm [26]. **Figure 1(b)** denotes the unit of AWE [24].

1.1.3 Solid oxide water electrolysis (SOWE)

The development of the SOWE systems that utilize a ceramic electrolyte and usually operate at elevated temperatures that exceed 500°C is gaining some research interest [23]. The SOWE is typical of operating at high cell efficiencies with the cell potential almost closer to the thermoneutral potential but their degradation possibility is accelerated owing to high temperature operation [25]. The ceramics (oxide-ion conducting) participate as cell separators and solid electrolytes. The SOWE (high temperature) involves reduction of water vapor at the cathode (Eq. (7)) and migration of the resulting oxygen ions to the anode for O_2 evolution (Eq. (8)) [24].

$$\textit{Cathode reaction}: H_2O_{(g)} + 2e^- \rightarrow H_{2(g)} + O^{2-} \tag{7}$$

$$\textit{Anode reaction}: 2O^{2-} \rightarrow O_{2(g)} + 4e^- \tag{8}$$

The oxide ions transportation via the electrolyte involves the ionic diffusion process while the Ohmic loss is reduced by the ceramic membranes [24]. Porous nickel and porous perovskite materials are used as steam cathode and air anode respectively. The electrolytes that are used in SOWE, such as yttria-stabilized zirconia (YSZ) portray excellent ion conductivity at high operating temperatures which are above or equivalent to 700°C [24]. However, the YSZ is subjected to more degradation rates, electrode contamination as well as shorter system lifetime. Some minerals that makeup the SOWE cell are known to be abundant and affordable and they include Sr, Ni, Mn, Zr, Y, La along with some ceramic materials [24]. A SOWE is shown on **Figure 1(c)**. The comparison of PEMWE, AWE and SOWE has been established in **Table 2** where both the advantages and disadvantages are indicated [21, 24, 26, 27].

1.2 Power supply

The choice of power source is very crucial in determining the overall cost of hydrogen production [8]. Ideally, hydrogen production process via WE is regarded one of the costly method of hydrogen gas generation and research is focusing on renewable energy sources [28]. Thus far, the direct current has been utilized as the power source. Other alternative power sources include solar energy whereby the solar-driven electricity initiates the water splitting process [12, 14]. Although solar energy is a renewable energy source, one of the challenges is that it is affected by the weather conditions hence harnessing solar energy and storing it for longer has been resorted to [14, 29]. Other green energy sources are hydro and wind energy thus their contribution to PEMWE system may be an alternative [14]. Mokete et al. [14] utilized the incoming solar energy to initiate the electrolysis process and the WE efficiency was 61.8% at temperature 25.5°C, irradiance of 810.1 W/m^2 and relative humidity of 42.2% whereas the PV cell efficiency was 12.7%. Both hydrophobic and hydrophilic properties are necessary for enabling water-dissolved ions and gases' transport [16].

2. Thermodynamics and reaction kinetics of HER and OER

2.1 Thermodynamics

The electrochemical process during water decomposition is attributed to the thermodynamic parameters of heat energy and voltage. These parameters correspond to Gibb's free energy and they can be expressed based on reversible potential. With respect to the standard condition of electrode reactions, the required minimum voltage hence energy to decompose water at the electrode is referred to as reversible work or potential. The Gibb's free energy can be associated with reversible potential as shown in Eq. (9) [30].

$$\Delta G = nFV_{rev} \tag{9}$$

Therefore, the reversible potential is as follows (Eq. (10)) where *G*, *n* and *F* represent Gibb's free energy (237 kJ mol^{-1}), number of electrons and Faraday's constant (96,485 C mol^{-1}) respectively [30, 31]:

$$\mathrm{V}_{rev} = \frac{\Delta G}{nF} \tag{10}$$

The involved overpotentials on PEMWE include activation Overpotential and Ohmic overpotential therefore, these overpotentials along with the reversible potential relate to the operating (cell) voltage as follows:

$$\mathrm{V}_{\mathrm{cell}} = \mathrm{V}_{rev} + \eta_{act} + \eta_{Ohm} \tag{11}$$

The Ohmic overpotential (η_{Ohm}) is given as a product of membrane resistance (R_{Ohm}) and the related current density (i) (Eq. (12)) [31, 32].

$$\eta_{Ohm} = \mathrm{R}_{Ohm}\mathrm{i} \tag{12}$$

The activation overpotential is calculated according to Eq. (13) where V_{rev} is reversible overpotential while *F*, *R* and T_{elec}. represent the Faraday's constant, universal gas constant and electrolyzer's temperature respectively [31]. The $i \; and \; i_o$ imply the current density as well as exchange current density.

$$\eta_{act} = \mathrm{V}_{rev} + \frac{RT_{elec.}}{2F}\ln\left(\frac{i}{i_O}\right) \tag{13}$$

Typically, the gaseous products of WE are quantified by current density based on specific time [33]. Mokete et al. [31] has presented the calculation of Faradaic efficiency (Eq. (14)) where $V_{H_2(produced)}$ and $V_{H_2(theoretical)}$ are the volume of hydrogen gas produced experimentally and the theoretical H_2 gas respectively.

$$FE(\%) = \frac{V_{H_2(produced)}}{V_{H_2(theoretical)}} \times 100 \tag{14}$$

The relationship between several thermodynamics of WE which include Gibbs free energy (ΔG), change in entropy ($T\Delta S$) and enthalpy (ΔH) has been established with respect to Eq. (15) whereby ΔH and ΔG are equivalent to 286 kJ mol^{-1} and 237 kJ mol^{-1} respectively [31].

$$\Delta H = \Delta G + T\Delta S \tag{15}$$

2.2 HER and OER kinetics and mechanisms

2.2.1 Kinetics and mechanisms of HER

The industrial relevance and simplicity of the HER has made it to be one of the most explored electrochemical reactions, especially in WE applications. The HER kinetics on noble metals such as platinum-group metals electrodes have proven to be faster to an extent where there is a possibility of practical current densities (>1 A cm^{-2}) at just little overpotentials with the exception of alkaline media [24]. Nickel is one of the metal-surfaces that has been used during the HER mechanism investigation [24]. The activity of HER varies in acidic media where it is related to hydrogen adsorption whereas it is attributed to hydroxyl adsorption in alkaline medium. One of the earliest investigations revealed that in acidic conditions, the proton is adsorbed on the metal's surface (M-H$^+$) whereas in alkaline or neutral media, the water molecule is adsorbed (M-HOH). Subsequently, the reduction of the adsorbed proton or water molecule is exhibited, resulting in the formation of M-H* and possible release of OH^- during the reduction of chemisorbed water [24]. It has been further deduced that the chemisorbed hydrogen can bind with another chemisorbed H* (Tafel step/Volmer-Tafel mechanism) or the chemisorbed proton can electrochemically react with another proton (or water) after which H_2 will be further desorbed (Heyrovsky step/Volmer-Heyrovsky mechanism) [24]. **Table 3** shows the HER pathways in both the acidic and alkaline conditions [21, 24].

When the two mechanisms of HER are within the acidic medium, the reaction steps of concern are with respect to **Table 4**. Some factors which the HER mechanism is dependent on include pH, surface coverage for the intermediate species, applied overpotential as well as electrode potential, nature and structure [24, 34]. The metal's catalytic activity is influenced by the strength of the M-H* bond whereby the initial step that commences the reaction to favor the M-H* requires some strength. However, the excessive strength of the M-H bonding has some impacts since it may lead to the hindrance of the reaction species (intermediate or products) or be discharged from the surface thus blocking it [24]. The Volmer-Heyrovsky mechanism is evident in low surface coverage under increased overpotentials whereas the Volmer-Tafel mechanism is exhibited at high surface coverage and low overpotentials [34].

Step	Acidic	Alkaline
1	$H^+ + e^- \rightarrow H^*$ (16)	$H_2O + e^- \rightarrow OH^- + H^*$ (19)
2	$H^* + H^* \rightarrow H_2$ (17)	
3	$H^* + H^+ + e^- \rightarrow H_2$ (18)	$H_2O + e^- \rightarrow H_2 + OH^-$ (20)

Table 3.
HER mechanisms in acidic and alkaline media.

Step	Tafel	Heyrovsky
1	$M + H^+ \rightarrow M\text{-}H^+$ (21)	
2	$M\text{-}H^+ + e^- \rightarrow M\text{-}H$ (22)	
	↓*Volmer-Tafel mechanism*	↓*Volmer-Heyrovsky mechanism*
3	$2M\text{-}H \rightarrow H_{2(g)} + 2M$ (23)	$M\text{-}H + H^+ \rightarrow M\text{-}H\text{-}H^+$ (24)
4		$M\text{-}H\text{-}H^+ + e^- \rightarrow H_{2(g)} + 2M$ (25)

Table 4.
HER (Tafel and Heyvrovsky mechanisms) in acidic conditions.

Step	Acidic	Alkaline
1	$H_2O + M \rightarrow M\text{-}OH^* + H^+ + e^-$ (26)	$OH^- + M^+ \rightarrow M\text{-}OH^* + OH^- + e^-$ (30)
2	$M\text{-}OH^* \rightarrow M\text{-}O^* + H^+ + e^-$ (27)	$M\text{-}OH^* + OH^- \rightarrow M\text{-}O^* + H_2O + e^-$ (31)
3	$M\text{-}O^* + H_2O \rightarrow M\text{-}OOH^* + H^+ + e^-$ (28)	$M\text{-}O^* + OH^- \rightarrow M\text{-}OOH^* + e^-$ (32)
4	$M\text{-}OOH^* + OH^- \rightarrow M + O_2 + H^+ + e^-$ (29)	$M\text{-}OOH^* + OH^- \rightarrow M + O_2 + H^+ + e^-$ (33)

Table 5.
OER mechanisms.

2.2.2 Kinetics and mechanisms of OER

The pH plays a key role during the reaction kinetics involved in WE since alterations in pH have a direct impact on the proton content, which is crucial for both OER as well as the HER. The involvement of protons in the HER is in the reduction reaction to produce H_2 gas thus the change in acidic to alkaline conditions affects the ions that participate in the catalytic mechanisms. The H^+ ions become reduced with increasing pH hence the rate of the reactions decreases. However, the OH^- ions partake in the oxidation process involved in the OER so the catalytic efficiency depends on the balance between protons and hydroxide ions [33]. It has been indicated that the OER differs depending on the conditions (acidic versus alkaline) [21]. First, the active species (OH^*) are produced by water on the electrode's surface. Secondly, the OH^* is transformed to O^* on the electrode's surface. Thirdly, the reaction of absorbed O^* and H_2O or OH^- to form OOH^* occurs. Lastly, desorption of OOH^* takes place then O_2 is produced. All the involved equations (Eqs. (26)–(33)) are shown on **Table 5**.

3. Electrodes catalysts

As indicated prior, there two electrodes in the PEMWE system are the anode and the cathode [14, 16]. Typically, the two electrochemical processes involved are oxygen evolution reaction (OER, at anode) and hydrogen evolution reaction (HER, at cathode) [35]. Some widely used effective materials for the cathode for HER are platinum (Pt), Ir or carbon felt whereas Ru, Ir (IrO_2), Pt or titanium are used as anode catalysts for OER [14, 21, 36, 37]. The core of WE is the employment of catalysts with high catalytic activity and reusability during water splitting process [38]. Usually, efficient catalysts for the HER and OER are regarded as materials associated with the platinum group [35]. Some affordable and abundant transitional metal-based materials have recently drawn research attention as electrode catalysts [35].

Wu et al. [35] successfully doped nickel hydroxide ($Ni(OH)_2$) nanosheets with iron (Fe) and cobalt (Co) then applied them as electrode catalysts. Both the Fe-doped and Co-doped $Ni(OH)_2$ are applicable as anode and cathode catalysts during WE since a high HER activity resulted from use of Co-doped $Ni(OH)_2$ as a HER catalyst [35]. Also, the incorporation of both Fe and Co doped-$Ni(OH)_2$ in a WE cell resulted in the current density of 10 mA cm^{-2} at 1.59 V thus outperforming RuO_2 and Pt/C catalysts [35]. A membrane electrode assembly (MEA) was assembled by Siracusano et al. [39] where Iridium oxide (IrO_2) was sprayed on Nafion 115 PEM to form an anode backed up with Ti mesh while the cathode catalyst was Pt/C (carbon cloth backup) [39]. With the active area of 100 cm^2, a short stack made up of 3 cells connected in series was done. Upon application of 330 W power and 60 A current, the produced H_2 was 80 L h^{-1} [39]. In a study by Grigoriev et al. [40], cobalt and iron (II) hexachloroclathrochelates were used as part of the electrocatalysts that are clathrochelate-based during PEMWE. The obtained results of current-voltage performance indicated the successful utilization of these cage complexes as electrocatalyst for HER in PEMWE [40].

When dispersed on effective support materials, nano-scale (core shell) catalysts and corrosion resistant support are some of the effective electrode catalysts [21]. Metal-based oxides like Sn, W, Ti and In qualify as some good support materials that are resistant to corrosion. In acidic atmosphere, metal nitride such as titanium nitride (TiN) proved to be resistant to corrosion and electrochemically stable [21]. This was verified by Nan et al. [41] who found out that the incorporating TiN nanotube as a catalyst support resulted in 0.78 A/mg Pt mass activity which was 3.6 times more than that of Pt/C. The carbonized complexes of natural pectin polysaccharides were proven by Kholin et al. [42] to be some of the effective non-platinum catalyst in the energy applications. In their study, Kholin et al. [42] used the carbonization technique to prepare sodium pectate by interchanging 25% of sodium ion with nickel (Ni(II)) ions thus forming 25%Ni-NaPG complex within an inert condition at carbonization temperatures (280°C, 550°C and 800°C). It was found out that the sample carbonized at 550°C outperformed others in the hydrogen-oxidation with the corresponding current density and power of 56 mA cm^{-2} and 5.96 mW cm^{-2}, whereas the 800°C sample was the best for oxygen-reduction reaction with the highest current density and power of 71.5 mA cm^{-2} and 13 mW cm^{-2} respectively [42].

A Pt nanoparticle cryoaerogel (Pt-NP-CG) was applied on the gas diffusion layer of carbon support then used as a cathode catalyst for WE by Stein et al. [32]. The results revealed that in the presence of the cryoaerogel, the low Pt loading (0.15 mg cm^{-2}) was achieved due to the increased connectivity and porosity to the structure. The stable performance in terms of maximum current density of 2 A cm^{-2} was exhibited by the croaerogel-based catalyst [32]. In a study by Wang et al. [25], the performance of a PEMWE stacks and systems was analyzed via the development of a cell model. Both the OER and HER were characterized through respective application of the TiO_2-supported IrO_2 as well as the unsupported powder of IrO_2. Basically, an electrochemical PEMWE system containing an electrolyzer stack wherein the anode catalyst (IrO_2/TiO_2 with Ir loading of 2 mg cm^{-2}), cathode catalyst (Pt/C with Pt loading of 1 mg cm^{-2}) and the PEM (7 mil PFSA membrane (N117)) were incorporated. The exhibited efficiencies of the stack (65.3%) and system (60.3%) were attained at specific energy consumptions of 51.1 kWh/kg-H_2 and 55.3 kWh/kg-H_2 respectively. Also, the stack degradation rate of 3.3 mV h^{-1} was exhibited [25].

4. PEMs: Nature and structure

A PEM is the most significant component of the PEMWE that facilitates proton transfer and acts as a gas separator [39, 43]. An excellent PEM is characterized by improved proton conductivity [43]. The components of a PEM include the polymer matrix, filler and crosslinker (**Table 6**). The thin film or PEM (20–200 μm) achieves high efficiency due to reduced Ohmic losses [29]. The most extensively used

Category	Material	Structure	Reference
Polymer	Poly (vinyl alcohol) (PVA)	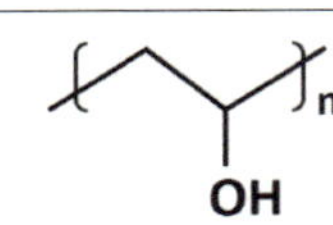	[44–46]
	Poly (vinylpyrrolidone) (PVP)	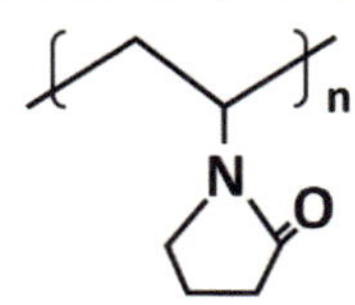	[44, 46]
	Poly (acrylamide) (PAM)	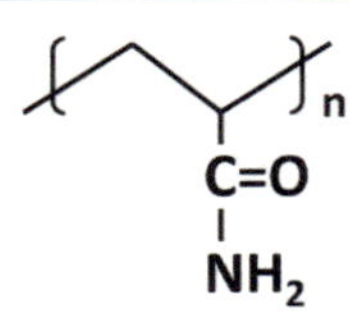	[44]
	Polyethylene oxide (PEO)		[44, 47]
	Polyvinyl carbonate (PVC)	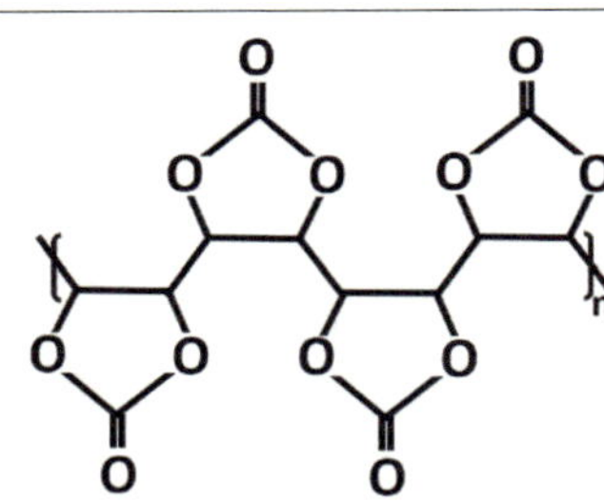	[47]
	Sulfonated poly(arylene ether sulfone) (SPAES)	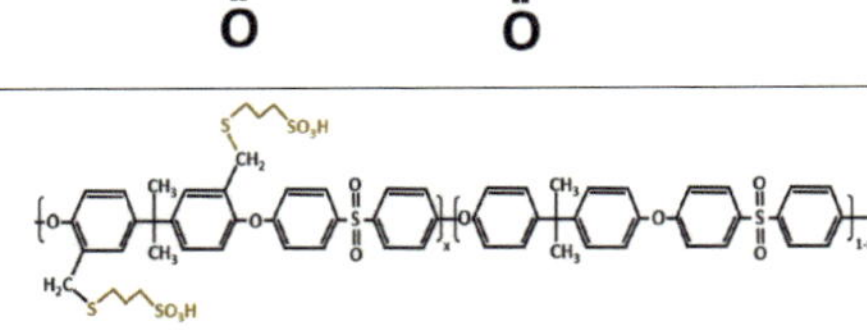	[48–51]
	Poly(ethylene glycol) dimethyl ether (PEGDE)		[46]
	Nafion	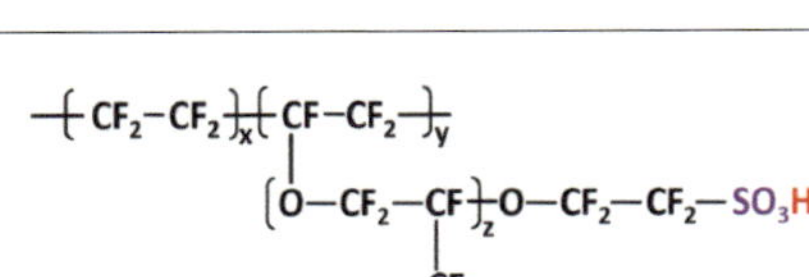	[18, 28, 52]

Category	Material	Structure	Reference
Crosslinker	Sulfosuccinic acid (SSA)	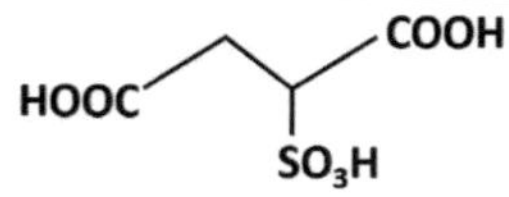	[53]
	Glutaldehyde (GA)	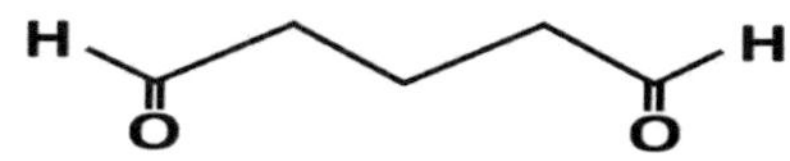	[45, 54]

Table 6.
Common polymers and crosslinkers utilized during PEM fabrication and their structures.

commercial PEM is Nafion PEM due to its excellent proton conductivity, stability (thermal and chemical), good water permeability, high mechanical strength, high chemical and electronic resistance [8, 18, 28]. In the mid-1960s, DuPont introduced Nafion, a flouropolymer that is composed of sulfonated polytetraflurorthylene [18]. In order for the proton conduction to be facilitated, there is a requirement for water contact with Nafion [18, 55]. The main transport mechanisms are either hopping (Grotthus) or diffusion (vehicle) mechanisms [20, 55]. Although Nafion is categorized as an excellent proton conducting PEM, its main challenges are scarcity and high cost [8, 28]. Therefore, this calls for readily available and affordable PEMs as alternatives. The shift of PEMWE research is the improvement of the electrodes or PEM. The operating conditions such as temperature, liquid state (such as acidic, neutral) among others, are some of the important factors to consider in the PEMWE system [56]. The development of the efficient advanced PEM has to take into account the proton conductivity, mechanical strength and durability [28]. The fabrication of the PEMs must be inclusive of the materials that will allow enhanced proton conductivity.

The improvement of proton conductivity while maintaining the mechanical strength has to be explored [44]. The properties of the PEMs can be improved using effective techniques such as crosslinking, grafting, co-polymerization and polymer blending [44]. Irrespective of the method of fabrication for the PEMs, the incorporated polymer in the PEM is significant since it influences the proton conductivity along with gas permeability [44]. Water soluble polymers are commonly utilized polymers in PEMs to facilitate proton conductivity. Most of the water soluble polymers are among the natural and biological polymers like polynucleotides, polysaccharides and polynucleotides. Such naturally and most abundant polymers are starch, cellulose and chitin which are modified to improve water solubility with derivatives of additional functional groups like sulfate and phosphate [57]. Some commercial water soluble polymers include polyvinyl alcohol (PVA), poly (vinylpyrrolidone) (PVP), poly hydroxyethyl acrylate (PHEA) [44]. **Table 6** indicates commonly utilized polymers as the PEM matrices. Among the polymers used in PEM fabrication, PVA has been widely used due to its availability, hydrophilicity and flexibility. The polymer's artwork determines its properties which include; swellability, thermal stability, chemical resistance, mechanical properties as well as the permeability [44]. Within the PEMs, the polymer comprises most of the PEM's surface which contributes as a matrix or holder [44]. All other constituents of the PEMs are incorporated and distributed within the polymer matrix [44].

Alternative to fluorocarbon-based polymers (Nafion), the hydrocarbon-based polymers such as SPAES have been used as proton conducting PEM materials [48].

Although the SPAES PEMs are mostly used in methanol fuel cells, they are still applicable in PEMWE [18]. SPAES is categorized as a hydrocarbon polymer that is extensively used in PEM applications since it is easy to fabricate and it can be modified to attain great membrane forming properties, however, it has low hydrophobicity [48, 49]. Also, its capability to form thinner membrane is a benefit in the reduction of Ohmic resistance thus marvelous mechanical properties. Even though hydrocarbon-based membranes like SPAES exhibit good proton conductivity, the physical and chemical stabilities they acquire are very poor [49]. Studies on hydrocarbon based-PEMs as both ionomer and membrane are not numerous hence research focus much continue to investigate them as alternative PEMs [49].

4.1 Crosslinking

Crosslinking in polymers, which comprise a larger portion of the PEM, is known as the connection of polymers chains to other materials by introducing one or more interlaced networks which are bonded covalently or non-covalently [44, 58]. The polymer chains' mobility will also determine its state as a gel or solid [44]. The cross-linked polymers are considered as engineering materials that are useful in various applications [17]. Also, the cross-linking approach help to modify and enhance the PEM's performance thus it improves the overall properties and performance of the PEM [44]. One important thing to consider during cross-linking is the degree of crosslinking because it also influences the nature and performance of the PEM [17]. Crosslinking is conducted under specific conditions such as temperature and using some chemicals referred to as cross-linkers [17, 59]. Chemical, thermal and photo cross-linking are employed during the PEMs' fabrication process depending on the desired application [17, 44]. During the PEM fabrication processes, crosslinking is sequential whereby chemical crosslinking agents are added followed by the thermal treatment [17]. The chemical and thermal treatment during the PEM fabrication also control the crosslinking degree. Typical crosslinkers that have been used include sulfosuccininc acid (SSA) [17] and glutaraldehyde (GA) [60]. The chemical crosslinkers that have sulfonic acid groups determine the proton conduction ($-SO_3^-H^+$). While crosslinking is considered to facilitate the proton conductivity, it sometimes plays a role in the hindrance of mobility of ion thus affecting the ultimate conductivity of the PEM [44]. When a chemical crosslinking agent is added to the polymer mixture, it becomes crosslinked to the alcohol groups of the polymer (PVA) thus creating the morphological stability within the polymer while preventing the dissolution of the polymer into the solvent and possibly creating the hydrophobic barrier which limits the gases diffusion [17]. Upon the thermal treatment, −OH from polymer (PVA) plus the $-SO_3H$ or COOH from the crosslinker (SSA) form hydrogen bonds owing to the decrement in the polymer chains' distance [60]. **Table 6** shows the crosslinkers (SSA and GA) usually used during the PEMs' fabrication along with their structures [47].

4.2 PEM modification

The main constituents of a PEM include a matrix and crosslinker. Further modifications of the PEMs give rise to mixed matrix membranes (MMM) or blended polymers (BP) wherein the additive (filler) is dispersed within the matrix [61]. Typical matrix materials are polymers such as PVA and cross-linkers that include sulfosuccinic acid (**Table 6**), whereas fillers can be nanoparticles or bio-materials among others [31, 61]. Various attempts to improve the PEMs' proton conductivity under suitable conditions like temperature and humidity have been considered [62]. Such include synthesizing the PEMs that bear protogenic groups [62]. Among the newly fabricated PEM materials, each usually has its unique properties based on the precursor materials; therefore, sometimes they may not have all the necessary requirements for hydrogen generation in PEMWE [62]. The thorough insight on proton transfer is helpful in assisting them to develop effective PEMs [62]. Since hydrogen is crucial during the proton transfer, the design of the new PEMs has to be centered on H-bonds with various protogenic groups in the PEMs [62]. The development of new PEMs involves polymer blending technology as a versatile approach. Blended polymers are deemed to have enhanced properties like proton conductivity and mechanical strength [63]. Most polymer blends are immiscible [44]. The introduction of the additional materials within the polymer matrix is another important form of blending. These materials referred to as fillers have to be compatible, while the proton conductivity and mechanical properties are maintained and improved [44]. The representation of a cross-linked PEM with filler additive within the polymer matrix is shown in **Figure 2** [17, 31, 44, 53, 58].

4.3 PEM properties

The PEMs' properties depend on its make-up since additives can alter the original properties by improving them [58]. The properties of concern in the PEM are the following:

i. *Morphological properties*: The surface morphological observation of the PEMs is done using the scanning electron microscope (SEM) and/or atomic force microscope (AFM) [64]. For the surface area measurement of the additives, if present in a PEM, the Brunauer-Emmett-Teller (BET) surface area analyzer is utilized [65]. Furthermore, the crystal structure of the PEMs is characterized using the X-ray diffraction (XRD). Functional groups available on the surface are determined using the Fourier transform infrared spectroscopy (FT-IR) spectrometer [64].

ii. *Water uptake, adsorption, and swelling properties*: The contact angle meter is used to study the PEMs' hydrophilicity [65]. For the water uptake (WU), the

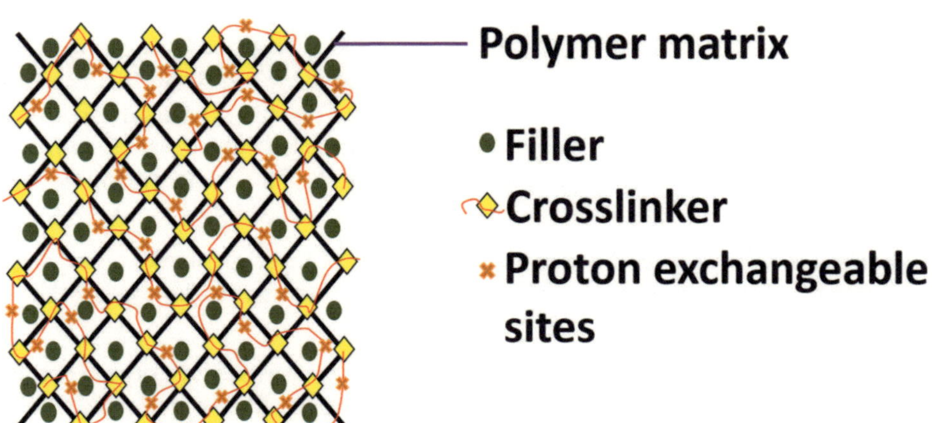

Figure 2.
Schematic diagram of a crosslinked filler-modified PEM.

PEMs become dried in an oven at the specific temperature (120°C) until a constant weight is reached; then, it is determined using the microbalance and recorded as dry weight (W_{dry}) [53]. Afterward, the PEMs get immersed in water for a specific time (24 h); then, they are weighed and recorded as wet weight (W_{wet}). The percentage of water uptake is calculated according to Eq. (34) [48, 53, 61].

$$\text{Water uptake}(\%) = \frac{W_{wet} - W_{dry}}{W_{dry}} \times 100 \tag{34}$$

For the PEMs' dimensional stability assessment, the swelling ratio (SR) is used (Eq. (35)). The wet and dry lengths of the PEM are measured during the water uptake procedure above and used as L_{wet} and L_{dry}, respectively [48, 64].

$$\text{Swelling ratio}(\%) = \frac{L_{wet} - L_{dry}}{L_{dry}} \times 100 \tag{35}$$

iii. *Ion exchange capacity (IEC)*: The ion exchange capacity is carried out using the titration method. The known weight of the dried membrane according to the water uptake procedure is put into the NaCl solution so as to exchange the sulfonic acid groups with sodium ions; then, the titration procedure with NaOH is conducted whereby phenolphthalein is utilized as an indicator. The moles number of protons is equivalent to the moles number of sulfonic groups. The IEC is as follows (Eq. (36)) [48, 53, 66]:

$$IEC\left(\frac{meq.}{g}\right) = \frac{\Delta V_{NaOH} \times C_{NaOH}}{W_S} \tag{36}$$

Where the volume change and concentration of NaOH solution are indicated by ΔV_{NaOH} and C_{NaOH}, while the PEM weight is represented by W_S.

iv. *Thermal properties*: The thermogravimetric analyzer (TGA) is utilized for determining the thermal properties [53]. During this TGA analysis, the water vapor adsorption is also carried out under various thermal conditions using the isotherms in order to observe the PEMs adsorption behavior [16, 31, 37, 67].

v. *Mechanical properties*: The tensile strength and elongation break are some mechanical properties employed. These measurements are done using the universal testing machine [64].

vi. *Proton conduction properties*: The materials' properties such as proton conductivity are usually attributed to the surface nature including the functionalities. The presence of water is important in facilitating the proton conductivity [68]. The Impedance analyzer is used to measure the proton conductivities at varying temperatures (25–80°C). The equation for proton conductivity calculation is Eq. (37). L, R, and A represent the respective distance between the electrodes, PEM resistance, and PEM surface area.

$$\delta\left(\frac{S}{cm}\right)=\frac{L}{R\times A} \tag{37}$$

vii. *Electrochemical properties*: by applying a known voltage to the electrolyzer and measuring either current density or efficiency [16, 31, 64]. The proton conductivity is a factor that is related to the performance and efficiency of PEMWE since enhanced proton conduct may indicate good performance [69]. The efficiency of the electrolyzer can be referred to as hydrogen efficiency and it is obtained by dividing experimentally produce H_2 by theoretically produced H_2 (Eq. (14)) [2, 31, 70].

4.4 PEM fabrication and conditions

Different PEMs as important components of PEMWE have been fabricated and reported in **Table 7**. As the most commonly polymer for PEM matrix, the PVA-based PEMs (and other PEMs) that have been crosslinked and modified with additives are indicated in this subsection (**Table 7**). A PVA/GA PEM with a proton conducting nano-zeolite or H-MFI-24 ($H_2Si_{25}Al_2O_{54}{\cdot}nH_2O$) filler was prepared (thickness = 40 μm) by Nishihara et al. [16] and then applied in WE. The through-plane proton conductivity (PC) measured at 100°C, and relative humidity (RH) of 95% was found to be 2.8 mS/cm. Accordingly, this PC was not high enough for practical applications due to the low PC of zeolite-PVA. The hydrogen yielded was 66% from the theoretical value at 1.8 V potential and current density of 18 mA/cm^2 [16]. The 50% WU indicates that the zeolite filler has good water absorbing capacity [16]. The OH stretching was displayed on the IR spectra between 3600 and 3000 cm^{-1} wavenumber [16].

Tasarin et al. [61] double crosslinked PVA (5 wt%) with both SSA and GA as double crosslinkers to form PVA/SSA-GA PEM. Accordingly, the SSA was added as a crosslinker and proton source, whereas the GA was for the improvement of water resistance. For this PVA/SSA-GA PEM, first, the PVA/SSA with the PVA:SSA molar ratio of 1:0.2 was prepared then soaked in GA (0.5 M) solution. The ZIF-8 filler (5%) was introduced immediately after the SSA addition (PVA:SSA molar ratio 1:0.2); then, the PVA/SSA/ZIF-8 PEM was casted. Another additive of imidazole (Im) was added as a filler that facilitates proton transfer and the PEM cast thereof was referred to as PVA/SSA/ZIF/Im. In this PVA/SSA/ZIF/Im PEM, the molar ratio for Im to sulfonic acid group (SAG) that proved to show better performance was 3. The respective WU and PC of the PVA/SSA-GA PEM at 27°C were 47.8% and the PC of 2.7 × 10^{-3} S/cm. The consideration of this PEM's PVA:SSA molar ratio (1:0.2) for further modification with additives was based on its hydrolytic stability as opposed to PC only which increased with increasing SSA content. Therefore, at 27°C, the 5% loading of ZIF-8 on PVA/SSA PEM (PVA:SSA molar ratio 1:0.2) resulted in the proton conductivity (highest), IEC and WU of 6.2 × 10^{-3} S/cm, 2.17 meq/g, and 58.5% respectively. After portraying the best PC, the PVA/SSA/ZIF-8 PEM with the 5% ZIF-8 loading was considered and Im was doped implying modification with two additives. The highest PC (2.3 × 10^{-4} S/cm) under dry conditions was achieved at 80°C by PVA/SSA/ZIF/Im PEM with Im to sulfonic ratio of 3; hence, the increase in PC was obtained with

PEM composite			**Properties**				**Refs.**
Polymer/crosslinker/filler	Thickness (μm)	WU (%)	SR (%)	IEC ($\frac{meq.}{g}$)	M-TS (MPa)	PC (mS/cm)	
PVA/GA/*H-MFI-24*	40	50	—	—	—	2.8	[16]
PVA/SSA-GA	~0.3 (mm)	47.8		—	—	2.7	[61]
PVA/SSA/*ZIF-8*	~0.3 mm	58.5	—	2.17	—	6.2	[61]
PVA/SSA/*ZIF-8, Im*	~0.3 mm	—	—	—	—	2.3×10^{-1}	[61]
PVA-PVP/SSA/*PB*	90	133.7	34.4	4.1	1	1.33	[62]
SPAES/DMAc/*PIN*	21	35	19	2.3	>40	170–200	[48]
PVA/GA/*BYb*	—	9	—	—	106.9	64	[37]
AQV-IR	150	24.3	—	—	—	103	[71]

Legend: WU, water uptake (%); SR, swelling ratio (%); IEC, ion exchange capacity ($\frac{meq.}{g}$); M-TS, mechanical property-tensile strength (MPa); PC, proton conductivity (mS/cm); Ref., references.

Table 7.
Modified PEMs and their properties.

increasing Im content and temperature. The XRD results of the PVA crosslinked PEMs displayed the crystalline and amorphous peaks at 2θ equal to 19.9° and 40.4°, respectively [61].

Park et al. [48] fabricated a PEM using a sulfonated poly(arylene ether sulfone) (SPAES) polymer with N,N-dimethylacetamide (DMAc) to incorporate HSO_3 groups, and then, the polyimide nanofiber (PIN) was used as a filler. This newly developed PEM is referred to as SPAES/DMAc and its properties are shown in **Table 7**. The SPAES/DMAc PEM is regarded as non-fluorine PEM and it is one of the promising PEMs to be widely applied in PEM technology [48]. Furthermore, the mixed metal oxide nanoparticles ($BaCe_{0.85}Yb_{0.15}O_{3-\delta}$ *or BYb*) were used for the first time as fillers in the GA crosslinked PVA to prepare a PVA/BYb PEM [37]. The 1.5 wt% of BYb showed good performance as shown in **Table 7**. The proton conductivity of the PEM at 70°C was 64 mS/cm, whereas the tensile strength (TS) was 106.9 MPa. The peaks of 2850 cm^{-1} indicated the aldehyde group, at wavelengths 1000, 1200–1250 cm^{-1} was the C−O group of PVA, the hydrogen bonding and OH vibrations were at 3200–3600 cm^{-1}, the C=O group was identified at 1720–1730 cm^{-1}, whereas the C−H bands were at 2907 and 324–1422 cm^{-1}. The H_2/O_2 evolution experiments were conducted and the power performance of 9.48 and 29 mW/cm^2 was obtained at respective current densities of 59 and 210 mA/cm^2 as well as operational temperatures of 25°C and 70°C. These electrochemical results reveal that the operational temperature and current density increase led to good performance thus indicating the higher proton conductivity of the PEM [37].

The pine bark (PB) filler was incorporated as filler by Mokete et al. [31] to the SSA crosslinked PVA and PVP polymers to form the PVA-PVP/SSA/PB PEM. The respective PVA-PVP/SSA/PB PEM's IEC, WU, SR, and PC were 4.1 meq/g, 133.7%, 34.4%, and 1.33 mS/cm. This PEM exhibited the highest water vapor adsorption following the type III adsorption isotherm. The performance of this PEM was in terms of Faradaic efficiency, which was 83.7% at 10 V application. Also, this PEM proved to be one of the effective filler-modified PEMs, which incorporated two polymers. Other PEMs are discussed thoroughly elsewhere in the literature [31]. Recently, a study investigated the ionizing irradiation effect on the PEM's structural properties and then utilized it in an electrochemical hydrogen pump [71]. The PEM was Aquivion® E98-15S, and it was transformed to H^+ form and was irradiated with RS-20MR installation, which had the respective electron beam current and electron energy of 75 kA and 1–1.5 MeV. The PEM of concern from this study shall be referred to AQV-IR. The WU at 20°C was 24.3%, whereas its conductivity and resistivity were 103 mS cm^{-1} and 9.7 Ohm cm, respectively [71].

PEM limitations as a result of low ion conductivity have been countermeasured by the development of polymer nanofiber framework. Composite PEMs of sulfonated polyimide nanofibers that show improved proton conductivity (10 times more than PEM without nanofibers) have been reported by Kawakami et al. [72]. Other effective nanofibers that were developed and incorporated in the PEM include phytic acid-doped polybenzimidazole nanofibers [72].

5. Hydrogen gas generation: Production, storage, and transport

The existence of H_2 involves three phases that are liquid, gas, and solid [73]. There are three stages attributed to H_2, which are production, storage, and utilization [74, 75]. Again, the significance of hydrogen as an energy carrier is evident in energy storage, and chemical and transportation applications [5]. The hydrogen fuel demand is expected to increase owing to its application in both heavy and light duty transportation. Although it is essential to develop the green, affordable, effective and safe systems that are practical for H_2 fuel applications, the apparent challenge has been the H_2 storage system establishment [9, 75]. The H_2 storage can be subdivided into physical methods such as liquid H_2, cryo-compressed H_2 and compressed H_2, and chemical methods that include chemical hydrides (borohydrides), metal hydrides (simple and complex metal hydrides), and sorbents (metal-organic framework) [73]. Thus far, one of the common ways of H_2 storage is high pressure (cylindrical) storage of compressed H_2 but the low densities and intense safety risks make it less appropriate [9, 75, 76]. Liquid H_2 storage is still not 100% effective as it may be subjected to evaporation in some instances hence H_2 loss [75]. Also, the energy that is lost during the liquid H_2 production is about 25% [9, 77]. Regarding the large-scale applications, the chemical storage of hydrogen through H_2 bonding has proven to be one of the promising alternatives for H_2 storage. To date, the solid-state storage of H_2 has gained significant interest because of good storage efficiency, safety, and good stability. Some important benefits of the solid-state H_2 storage include demonstration of low absorption and desorption temperature, good reversibility especially after repeated cycles of absorption and desorption, high H_2 capacity, and reaction kinetics retention. As part of the solid-state storage, the metal hydrides (including chemical and complex hydrides) have proven to successfully store H_2 since they exhibit high storage capacities while operating at low or pressures and non-cryogenic temperatures [9, 73, 75]. Such metal hydrides include MgH_2, AlH_3, and TiH_2, whereas the complex hydrides include $NaBH_4$, $NaALH_4$, $LiBH_4$, and $LiAlH_4$. Among the light metal hydrides, MgH_2 has surpassed others owing to its properties like high hydrogen capacity (7.6 wt%) [78, 79], the highest energy density (9 MJ/kg) [80] as well as low cost with the excellent reversibility [9, 81, 82].

As the universe's most abundant element, H_2 can form H-bearing compounds upon reaction with other elements [83]. Upon reaction with more electropositive elements including the alkali, alkaline as well as rare Earth metals, an electron is gained by hydrogen thus bearing a negative charge, which is referred to as a hydride ion (H^-). Some properties of H^- are strong reducibility and high polarization ability along with high redox potential (H^-/H_2: −2.3 V). These properties of H^- enable it to be a carrier of energy and a reactive H_2 species [68]. The H^--bearing materials have been identified and explored as hydrogen storage, thermal storage, chemical synthesis, ion conduction, and hydrogen separation materials [83]. Rb_2CaH_4 and Cs_2CaH_4 were synthesized by Rodenburg et al. [84] and it was discovered that both their conductivities increased due to the release of hydrogen at elevated temperatures; thus, both materials represented mixed hydride ion conductors which displayed the potential of chemical and energy conversion applications. In a study by Novak et al. [85], neutron scattering was employed to investigate low- and high-temperature phases of barium hydride (BaH_2) thus studying the mechanism of hydride ion diffusion. The results revealed that hydrogen allowance of hydride ion conduction is *via* hydrogen jumps of 3.2 Å length and owing to the phase transition, there is an evident reduction of hydrogen jump distances, energy barriers as well as resident times [85]. Accordingly, the phase transition partakes in the transformation of solid-state material with kinetics to a rapid H_2 conductor [85]. In another study by Ubukata et al. [86], a layered $Ba_{2-\delta}H_{3-2\delta}X$ (X = Cl, Br, and I) with structure of anti-Li_3LaSb_2 exhibited apparent conductivity which reached up to 1 mS/cm thus enabling the H^- conduction at room temperature. Kobayashi et al. [87] prepared the K_2NiF_4-type oxyhydrides of La_{2-x-y}

$Sr_{x+y}LiH_{1-x+y}O_{3-y}$ $(0 \leq x \leq 1, 0 \leq y \leq 2, 0 \leq x + y \leq 2)$, which portrayed the flexible storage of H^- and O^{2-} due to the presence of cation sublattices that incorporate more electron-donating cations as opposed to the sublattices of H^- as well as anions.

6. Summary and future outlook

This chapter is an overview of WE as a process that produces renewable hydrogen energy. Among the WE techniques are PEMWE, AWE, and SOWE. The water-splitting process is attributed to some thermodynamics of heat energy and voltage. Additionally, the OER and HER are governed by the respective reaction kinetics which depend on the conditions such as acidic or alkaline media. Within the WE system, the power supply to initiate the WE is key and renewable energy alternatives are highly recommended. The electrodes' catalysts that are highly effective have to be given more research attention. One of the promising H_2 generation techniques, which is the PEMWE, is also discussed in depth including the PEM, electrodes and their enhancement with catalysts, and power source that drives the PEMWE process. Also, the PEM nature and structure, crosslinking, properties as well as modification with additives are addressed. Some filler-modified PEMs have been investigated and their properties have been highlighted. Furthermore, this chapter includes the H_2 generation, storage, and transport. Solid-state H_2 storage is one of the storage techniques that has drawn a lot of research interest owing to its safety, good storage efficiency, and stability. Research attention has to be largely based on the development of the PEMWE system that includes the electrodes, PEMs, and renewable power sources. Specifically for PEMs, new materials that will contribute toward the formation of cost-effective and readily available PEMs must be explored. Again, emphasis on the study and application of renewable energy has to be made in order to promote the zero-emissions atmosphere. The design of future improvements of the PEMWE system is expected to consider reduction in material utilization, particularly the costly and hard to manufacture materials like titanium bipolar plates. This can be achieved through the reduction in catalyst loadings, PEM thickness and bipolar plates' thickness amongst others [8]. The shift of energy utilization has to be from carbon-emitting energy sources to renewable energy.

Acknowledgements

Acknowledgments go to the editors, and publishing manager and all the funding bodies.

Conflict of interest

The authors declare no conflict of interest.

Notes/thanks/other declarations

Gratitude goes to everyone who supported me throughout my research.

Nomenclature

PEM	polymer electrolyte membrane
WE	water electrolysis
PEMWE	polymer electrolyte membrane water electrolysis
OER	oxygen evolution reaction
HER	hydrogen evolution reaction
MEA	membrane electrode assembly
AWE	alkaline water electrolysis
SOWE	solid oxide water electrolysis
YSZ	yttria-stabilized zirconia
TiN	titanium nitride
PVA	poly vinyl alcohol
PVP	poly(vinylpyrrolidone)
PHEA	poly hydroxyethyl acrylate
SPAES	sulfonated poly(arylene ether sulfone)
SSA	sulfosuccininc acid
GA	glutaraldehyde
PAM	poly(acrylamide)
PEO	polyethylene oxide
PVC	polyvinyl carbonate
PEGDE	poly(ethylene glycol) dimethyl ether
MMM	mixed matrix membranes
BP	blended polymer
SEM	scanning electron microscope
AFM	atomic force microscope
BET	Brunauer-Emmett-Teller
XRD	X-ray diffraction
FT-IR	Fourier transform infrared spectroscopy
WU	water uptake
SR	swelling ratio
IEC	ion exchange capacity
TGA	thermogravimetric analyzer
H-MFI-24	proton-conducting nano-zeolite
RH	relative humidity
PC	proton conductivity
ZIF-8	zeolitic imidazolate framework-8
Im	imidazole
SAG	sulfonic acid group
SPAES	sulfonated poly(arylene ether sulfone)
DMAc	N,N-dimethylacetamide
PIN	polyimide nanofiber
BYb	mixed metal oxide nanoparticles ($BaCe_{0.85}Yb_{0.15}O_{3-\delta}$)
TS	tensile strength
PB	pine bark

Author details

Relebohile Mokete
JoVe, Onojo, Japan

*Address all correspondence to: relemokete@kyudai.com

References

[1] Sharma M, Das PP, Sood T, Chakraborty A, Purkait MK. Ameliorated polyvinylidene fluoride based proton exchange membrane impregnated with graphene oxide, and cellulose acetate obtained from sugarcane bagasse for application in microbial fuel cell. Journal of Environmental Chemical Engineering. 2021;**9**:106681. DOI: 10.1016/J. Jece.2021.106681

[2] Kamaroddin MFA, Sabli N, Nia PM, Abdullah TAT, Abdullah LC, Izhar S, et al. Phosphoric acid doped composite proton exchange membrane for hydrogen production in mediumtemperature copper chloride electrolysis. International Journal of Hydrogen Energy. 2020;**45**:22209-22222. DOI: 10.1016./j.ijhydene.2019.10.030

[3] Kiran Kumar V, Man Mohan K, Gunaseelan K, Gajalakshmi S. Insights into the role of bioinspiration, photosynthetic organisms, and biomass-derived electrodes/membranes in the development of bioelectrochemical systems. Sustainable Energy Technologies and Assessments. 2021;**48**:101570. DOI: 10.1016/j.seta.2021.101570

[4] Pei-Kang SHEN, Sheng-Long WANG, Zhi-Yi HU, Yong-Liang LI, Rong Z, Yue-Qiang H. Hydrogen production by alcohol electrolysis. Acta Physico-Chimica Sinica. 2007;**23**:107-110. DOI: 10.3866/PKU.WHXB20070122

[5] Holm T, Borsboom-Hanson T, Herrera OE, Mérida W. Hydrogen costs from water electrolysis at high temperature and pressure. Energy Conversion and Management. 2021;**237**:114106. DOI: 10.1016/J. enconman.2021.114106

[6] Demirdaöven N, Deutch J. Hybrid cars now, fuel cells later. Journal of Science. 2004;**305**:974-976. DOI: 10.1126/ science.1093965

[7] Schlapbach L, Züttel A. Hydrogen-storage materials for mobile applications. Nature. 2001;**414**:353. DOI: 10.1038/35104634

[8] Bareiß K, de la Rua C, Moöckl M, Hamacher T. Life cycle assessment of hydrogen from proton exchange membrane water electrolysis in future energy systems. Applied Energy. 2019;**237**:862-872. DOI: 10.1016/j. apenergy.2019.01.001

[9] Ali NA, Ismail M. Advanced hydrogen storage of the Mg–Na–Al system: A review. Journal of Magnesium and Alloys. 2021;**9**:1111-1122. DOI: 10.1016/j. jma.2021.03.031

[10] Dincer I, Acar C. Review and evaluation of hydrogen production methods for better sustainability. International Journal of Hydrogen Energy. 2015;**40**:11094-11111. DOI: 10.1016/j.ijhydene.2014.12.035

[11] Barbir F. PEM electrolysis for production of hydrogen from renewable energy sources. Solar Energy. 2005;**78**:661-669. DOI: 10.1016/j. solener.2004.09.003

[12] Coridan RH, Nielander AC, Francis SA, McDowell MT, Dix V, Chatman SM, et al. Methods for comparing the performance of energy-conversion systems for use in solar fuels and solar electricity generation. Energy & Environmental Science. 2015;**8**:2886. DOI: 10.1039/c5ee00777a

[13] Soni R, Miyanishi S, Kuroki H, Yamaguchi T. Pure water solid alkaline water electrolyzer using fully

aromatic and high-molecular-weight poly(fluorene-alttetrafluorophenylene)-trimethyl ammonium anion exchange membranes and ionomers. ACS Applied Energy Materials. 2021;**4**:1053-1058. DOI: 10.1021/acsaem.0c01938

[14] Mokete R, Mikšík F, Takata N, Thu K, Miyazaki T. A precursor study for modification of hydrogen production by electrolysis. In: International Conference on Applied Energy (ICAE). Vol. 20. Energy Proceedings; 2021. Available from: https://www.enerarxiv.org/page/thesis.html?id=3477. ISSN 2004-2965

[15] Ito H, Miyazaki N, Ishida M, Nakano A. Stack performance of proton exchange membrane based unitized reversible fuel cells. International Journal of Microgravity Science and Application. 2016;**33**:330305. DOI: 10.15011/jasma.33.330305

[16] Nishihara M, Terayama Y, Haji T, Lyth SM, Satokawa S, Matsumoto H. Proton-conductive nano zeolite-PVA composite film as a new water-absorbing electrolyte for water electrolysis. eXPRESS Polymer Letters. 2018;**12**:256-264. DOI: 10.3144/expresspolymlett.2018.23

[17] Kang M, Kim JH, Won J, Moon S, Kang YS. Highly charged proton exchange membranes prepared by using water soluble polymer blends for fuel cells. Journal of Membrane Science. 2005;**247**:127-135. DOI: 10.1016/j.memsci.2004.09.017

[18] Gagliardi GG, Ibrahim A, Borello D, El-Kharouf A. Composite polymers development and application for polymer electrolyte membrane technologies—A review. Molecules. 2020;**25**:1712. DOI: 10.3390/molecules25071712

[19] Grigoriev SA, Dzhus KA, Bessarabov DG, Millet P. Failure of PEM water electrolysis cells: Case study involving anode dissolution and membrane thinning. International Journal of Hydrogen Energy. 2014;**39**:20440-20446. DOI: 10.1016/j.ijhydene.2014.05.043

[20] Yodwong B, Guilbert D, Phattanasak M, Kaewmanee W, Hinaje M, Vitale G. Proton exchange membrane electrolyzer modeling for power electronics control: A short review. Journal of Carbon Research. 2020;**6**:29. DOI: 10.3390/c6020029

[21] Ababao PM, Oh I. Recent advances in catalyst materials for PEM water electrolysis. Journal of the Korean Electrochemical Society. 2023;**26**:19-34. DOI: 10.5229/JKES.2023.26.2.19

[22] Bernt M, Armin Siebel A, Gasteiger HA. Analysis of voltage losses in PEM water electrolyzers with low platinum group metal loadings. Journal of the Electrochemical Society. 2018;**165**:F305-F314. DOI: 10.1149/2.0641805jes

[23] Wang CR, Stansberry JM, Mukundan R, Chang HJ, Kulkarni D, Park AM, et al. Proton exchange membrane (PEM) water electrolysis: Cell-level considerations for gigawatt-scale deployment. Chemical Reviews. 2025;**125**:1257-1302. DOI: 10.1021/acs.chemrev.3c00904

[24] Chatenet M, Pollet BG, Dekel DR, Dionigi F, Deseure J, Millet P, et al. Water electrolysis: From textbook knowledge to the latest scientific strategies and industrial developments. Chemical Society Reviews. 2022;**51**:4583. DOI: 10.1039/d0cs01079k

[25] Wang X, Star AG, Ahluwalia RK. Performance of polymer electrolyte membrane water electrolysis systems: Configuration, stack materials,

turndown and efficiency. Energies. 2023;**16**:4964. DOI: 10.3390/en16134964

[26] Kuleshov VN, Kurochkin SV, Kuleshov NV, Gavriluk AA, Klimova MA, Smirnov SE. Hydrophilic fillers for anione exchange membranes of alkaline water electrolyzers. E3S Web of Conferences. 2023;**389**:02030. DOI: 10.1051/e3sconf/202338902030

[27] Wallnöfer-Ogris E, Grimmer I, Ranz M, Höglinger M, Kartusch S, Rauh J, et al. A review on understanding and identifying degradation mechanisms in PEM water electrolysis cells: Insights for stack application, development, and research. International Journal of Hydrogen Energy. 2024;**65**:381-397. DOI: 10.1016/j.ijhydene.2024.04.017

[28] Jinjun S. Composite membranes for proton exchange membrane fuel cells [browse all theses and dissertations]. Vol. 8362008. Ohio, United States: Wright State University and OhioLink; 2008. Available from: https://corescholar.libraries.wright.edu/etd_all/836

[29] Martinez R, Forero E. Estimation of energy efficiency in solar photovoltaic panels considering environmental variables. IOP Conference Series: Materials Science and Engineering. 2018;**437**:012008. DOI: 10.1088/1757-899X/437/1/012008

[30] Abdol Rahim AHA, Tijani AS, Kamarudin HS. An overview of polymer electrolyte membrane electrolyzer for hydrogen production: Modeling and mass transport. Journal of Power Sources. 2016;**309**:56-65. DOI: 10.1016/j.jpowsour.2016.01.012

[31] Mokete R, Mikšík F, Selyanchyn R, Takata N, Thu K, Miyazaki T. Fabrication of the novel mixed matrix polymer electrolyte membranes (PEMs) intended for renewable hydrogen production via electrolysis application. Energy Advances. 2024;**3**:1019-1036. DOI: 10.1039/d3ya00503h

[32] Stein L, Borg H, Wesemann C, Zhao Z, Moß C, Trinke P, et al. Platinum cryoaerogel as a low loading cathode catalyst in PEM water electrolysis: An initial concept evaluation. ACS Applied Energy Materials. 2025;**8**:194-207. DOI: 10.1021/acsaem.4c02255

[33] Irshad H, Zia M, Al-Hajri R, Khattak ZAK, Al-Abri M, Ahmad N, et al. Electrocatalysts for hydrogen and oxygen evolution reactions under neutral/near-neutral conditions: Summary and challenges. International Journal of Hydrogen Energy. 2025;**137**:1009-1041. DOI: 10.1016/j.ijhydene.2025.02.246

[34] Gebremariam GK, Jovanović AZ, Pašti IA. The effect of electrolytes on the kinetics of the hydrogen evolution reaction. Hydrogen. 2023;**4**:776-806. DOI: 10.3390/hydrogen4040049

[35] Wu Y, Ji S, Wang H, Pollet BG, Wang X, Wang R. A highly efficient water electrolyser cell assembled by asymmetric array electrodes based on Co, Fe-doped $Ni(OH)_2$ nanosheets. Applied Surface Science. 2020;**528**:146972. DOI: 10.1016/j.apsusc.2020.146972

[36] Zhao S, Yu H, Maric R, Danilovic N, Capuano CB, Ayers KE, et al. Calculating the electrochemically active surface area of iridium oxide in operating proton exchange membrane electrolyzers. Journal of the Electrochemical Society. 2015;**162**:F1292-F1298. DOI: 10.1149/2/0211512jes

[37] Salehi Artimani J, Arjomand M, Enhessari M, Javanbakht M. Proton conducting nanocomposite membranes based on poly vinyl alcohol (PVA)/glutaraldehyde (GA) for proton exchange membrane fuel cells. Iranian journal of

chemistry and chemical engineering. 2021;**40**:69-82. DOI: 10.30492/ijcce.2019.36828

[38] Hamaloğglu KÖ, Tosun RB, Kayı H, Kavaklı PA, Kavaklı C, Tuncel A. Monodisperse-porous cerium oxide microspheres carrying iridium oxide nanoparticles as a heterogeneous catalyst for water oxidation. Applied Surface Science. 2021;**547**:149219. DOI: 10.1016/j.apsusc.2021.149219

[39] Siracusano S, Baglio V, Di Blasi A, Briguglio N, Stassi A, Ornelas R, et al. Electrochemical characterization of single cell and short stack PEM electrolyzers based on a nanosized IrO_2 anode electrocatalyst. International Journal of Hydrogen Energy. 2010;**35**:558-5568. DOI: 10.1016/j.ijhydene.2010.03.102

[40] Grigoriev SA, Pushkarev AS, Pushkareva IV, Millet P, Belov AS, Novikov VV, et al. Hydrogen production by proton exchange membrane water electrolysis using cobalt and iron hexachloroclathrochelates as efficient hydrogen-evolving electrocatalysts. International Journal of Hydrogen Energy. 2017;**42**:27845-27850. DOI: 10.1016/j.ijhydene.2017.05.048

[41] Nan H, Dang D, Tian XL. Structural engineering of robust titanium nitride as effective platinum support for the oxygen reduction reaction. Journal of Materials Chemistry A. 2018;**6**:6065. DOI: 10.1039/C8TA00326B

[42] Kholin KV, Sabirova AF, Kadirov DM, Khamatgalimov AR, Khrizanforov MN, Nizameev IR, et al. Carbonized nickel complex of sodium pectate as catalyst for proton-exchange membrane fuel cells. Membranes. 2023;**13**:635. DOI: 10.3390/membranes13070635

[43] Yu J, Jung G, Su Y, Yeh C, Kan M, Lee C, et al. Proton exchange membrane water electrolysis system-membrane electrode assembly with additive. International Journal of Hydrogen Energy. 2019;**44**:15721-15726. DOI: 10.1016/j.ijhydene.2018.11.192

[44] Ye Y, Rick J, Hwang B. Water soluble polymers as proton exchange membranes for fuel cells. Polymers. 2012;**4**:913-963. DOI: 10.3390/polym4020913

[45] Liu C, Dai C, Chao C, Chang S. Novel proton exchange membrane based on crosslinked poly(vinyl alcohol) for direct methanol fuel cells. Journal of Power Sources. 2014;**249**:285-298

[46] Fu J, Qiao J, Lv H, Ma J, Yuan X, Wang H. Alkali doped poly(vinyl alcohol) (PVA) for anion-exchange membrane fuel cells—Ionic conductivity, chemical stability and FT-IR characterizations. ECS Transactions. 2010;**25**:15-23. DOI: 10.1149/1.3315169

[47] Yang X, Liu J, Pei N, Chen Z, Li R, Fu L, et al. The critical role of fillers in composite polymer electrolytes for lithium battery. Nano-Micro Letters. 2023;**15**:74. DOI: 10.1007/s40820-023-01051-3

[48] Park S, Chae K, Lee M. A sulfonated poly(arylene ether sulfone)/polyimide nanofiber composite proton exchange membrane for microbial electrolysis cell application under the coexistence of diverse competitive cations and protons. Journal of Membrane Science. 2017;**540**:165-173. DOI: 10.1016/j.memsci.2017.06.048

[49] Park JE, Kim J, Han J, Kim K, Park S, Kim S, et al. High-performance proton-exchange membrane water electrolysis using a sulfonated poly(arylene ether sulfone) membrane and ionomer. Journal of Membrane Science. 2021;**620**:118871. DOI: 10.1016/j.memsci.2020.118871

[50] Tinh VDC, Thuc VD, Kim D. Chemically sustainable fuel cells via layer-by-layer fabrication of sulfonated poly(arylene ether sulfone) membranes containing cerium oxide nanoparticles. Journal of Membrane Science. 2021;**634**:119430. DOI: 10.1016/j.memsci.2021.119430

[51] Deivanayagam P, Ramamoorthy AR, Jaisankar SN. Synthesis and characterization of sulfonated poly (arylene ether sulfone)/silicotungstic acid composite membranes for fuel cells. Polymer Journal. 2013;**45**:166-172. DOI: 10.1038/pj.2012.102

[52] Kreuer K, Paddison SJ, Spohr E, Schuster M. Transport in proton conductors for fuel-cell applications: Simulations, elementary reactions, and phenomenology. Molecules. 2020;**25**:1712. DOI: 10.1021/cr020715f

[53] Rhim J, Park HB, Lee C, Jun J, Kim DS, Lee YM. Crosslinked poly(vinyl alcohol) membranes containing sulfonic acid group: Proton and methanol transport through membranes. Journal of Membrane Science. 2004;**238**:143-151. DOI: 10.1016/j.memsci.2004.03.030

[54] Sapalidis AA. Porous polyvinyl alcohol membranes: Preparation methods and applications. Symmetry. 2020;**12**:960. DOI: 10.3390/sym12060960

[55] Agmon N. The Grotthuss mechanism. Chemical Physics Letters. 1995;**244**:456-462. DOI: 10.1016/009-2614(95)00905-J

[56] Alia SM. Current research in low temperature proton exchange membrane-based electrolysis and a necessary shift in focus. Current Opinion in Chemical Engineering. 2021;**33**:100703. DOI: 10.1016/j.coche.2021.100703

[57] Shalaby SW, McCormick CL, Buttler GB. Water-soluble polymers. Synthesis, solution properties and applications. In: ACS Symposium Series. Vol. 467. Washington, DC, USA: American Chemical Society; 1991. DOI: 10.1021/bk-1991-0467

[58] Shaari N, Kamarudin SK. Recent advances in additive-enhanced polymer electrolyte membrane properties in fuel cell applications: An overview. International Journal of Energy Research. 2019;**43**:2756-2794. DOI: 10.1002/er.4348

[59] Mokete R, Mikšík F, Selyanchyn R, Takata N, Thu K, Miyazaki T. The future of modified polymer electrolyte membranes to enhance green hydrogen gas generation by electrolysis. In: Proceedings of 11th International Young Researchers Conference on Industrial Engineering 2024—From Zero to Hero. Kaunas, Lithuania: Kaunas University of Technology (KTU); 2024. pp. 161-164. DOI: 10.5755/e01.2783-6363.2024

[60] Smitha B, Sridhar S, Khan AA. Polyelectrolyte complexes of chitosan and poly(acrylic acid) as proton exchange membranes for fuel cells. Macromolecules. 2004;**37**:2233-2239. DOI: 10.1021/ma0355913

[61] Tasarin S, Panawong C, Sumranjit J, Budsombat S. Enhancement of proton conductivity of crosslinked poly(vinyl alcohol) through introduction of zeolitic imidazolate framework-8 and imidazole. International Journal of Hydrogen Energy. 2021;**46**:36969-36981. DOI: 10.1016/j.ijhydene.2021.08.199

[62] Abdolmaleki A, Eskandari K, Molavian MR. Sulfonated or phosphonated membranes? DFT investigation of proton exchange in poly(oxadiazole) membranes. Polymer. 2016;**87**:181-193. DOI: 10.1016//j.polymer.2016.02.011

[63] Paul DR, Newman S. Polymer Blends. New York, NY, USA: Academic Press; 1978

[64] Munavalli BB, Kariduraganavar MY. Development of novel sulfonic acid functionalized zeolites incorporated composite proton exchange membranes for fuel cell application. Electrochimica Acta. 2019;**296**:294-307. DOI: 10.1016/j.electacta.2018.11.056

[65] Yin Y, Shao J, Zhang L, Cui Q, Wang H. Experimental study on heat transfer and adsorption cooling performance of MIL-101/few layer graphene composite. Energies. 2021;**14**:4970. DOI: 10.3390/en14164970

[66] Gouda MH, Tamer TM, MSM E. A highly selective novel green cation exchange membrane doped with ceramic nanotubes material for direct methanol fuel cells. Energies. 2021;**14**:5664. DOI: 10.3390/en14185664

[67] Park GM, Vargas RA, Bueno PR. Protonic conductivity and thermal properties of cross-linked PVA/TiO_2 nanocomposite polymer membranes. Journal of Non-Crystalline Solids. 2019;**522**:119520. DOI: 10.1016/j.jnoncrysol.2019.119520

[68] Furtmair M, Timm J, Marschall R. Sulfonation of porous materials and their proton conductivity. Microporous and Mesoporous Materials. 2021;**312**:110745. DOI: 10.1016/j.micromeso.2020.110745

[69] Kulasekaran P, Mahimai BM, Deivanayagam P. Novel cross-linked poly(vinyl alcohol)-based electrolyte membranes for fuel cell applications. RSC Advances. 2020;**10**:26521-26527. DOI: 10.1039/d0ra04360e

[70] Kumar SS, Himabindu V. Hydrogen production by PEM water electrolysis—A review. Materials Science for Energy Technologies. 2019;**2**:442-454. DOI: 10.1016/j.mset.2019.03.002

[71] Ivanova NA, Ivanov BV, Mensharapov RM, Spasov DD, Sinyakov MV, Nagorny SV, et al. Features of electrochemical hydrogen pump based on irradiated proton exchange membrane. Membranes. 2023;**13**:885. DOI: 10.3390/membranes13110885

[72] Kawakami H. Development of composite electrolyte membranes with functional polymer nanofiber frameworks. Polymer Journal. 2025;**57**:623-633. DOI: 10.1038/s41428-024-01007-3

[73] Wijayanta AT, Oda T, Purnomo CW, Kashiwagi T, Aziz M. Liquid hydrogen, methylcyclohexane, and ammonia as potential hydrogen storage: Comparison review. International Journal of Hydrogen Energy. 2019;**44**:15026-15044. DOI: 10.1016/j.ijhydene.2019.04.112

[74] Gabrielli P, Poluzzi A, Kramer GJ, Spiers C, Mazzotti M, Gazzani M. Seasonal energy storage for zero-emissions multi-energy systems via underground hydrogen storage. Renewable and Sustainable Energy Reviews. 2020;**121**:109629. DOI: 10.1016/j.rser.2019.109629

[75] Hitam CNC, Aziz MAA, Ruhaimi AH, Taib MR. Magnesium-based alloys for solid-state hydrogen storage applications: A review. International Journal of Hydrogen Energy. 2021;**46**:31067-31083 DOI: 10.1016/j.ijhydene.2021.03.153

[76] Liu H, Li Y. Modified carbon nanotubes for hydrogen storage at moderate pressure and room temperature. Fullerenes, Nanotubes, and Carbon Nanostructures. 2020;**28**:663-670. DOI: 10.1080/1536383X.2020.1738396

[77] Dalebrook AF, Gan W, Grasemann M, Moret S, Laurenczy G. Hydrogen storage: Beyond conventional methods. Chemical Communications. 2013;**49**:8735-8751. DOI: 10.1039/C3CC43836H

[78] Yartys VA, Lototskyy MV, Akiba E, Albert R, Antonov VE, Ares JR, et al. Magnesium based materials for hydrogen based energy storage: Past, present and future. International Journal of Hydrogen Energy. 2019;**44**:7809-7859. DOI: 10.1016/j.ijhydene.2018.12.212

[79] Ismail M. Effect of adding different percentages of $HfCl_4$ on hydrogen storage properties of MgH_2. International Journal of Hydrogen Energy. 2021;**46**:8621-8628. DOI: 10.1016/j.ijhydene.2020.12.068

[80] Bogdanović B, Bohmhammel K, Christ B, Reiser A, Schlichte K, Vehlen R, et al. Thermodynamic investigation of the magnesium-hydrogen system. Journal of Alloys and Compounds. 1999;**282**:84-92. DOI: 10.1016/S0925-8388(98)00829-9

[81] Imamura H, Masanari K, Kusuhara M, Katsumoto H, Sumi T, Sakata Y. High hydrogen capacity of nanosized magnesium synthesized by high energy ball-milling. Journal of Alloys and Compounds. 2005;**386**:211-216. DOI: 10.1016/j.jallcom.2004.04.145

[82] Zaluski L, Zaluska A, Ström-Olsen JO. Nanocrystalline metal hydrides. Journal of Alloys and Compounds. 1997;**253**:70-79. DOI: 10.1016/S0925-8388(96)02985-4

[83] Zhang W, Cao H, Chen P. Hydride ion conductor: A key material for innovative energy storage and conversion. The Innovation Materials. 2023;**1**:100006

[84] Rodenburg HP, Mutschke A, Ngamwongwan L, Gulino V, Kyriakou V, Kunkel N, et al. Mixed hydride-electronic conductivity in Rb2CaH4 and Cs2CaH4. Solid State Ionics. 2023;**403**:116384. DOI: 10.1016/j.ssi.2023.116384

[85] Novak E, Daemen L, Ramirez-Cuesta AJ, Cheng Y, Smith R, Egami T, et al. Uncovering the hydride ion diffusion pathway in barium hydride via neutron spectroscopy. Scientific Reports. 2022;**12**:6194. DOI: 10.1038/s41598-022-10199-8

[86] Ubukata H, Takeiri F, Shitara K, Tassel C, Saito T, Kamiyama T, et al. Anion ordering enables fast H^- conduction at low temperatures. Science Advances. 2021;7:eabf7883. DOI: 10.1126/sciadv.abf7883

[87] Kobayashi G, Kanno R. Development of H^- Conductive Oxyhydrides. New York: Research Frontiers; 2016. Available from: http://www.spring8.or.jp